In Memory of
Akira Tonomura
Physicist and Electron Microscopist

In Memory of
Akira Tonomura
Physicist and Electron Microscopist

Editors

Kazuo Fujikawa
Nishina Center
RIKEN, Japan

Yoshimasa A. Ono
FIRST Tonomura Project
Japan Science and Technology Agency, Japan

NEW JERSEY • LONDON • SINGAPORE • BEIJING • SHANGHAI • HONG KONG • TAIPEI • CHENNAI

Published by

World Scientific Publishing Co. Pte. Ltd.

5 Toh Tuck Link, Singapore 596224

USA office: 27 Warren Street, Suite 401-402, Hackensack, NJ 07601

UK office: 57 Shelton Street, Covent Garden, London WC2H 9HE

British Library Cataloguing-in-Publication Data
A catalogue record for this book is available from the British Library.

IN MEMORY OF AKIRA TONOMURA
Physicist and Electron Microscopist
(With DVD-ROM)

ISBN 978-981-4472-88-3
ISBN 978-981-4472-89-0 (pbk)

Printed in Singapore by Mainland Press Pte Ltd.

Preface

This memorial book in honor of Dr. Akira Tonomura, experimental physicist and electron microscopist, is to commemorate his enormous contributions to fundamental physics in addition to the basic technology of electron microscopy. Dr. Tonomura passed away on May 2, 2012 at the age of 70 during the course of a medical treatment on pancreatic cancer. He was Fellow of Hitachi, Ltd., Group Director of Single Quantum Dynamics Group of RIKEN, and Professor of Okinawa Institute of Science and Technology Graduate University.

The book consists of contributions from distinguished physicists who participated in international conferences planned by Tonomura himself and reprints of key papers by Tonomura and his team. In particular, representative invited speakers at the Memorial Symposium held in Tokyo on May 9-10, 2012, "Tonomura FIRST International Symposium on Electron Microscopy and Gauge Fields," includes Chen Ning Yang and other distinguished physicists such as Yakir Aharonov, Gordon Baym, Christian Colliex, Anthony J. Leggett, Naoto Nagaosa, Nobuyuki Osakabe, Masahito Ueda, and Yimei Zhu. This Symposium was originally planned to commemorate the start of the FIRST Tonomura Project to construct the 1.2 MV holography electron microscope, his longtime dream of the ultimate electron microscope, capable of observing quantum phenomena in the microscopic world. Here FIRST stands for "Funding Program for World-leading Innovative R&D on Science and Technology" sponsored by the Japanese government. Other contributions are from participants of the past ISQM-Tokyo symposia held at Hitachi, International Symposium on Foundations of Quantum Mechanics in the Light of New Technology, including Hidetoshi Fukuyama, Joseph Imry, Yoshinori Tokura, Jaw-Shen Tsai, and Anton Zeilinger, in addition to Tonomura's longtime friends such as Michael Berry and Jerome Friedman.

Dr. Tonomura made outstanding contributions to the development of fundamental quantum physics, and its application to technology such as superconductors and magnetism. He developed his own techniques of electron holography and coherent-beam Lorentz microscopy in order to precisely detect the phase information of electron beams and directly image quantum objects. For more than four decades, Tonomura developed bright and monochromatic field-emission electron beams for electron microscopes and applied them to the development of electron holography and coherent-beam Lorentz microscopy. He used his methods to clarify the fundamental questions of quantum mechanics and to observe hitherto-unobservable quantum objects. He demonstrated the

Aharonov-Bohm effect decisively and brought to light how the wave-particle duality of electrons is realized in Nature. Recently he applied these methods extensively to make visible microscopic structures of matter that had been so far inaccessible, including the space-time behavior of the quanta (so-called vortices) of magnetic flux lines in both metal and high-temperature superconductors, and magnetic behavior of spintronics materials, such as magnetic recording media and heads.

We have always been impressed by Dr. Tonomura's youthful enthusiasm for new sciences, which he pursued until the end of his life. All of us who have known him well will deeply miss him.

The book is edited by Kazuo Fujikawa of RIKEN, Tonomura's longtime friend from college, and Yoshimasa A. Ono of Japan Science and Technology Agency, Tonomura's associate at the Hitachi Advanced Research Laboratory and now in the FIRST Tonomura Project. It is indeed our greatest honor to edit this book in memory of Dr. Tonomura and we would like to share Dr. Tonomura's amazing achievements with readers of the book.

We would like to thank Ms. Kimi Matsuyama, Tonomura's secretary, for collecting the most precious photos with the help of Dr. Tonomura's family. We are particularly indebted to Dr. Nobuyuki Osakabe, Tonomura's longtime collaborator at Hitachi and presently General Manager of Hitachi Central Research Laboratory, for his guidance and comments during the preparation of this book. We also would like to thank Professor Kok Khoo Phua of World Scientific for suggesting the publication of this book.

June 2013   Kazuo Fujikawa
                                                       Yoshimasa A. Ono

Contents

Quantum Physics

Photograph Collection of Akira Tonomura

Dr. Akira Tonomura (April 25, 1942 – May 2, 2012)
(Photo taken in October 2009) (courtesy of Kimi Matsuyama)

Photo 1. Dr. A. Tonomura showing his electron microscope to Prof. C. N. Yang during the FIRST
ISQM-Tokyo at Hitachi Central Research Laboratory (August 1983)
(courtesy of Hitachi, Ltd.)

Photo 2. Prof. C. N. Yang and Prof. Y. Aharonov discussing with Dr. A. Tonomura during the
First ISQM at Hitachi Central Research Laboratory (August 1983)
(courtesy of Hitachi, Ltd.)

Photo 3. Dr. Akira Tonomura with his colleagues in front of the 350-kV holography electron
microscope at Hitachi Advanced Research Laboratory at Hatoyama, Saitama (1990)
(courtesy of Hitachi, Ltd.)

Photo 4. Dr. Akira Tonomura in front of his holography electron microscope at Hitachi Advanced
Research Laboratory in Hatoyama, Saitama (January 27, 2010)
(courtesy of RIKEN)

Photo 5. Photographer Akira Tonomura at Lake Nozori in Gunma, Japan (July 17, 2009)
(courtesy of Osamu Tonomura)

Photo 6. Dr. Akira Tonomura with his wife, Miwako, and his two sons, Masaru and Osamu (Winter 2010)
(courtesy of Miwako Tonomura)

My Dream of Ultimate Holography Electron Microscope

Akira Tonomura (deceased)

Central Research Laboratory, Hitachi, Ltd., Hatoyama, Saitama 350-0395, Japan
Advanced Science Institute, RIKEN, Wako, Saitama 351-0198, Japan
Okinawa Institute of Science and Technology Graduate University,
Onna-son, Okinawa 904-0495, Japan

This note was prepared by Tonomura himself on August 25, 2010 for the feature article and interview of IOP Asia-Pacific (http://asia.iop.org/cws/article/news/43723). This is the most recent narrative of Tonomura on his research starting from his joining Hitachi Central Research Laboratory to the aims of the FIRST Tonomura Project that started in 2010 (Editor).

Dr. Akira Tonomura in his office (October 2009)
(courtesy of Kimi Matsuyama)

1. Background

I joined the Hitachi Central Research Laboratory's Electron Microscope Group in 1965, just when the research on electron microscopes was in full bloom. I was particularly captivated by the experimental work[1] of Hitachi's Hiroshi Watanabe, who with a single electron micrograph proved the Bohm-Pines theory[2] that had only recently been postulated. Looking back even now, at that time in Japan we had a stellar cast of the best researchers of the world, including Hiroshi Watanabe and others in Hitachi. Later, I was privileged to receive advice and guidance from a number of exceptional mentors outside of Hitachi including Professor Ryoji Uyeda of Nagoya University, Professor Gottfried Möllenstedt of Tübingen University, and Professor Chen Ning Yang of State University of New York at Stony Brook, who earned the Nobel Prize in Physics in 1957; they instilled in me a fundamental attitude toward the basic research. Every time I came up with a result that I thought was impressive, I sent off the report to Professor Uyeda, but after the jubilation subsided, he would always add: "You should not be satisfied with this. It may not be imitative, but it will not compete against the results achieved by scientists in western countries, and really doesn't go beyond the level of *basic practice*. It is only when your results become a great trunk supporting myriad boughs and branches, and then your work would be considered *basic research*. Cultivating the roots of robust science and technology traditions in Japan will take 60 years of concerted effort. You should maintain confidence in yourself, keep plugging away, and try not to get bogged down in petty chores." I did not understand at that time why we needed 60 years, but I thought that great research results of Faraday and Maxwell were in Professor Uyeda's mind.

I was fresh out of college at that time. I would consider myself lucky that I was given an opportunity to work more than 40 years conducting research on electron microscopes. This is made possible by this supportive environment in Hitachi as well as support from the Japanese Government including RIKEN in recent years.

My research was concentrated on interference of electron beams throughout my life. I wanted to develop a new imaging method called *electron holography* by exploiting the wave nature of electrons.[3] When I started working on this, the greatest obstacle to my dream was that extraordinary long exposure times were necessary to take beautiful pictures of the interference patterns. Therefore, I immediately started to develop brighter and yet coherent electron beams, which can be compared with a laser beam in light optics. The electron beams are produced simply by applying a voltage to a sharp metal tip which draws out the electrons due to the tunneling effect.

The ten-year period up to 1978 saw dramatic improvements in the development of electron beams providing two-order-of-magnitude improvements in brightness. Electron interference fringes that previously could not be seen at all became directly viewable on fluorescent screens, and the number of interference fringes captured on films increased dramatically from 300 to 3,000. In the years from 1978 to present, we have undertaken a number of projects to enhance the brightness even further. Brightness was enhanced by increasing the acceleration voltage without causing an electric breakdown and by employing various other refinements, opening up a whole new range of potential applications. Our most powerful 1-MV electron microscope available today delivers coherent beams four orders of magnitude brighter than our initial electron microscope, resulting in visualization of more than 10,000 interference fringes.

By using coherent electron beams, we have been observing strange and paradoxical phenomena in quantum world since we are observing not only the intensity but also the phase of electrons, i.e., "wave functions".

In addition to the observation of quantum phenomena, we found that electromagnetic fields in the microscopic world can be visualized by electron holography.

In March 2010, I received a grant from the Japanese government with the six billion-yen, budgeted over five years, for the "FIRST Tonomura Project" which are conducted jointly by RIKEN and Hitachi, Ltd. Here "FIRST" stands for "funding program for world-leading innovative research and development on science and technology". In this project we are going to develop an atomic-resolution holography electron microscope by incorporating cutting-edge technologies to realize my dream of reconstructing the wave fronts of electrons scattered by atoms, first proposed by Dennis Gabor in 1948.

2. Aims of the research and main scientific problems to solve

The aims I want to attain in the FIRST Tonomura Project are to develop unprecedented technology to observe the microscopic and quantum phenomena by precisely measuring phase distribution of electrons at atomic dimensions. In order to realize this we have to develop a technique to spatially resolve atoms within 1/10 of the atomic scale and have to increase the phase detection limit to 1/1000 of electron wavelength from the present highest value of 1/100. In short, we have to develop an electron holography technique available in atomic dimensions, so that electron wavefronts from each atom can be reconstructed, enabling observation of atomic arrangements in three dimensions. This helps to elucidate quantum phenomena often appearing in microscopic regions.

3. Originality of our methodology/approach

Our method to achieve these aims is essentially based on electron holography, which is the technology we have been developing by ourselves. We also make full use of the technique of ultra-high voltage electron microscope we are good at. The present level of electron holography is, however, not sufficient at all for our purpose. We have to improve the spatial resolution from 0.3 nm to below 0.04 nm and phase precision from the 1/100 to 1/1000 of the electron wavelength. This super resolution can be realized by removing aberrations from the lens system; by stabilizing the high-voltage and lens current up to about 10^{-8}/min; by improving the mechanical stability of sample holders; and by putting the microscope in perfect isolation free from electromagnetic disturbances.

4. Experimental or theoretical procedures

This new super microscope must be developed in steps. We also must open new application fields with this electron interference microscopy, where the wavefunction of electrons is detected instead of the intensity.

5. Results and findings

Main findings of our research on electron holography for the past 40 years are as follows:

(1) The first step to electron holography: in 1968 we demonstrated[4] that holography is feasible by using electrons.

(2) The second step to electron holography: in 1979 we obtained[5] bright field-emission electron beams after ten years of efforts, and increased the spatial resolution of the holographically reconstructed images by more than one order of magnitudes to 3 Å.

(3) In 1980 we demonstrated[6] that phase contours of electrons transmitted through magnetic fields in a microscopic region show directly quantitative magnetic lines of force in h/e flux units. This is the most exciting result and is the basis for our research! Although this result can be derived from the Schrödinger equation easily, the phenomenon has not been observed experimentally until then. In 1986, we decisively demonstrated[7] the existence of the Aharonov-Bohm effect using electron holography with a tiny ring magnet covered with a superconducting film and with a copper film. In 1989 we explicitly showed[8] the wave-particle duality of electrons by demonstrating how, in the two-slit interference experiment (using actually electron biprism in place of the two slits), the interference pattern emerges from electrons detected one by one on the image plane.

(4) In 1990s, we observed for the first time static and dynamic behavior of magnetic flux lines (vortices) in metal superconductor thin films[9,10] by using electron holography and Lorentz microscopy with our 350-kV prototype holography electron microscope.

(5) In 2000, we succeeded in developing a 1-MV holography electron microscope[11] and with this microscope we investigated dynamic behavior of quantum vortices in high-temperature superconducting thin films.[12,13] This was our 20 years' dream and successful observation was the most joyful event. Since then we have observed many unexpected behavior of vortices' forms, arrangements and dynamics, which is now summarized[14] and will soon be submitted.

6. Expected industrial and social implications

The new microscope we are now developing in the FIRST Tonomura Project is expected to contribute to developments in a wide range of industrial fields ranging from fundamental physics to materials science and technology, and life sciences as indicated below:

(1) Ultra-high resolution quantum observation of materials will reveal behavior of atoms, lattice defects, vacancies, interstitials and others, not only in the semiconductors but in lithium-ion batteries, leading to the 3rd industrial revolution.

(2) Observation of local magnetic and electric fields in nanoscale devises[15,16] will accelerate the development of advanced technologies such as spintronics and quantum computers.

(3) Many key technologies developed in the project will promote the development of surrounding industries.

References

1. H. Watanabe, Experimental evidence for the collective nature of the characteristic energy loss of electrons in solids –Studies on the dispersion relation of plasma frequency–, *J. Phys. Soc. Japan* **11**(2), 112-119, (1956).
2. D. Bohm and D. Pines, A collective description of electron interactions: III Coulomb interactions in a degenerate electron gas, *Phys. Rev.* **92**(3), 609-625, (1953).
3. A. Tonomura, *Electron Holography*, 2nd edition (Springer-Verlag, Berlin, Heidelberg, 1999).
4. A. Tonomura, A. Fukuhara, H. Watanabe, and T. Komoda, Optical reconstruction of image from Fraunhofer electron-hologram, *Jpn. J. Appl. Phys.* **7**, 295, (1968).
5. A. Tonomura, T. Matsuda, J. Endo, H. Todokoro, and T. Komoda, Development of a field emission electron microscope, *J. Electron Microsc.* **28**(1), 1-11, (1979).
6. A. Tonomura, T. Matsuda, J. Endo, T. Arii, and K. Mihama, Direct observation of fine structure of magnetic domain walls by electron holography, *Phys. Rev. Lett.* **44**(21), 1430-1433, (1980).
7. A. Tonomura, N. Osakabe, T. Matsuda, T. Kawasaki, J. Endo, S. Yano, and H. Yamada, Evidence for Aharonov-Bohm effect with magnetic field completely shielded from electron wave, *Phys. Rev. Lett.* **56**(8), 792-795, (1986).
8. A. Tonomura, J. Endo, T. Matsuda, T. Kawasaki, and H. Ezawa, Demonstration of single-electron buildup of an interference pattern, *Am. J. Phys.* **57**(2), 117-120, (1989).
9. K. Harada, T. Matsuda, J. Bonevich, M. Igarashi, S. Kondo, G. Pozzi, U. Kawabe, and A. Tonomura, Real-time observation of vortex lattices in a superconductor by electron holography, *Nature* **360**, 51-53, (5 November 1992).
10. T. Matsuda, K. Harada, H. Kasai, O. Kamimura, and A. Tonomura, Observation of dynamic interaction of vortices with pinning centers by Lorentz microscopy, *Science* **271**, 1393-1395, (8 March 1996).
11. T. Kawasaki, T. Yoshida, T. Matsuda, N. Osakabe, A. Tonomura, I. Matsui, and K. Kitazawa, Fine crystal lattice fringes observed using a transmission electron microscope with 1 MeV coherent electron waves, *Appl. Phys. Lett.* **76**(10), 1342-1344, (2000).

12. A. Tonomura, H. Kasai, O. Kamimura, T. Matsuda, K. Harada, Y. Nakayama, J. Shimoyama, K. Kishio, T. Hanaguri, K. Kitazawa, M. Sasase, and S. Okayasu, Observation of individual vortices trapped along columnar defects in high-temperature superconductors, *Nature* **412**, 620-622, (9 August 2001).

13. T. Matsuda, O. Kamimura, H. Kasai, K. Harada, T. Yoshida, T. Akashi, A. Tonomura, Y. Nakayama, J. Shimoyama, K. Kishio, T. Hanaguri, and K. Kitazawa, Oscillating rows of vortices in superconductors, *Science* **294**, 2136-2138, (7 December, 2001).

14. A. Tonomura, Electron phase microscopy for observing superconductivity and magnetism, in *Proc. 9th International Symposium on Foundations of Quantum Mechanics in the Light of New Technology ISQM-Tokyo '08*, S. Ishioka and K. Fujikawa, Eds. (World Scientific, Singapore, 2009), pp. 301-306.

15. J. J. Kim, K. Hirata, Y. Ishida, D. Shindo, M. Takahashi, and A. Tonomura, Magnetic domain observation in writer pole tip for perpendicular recording head by electron holography, *Appl. Phys. Lett.* 92(16), 162501, (2008).

16. Y. Murakami, H. Kasai, J. J. Kim, S. Mamishin, D. Shindo, S. Mori, and A. Tonomura, Ferromagnetic domain nucleation and growth in colossal magnetoresistive manganite, *Nature Nanotechnology* 5(1), 37-41, (January 2010).

Biography of Akira Tonomura (April 1942 – May 2012)

Nobuyuki Osakabe

Central Research Laboratory, Hitachi, Ltd.
Kokubunji, Tokyo 185-8601, Japan
E-mail: nobuyuki.osakabe.hu@hitachi.com

Akira Tonomura was born on 25 April 1942 in Hyogo Prefecture, Japan. He graduated from the Department of Physics in the University of Tokyo in 1965 and immediately joined the Central Research Laboratory of Hitachi, Ltd. Tonomura was motivated to join the lab because of Hiroshi Watanabe,[1] who, with a single electron micrograph, proved the Bohm-Pines plasma oscillation theory[2] by using electron energy-loss spectroscopy. On the recommendation of Watanabe, Tonomura pursued research with electron beams to achieve holography, invented by Dennis Gabor[3] in 1948. Tonomura frequently communicated with Gabor, who encouraged him to realize aberration correction using electron holography, something that Gabor himself wanted to do, but he lacked a coherent electron source when he was doing his own research.

After demonstrating the basic principle of holography using a conventional electron microscope in 1968, Tonomura started developing an electron microscope with a field-emission electron source invented by Albert Crewe.[4] Tonomura intended the use of the source, which has brightness several orders of magnitude higher than that of a conventional thermionic-emission source, to generate a coherent electron wave. It was during that period (1973-1974) that Tonomura went to work as a research scientist with Gottfried Möllenstedt at the University of Tübingen in Germany; Möllenstedt had invented a device called an electron biprism[5] for interference experiments. The experience proved much inspiration for Tonomura's future work. After several years of technical development, in 1979 he achieved high coherence in his electron beams.[6] This brought holography electron microscopy into the practical realm, making it possible to observe phase information contained in the electron wave passing through specimens at a much higher resolution. Use of this phase information enabled measurement of minute magnetic fields[7] that cannot be directly observed. For this achievement, Tonomura received the Optics Paper Award from the Optical Society of Japan and the Setoh Prize from the Japanese Society of Electron Microscopy, both in 1980, and was commended as

Persons of Scientific and Technological Research Merit by the Minister of State for Science and Technology (Science and Technology Agency, Japan) in 1984.

Tonomura then went on to use electron holography to observe the paradoxical quantum world, and his 1986 verification[8-10] of the Aharonov-Bohm (AB) effect[11] is one of his singular achievements. Yakir Aharonov and David Bohm stated in 1959 that the vector potential is itself a fundamental physical entity and can affect a charged particle in a region where there is no magnetic field and therefore no force acting on the charged particle. The significance of the AB effect was renewed when Tai Tsun Wu and C. N. Yang[12] recognized the effect as an experimental manifestation of the nonintegrable phase factor; they introduced and generalized it to a non-Abelian gauge field to explain different interactions in a unified manner.

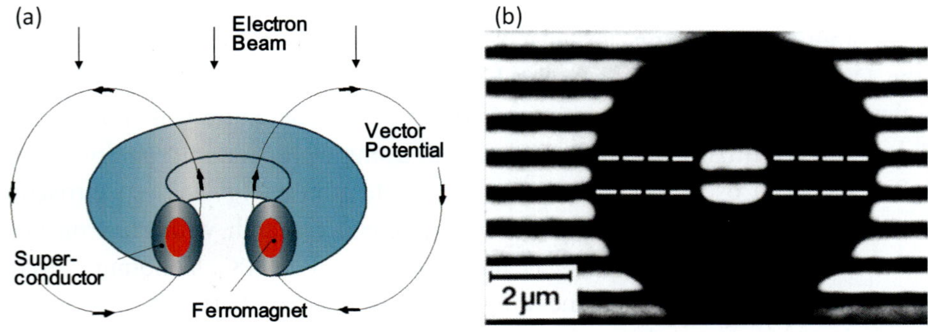

Fig. 1. Experimental verification of the Aharonov-Bohm (AB) effect. (a) Schematic diagram of the experimental device; (b) Interferogram showing the AB effect.

To verify the AB effect, Tonomura and his colleagues fabricated a tiny toroidal ferromagnet covered with a layer of superconducting niobium to perfectly shield the magnetic field (See Fig. 1a). They then measured a phase difference between the electrons that traveled through the central hole of the toroid and those outside it. Although the electrons had only progressed through regions free of electromagnetic fields, there was an observable effect produced by the existence of vector potentials (See Fig. 1b). For this achievement, Tonomura was awarded the Nishina Memorial Prize in 1982, the Asahi Prize in 1987, and the Japan Academy Prize and Imperial Prize in 1991.

To create a new trend of applying cutting-edge engineering to the various issues in quantum mechanics, an International Symposium on Foundations of Quantum Mechanics in the Light of New Technology (ISQM-Tokyo)[13] was established in 1983 at the Central Research Laboratory, Hitachi, Ltd., in Kokubunji, Tokyo, after Tonomura's first experiment to verify the AB effect.

Fig. 2. Poster of the First International Symposium on Foundations of Quantum Mechanics in the Light of New Technology (ISQM-Tokyo) with invited speakers included.

Fig. 3. Professor C. N. Yang and Professor Y. Aharonov discussing with Dr. A. Tonomura during the First ISQM-Tokyo (courtesy of Hitachi, Ltd.)

Many international conferences[14] with the same scope have since been held every three years at Hitachi Research Laboratories: Second to Fourth ISQMs were held at the Central Research Laboratory and Fifth to Ninth ISQMs were held at the Advanced Research Laboratory in Hatoyama, Saitama.

Tonomura's experiment[15] in 1989 showed single-electron buildups of electron wave interference fringe patterns. It revealed the dual nature of electrons and was referred to in a number of textbooks in quantum mechanics and even prominently featured in the first-year physics textbooks, e,g., in the classic "Fundamentals of Physics" by David Halliday and Robert Resnick. Furthermore this result was described by *Physics World* magazine[16] as the world's most beautiful physics experiment, ranking above the historical experiments of Galileo Galilei and Robert Millikan.

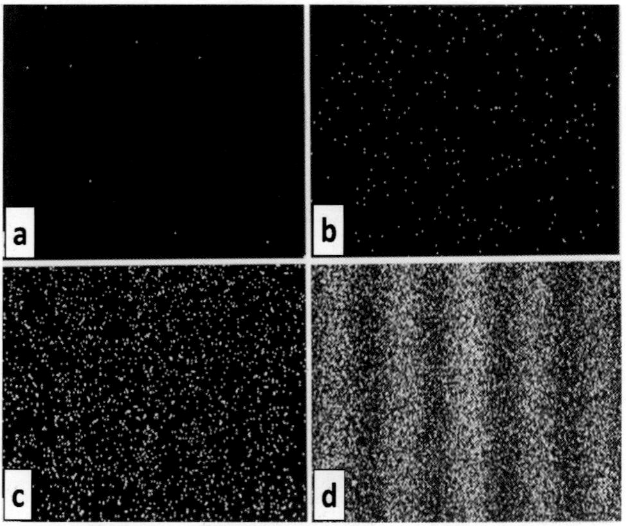

Fig. 4. Buildup of the electron interference pattern. (a) Number of electrons = 10; (b) Number of electrons = 100; (c) Number of electron = 3000; (d) Number of electrons = 70,000. (Ref. 15) (courtesy of American Association of Physics Teachers)

From 1989 to 1994, Akira Tonomura was the Research Director of ERATO Tonomura Electron Wavefront Project aiming at developing new methods of electron holography for real-time, high precision, and high-resolution measurement and applying them in a practical manner. Here ERATO stands for "Exploratory Research for Advanced Technology" sponsored by Research Development Corporation of Japan, which is now Japan Science and Technology Corporation. Research results include the following: Real-time electron holography[17] by using a liquid crystal panel, three-dimensional reconstruction[18,19] of electric potential and magnetic flux distribution, a phase-shifting method[20] for precise phase detection, and electron anti-bunching measurement[21] by developing a very fast electron counting technique.[22]

In 1992, Tonomura and his colleagues developed a method, a coherent-beam Lorentz microscopy,[23,24] to observe magnetic flux lines (vortices)[25] inside superconductors. Pinning down vortices inside superconductors is necessary in order to obtain practical superconducting magnets. For that purpose it is necessary to understand how vortices behave and how they are affected by changes in current, magnetic field, and temperature. Using a 350 kV holography electron microscope[26] installed at Hitachi Advanced Research Laboratory in Hatoyama, the researchers succeeded in achieving the first dynamic observations of vortex motion in metal superconductors.[23]

Fig. 5. Akira Tonomura with his colleagues in front of the 350-kV holography electron microscope installed at Hitachi Advanced Research Laboratory in 1990 (courtesy of Hitachi, Ltd.)

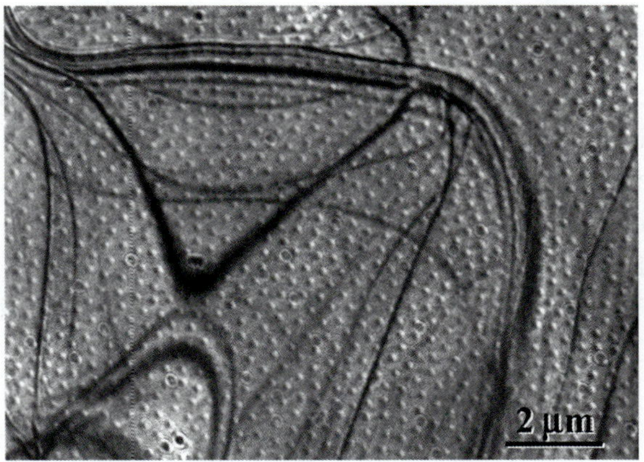

Fig. 6. Lorentz micrograph of Nb thin film. $B = 10$ mT; $T = 4.5$ K.

In 1993, they used the same technique to observe vortices and their dynamics in high-temperature superconducting thin films of $Bi_2Sr_2CaCu_2O_\chi$.[27] For these results Bishop wrote in his 1993 Nature article[28] entitled "Heroic holograms" as follows:

> They [Tonomura and his colleagues] are the first to generate real-space, real-time images of a melting magnetic flux-line lattice in a type-II superconductor. It is an experimental tour de force.

Many discoveries based on Lorentz microscopy observation followed: In 1996, dynamic interactions of vortices[29] with artificial pinning centers in the Nb thin film, called "Vortex Rivers," were reported. In 1997, they discovered vortex-antivortex pair annihilation[30] in the Nb thin film when the applied magnetic field was reversed.

Based on the achievements described above, in 1999 Akira Tonomura was awarded the Benjamin Franklin Medal in Physics from the Franklin Institute in the USA with the citation "for his contributions in the development of an electron beam and high-resolution microscopic devices."

In 2000, Tonomura and his colleagues succeeded in developing a 1-MV holography electron microscope[31,32] with a beam brightness of 2×10^{10} $Acm^{-2}str^{-1}$ and lattice resolution of 49.5 pm, both world records. Using this machine they clarified unconventional arrangements of vortices[33-38] in layered high-temperature superconductors $Bi_2Sr_2CaCu_2O_{8+\delta}$ and $YBa_2Cu_3O_{7-\delta}$. The development of this 1 MV electron microscope and its application to the observation of vortices in high-*Tc* superconductors were financially supported by Japan Science and Technology Agency through Core Research and for Evolutional Science and Technology (CREST) program (1995-2000) and Solution-Oriented Research for Science and Technology (SORST) program (2001-2004). The series of experiments he conducted generated a wealth of new information concerning the unconventional behavior of vortices in high-temperature superconductors, such as the mechanism of flux pinning and the artificial control of vortices using microfabricated channels.

Throughout his career, Tonomura worked in an industrial research laboratory. He also held concurrent appointments as the group director at RIKEN and principal investigator at Okinawa Institute of Science and Technology. At both institutions, he led national projects in electron holography aimed at a deeper understanding of quantum effects and explored field distributions in magnetic and superconducting matter. In 2010 he received a Japanese government grant of six billion yen (~ 50 million US dollars), budgeted over five years, to develop an atomic-resolution holography electron microscope. In this FIRST Tonomura Project,[39] where FIRST stands for "Funding Program for World-Leading Innovative R&D on Science and Technology," Tonomura aimed at building a 1.2 MV holography electron microscope that would enable him to realize his dream of reconstructing the wave fronts of electrons scattered by atoms, first proposed by Gabor in the 1940s. However, he passed away before he could fulfill his dream.

Among many recognitions received in Japan, Tonomura was awarded with the Cultural Merit Award from the Ministry of Education, Culture, Sports, Science and Technology in 2002 and with the Order of the Rising Sun, Gold and Silver Star from the Japanese government in 2012. He was elected as a Member of the Science Council of Japan in 2005, and the Japan Academy in 2007. Overseas, he was honored to serve as Foreign Associate of the National Academy of Sciences in the United States of America since 2000, of Royal Society of Sciences in Uppsala since 2003, and of the Royal Swedish Academy of Engineering Sciences since 2006. He was conferred with a Laurea Honoris Causa from the University of Camerino, Italy, in 2005, and he was elected as a Fellow of the American Physical Society in 1999, of the American Association for the Advancement of Science in 2007, and of the Institute of Physics in 2011.

One notable attribute of Tonomura was his strong will to complete his research. In his experiment[8] to verify the AB effect in 1982, many other researchers pointed out problems. But with advice from Yang, he conceived of the experiment using the toroidal ferromagnet.[9] Despite its being a difficult experiment, Tonomura allowed no compromise, and he finally succeeded. I remember him saying a scientist should aim for a research target so high that if he cannot succeed, then he can confidently say, "No one else in the world can." In just that way, he was unrelenting in his research, but to the people around him, he was a kind and thoughtful man. Officially and personally, he looked after researchers to ensure they had a good research environment.

Other quotable comments of Tonomura on R&D are listed below:

"You have to develop new equipment when you attack a new problem."

"You must have 'new ideas' and 'persistence' to attain your goal."

"Finding the cause and solving riddles are indeed the real thrill of R&D."

"We started with a strong determination that 'We will try everything and anything to prove the Aharonov-Bohm effect.

"Technical difficulty is no excuse."

"Research results must be 'the first in the world"—imitations have no value.

When he was first diagnosed in the spring of 2011, even as he sought the best treatment for his illness, Tonomura continued to think about his research and maintained his pride as a researcher until the end. Tonomura passed away in Hidaka, Saitama, Japan, on 2 May 2012 after his fight against pancreatic cancer.

I am deeply saddened for the premature departure of such a warm and kind colleague and such a great scientist. I have the deepest respect for the man and his many achievements.

Fig. 7. Akira Tonomura in front of his holography electron microscope at the Advanced Research Laboratory, Hitachi, Ltd., in Hatoyama, Saitama, taken on January 27, 2010 (courtesy of RIKEN)

References

1. H. Watanabe, Experimental evidence for the collective nature of the characteristic energy loss of electrons in solids –Studies on the dispersion relation of plasma frequency–, *J. Phys. Soc. Japan* **11**(2), 112-119, (1956).
2. D. Bohm and D. Pines, A collective description of electron interactions: III Coulomb interactions in a degenerate electron gas, *Phys. Rev.* **92**(3), 609-625, (1953).
3. D. Gabor, A new microscopic principle, *Nature* **161**(4098), 777-778, (1948).
4. A. V. Crewe, D. N. Eggenberger, J. Wall, and L. M. Welter, Electron gun using field emission source, *Rev. Sci. Instr.* **39**(4), 576–583, (1968); A. V. Crewe, J. Wall, and L. M. Welter, A high-resolution scanning transmission electron microscope, *J. Appl. Phys.* **39**(13), 5861–5868, (1968).
5. G. Möllenstedt and H. Düker, Beobachtungen und Messungen an Biprisma-Interferenzen mit Elektronenwellen, *Z. Phys.* **145**, 377-397, (1956).
6. A. Tonomura, T. Matsuda, J. Endo, H. Todokoro, and T. Komoda, Development of a field emission electron microscope, *J. Electron Microsc.* **28**(1), 1-11, (1979).
7. A. Tonomura, T. Matsuda, J. Endo, T. Arii, and K. Mihama, Direct observation of fine structure of magnetic domain walls by electron holography, *Phys. Rev. Lett.* **44**(21), 1430-1433, (1980).
8. A. Tonomura, T. Matsuda, R. Suzuki, A. Fukuhara, N. Osakabe, H. Umezaki, J. Endo, K. Shinagawa, Y. Sugita and H. Fujiwara, Observation of Aharonov-Bohm effect by electron holography, *Phys. Rev. Lett.* **48**(21), 1443-1446, (1982).
9. A. Tonomura, N. Osakabe, T. Matsuda, T. Kawasaki, J. Endo, S. Yano and H. Yamada, Evidence for Aharonov-Bohm effect with magnetic field completely shielded from electron wave, *Phys. Rev. Lett.* **56**(8), 792-795, (1986).

10. N. Osakabe, T. Matsuda, T. Kawasaki, J. Endo, A. Tonomura, S. Yano, and H. Yamada, Experimental confirmation of the Aharonov-Bohm effect using a toroidal magnetic field confined by a superconductor, *Phys. Rev. A* **34**(2), 815-822, (1986).

11. Y. Aharonov and D. Bohm, Significance of electromagnetic potentials in the quantum theory, *Phys. Rev.* **115**(3), 485-491, (1959).

12. T. T. Wu and C. N. Yang, Concept of nonintegrable phase factors and global formation of gauge fields, *Phys. Rev. D* **12**(12), 3845-3857, (1975).

13. *Proceedings of the International Symposium—Foundations of Quantum Mechanics in the Light of New Technology*, Eds. S. Kamefuchi, H. Ezawa, Y. Murayama, M. Namiki, S. Nomura, Y. Ohnuki, and T. Yajima, Tokyo, Japan (August 1983) (Physical Society of Japan, Tokyo, 1984).

14. See, for example, *Foundations of Quantum Mechanics in the Light of New Technology—Selected Papers from the Proceedings of the First through Fourth International Symposia on Foundations of Quantum Mechanics*. Eds. S. Nakajima, Y. Murayama, and A. Tonomura (World Scientific, Singapore, 1996).

15. A. Tonomura, J. Endo, T. Matsuda, T. Kawasaki, and H. Ezawa, Demonstration of single-electron buildup of interference pattern, *Amer. J. Phys.* **57**(2), 117-120, (1989).

16. R. P. Crease, The most beautiful experiment, *Physics World* September 2002.

17. J. Chen, T. Hirayama, G. Lai, T. Tanji, K. Ishizuka, and A. Tonomura, Real-time electron-holographic interference microscopy with a liquid-crystal spatial light modulator, *Opt. Lett.* **18**(22), 1887-1889, (1993)

18. G. Lai, T. Hirayama, K. Ishizuka, T. Tanji, and A. Tonomura, Three-dimensional reconstruction of electric-potential distribution in electron-holographic interferometry, *Appl. Opt.* **33**(5), 829-833, (1994).

19. G. Lai, T. Hirayama, A. Fukuhara, K. Ishizuka, T. Tanji, and A. Tonomura, Three-dimensional reconstruction of magnetic vector fields using electron-holographic interferometry, *J. Appl. Phys.* **75**(9), 4593-4598, (1994).

20. Q. Ru, G. Lai, K. Aoyama, J. Endo, and A. Tonomura, Principle and application of phase-shifting electron holography, *Ultramicroscopy* **55**(2), 209-220, (1994).

21. N. Osakabe, T. Kodama, J. Endo, A. Tonomura, K. Ohbayashi, T. Urakami, H. Tsuchiya, and Y. Tsuchiya, Fast and precise electron counting system for the observation of quantum mechanical electron intensity correlation, *Nucl. Instrum. & Methods* **A365**, 585-587, (1995).

22. S. Saito, J. Endo, T. Kodama, A. Tonomura, A. Fukuhara, and K. Ohbayashi, Electron counting theory, *Phys. Lett.* **A162**, 442-448, (1992).

23. K. Harada, T. Matsuda, J. Bonevich, M. Igarashi, S. Kondo, G. Pozzi, U. Kawabe, and A. Tonomura, Real-time observation of vortex lattices in a superconductor by electron microscopy, *Nature* **360**(6399), 51-53, (1992).

24. For details of Tonomura's electron microscopy experiments on vortex motions in superconductors, see K. Harada, N. Osakabe, and Y. A. Ono, Electron microscopy study on magnetic flux lines in superconductors: Memorial to Akira Tonomura, *IEEE Trans. Appl. Superconductivity* **20**(1), 8000507, (2013).

25. A. A. Abrikosov, On the magnetic properties of superconductors of the second group, *Soviet Phys. JETP* **5**(6), 1174–1182, (1957).

26. T. Kawasaki, J. Endo, T. Matsuda, and A. Tonomura, Development and application of a 350 kV transmission electron microscope with a magnetic field superimposed field emission gun, *Microbeam Analysis* **3**, 287-291, (1994).

27. K. Harada, T. Matsuda, H. Kasai, J. E. Bonevich, T. Yoshida, and A. Tonomura, Vortex configuration and dynamics in $Bi_2Sr_{1.8}CaCu_2O_z$ thin film by Lorentz microscopy, *Phys. Rev. Lett.* **71**(20), 3371-3374, (1993).

28. D. J. Bishop, Heroic holograms, *Nature* **366**(6452), 209, (1993).

29. T. Matsuda, K. Harada, H. Kasai, O. Kamimura, and A. Tonomura, Observation of dynamic interaction of vortices with pinning centers by Lorentz microscopy, *Science* **271**(5254), 1393-1395, (1996).

30. K. Harada, H. Kasai, T. Matsuda, M. Yamasaki, and A. Tonomura, Direct observation of interaction of vortices and antivortices in a superconductor by Lorentz microscopy, *J. Electron Microsc.* **46**(3), 227-232, (1997).

31. T. Kawasaki, I. Matsui, T. Yoshida, T. Katsura, S. Hayashi, T. Onai, T. Furutsu, K. Myochinm, M. Numata, H. Mogaki, M. Gorai, T. Akashi, O. Kamimura, T. Matsuda, N. Osakabe, A. Tonomura, and K. Kitazawa, Development of a 1MV field-emission transmission electron microscope, *J. Electron Microsc.* **49**(6), 711-718, (2000).

32. K. Kawasaki, T. Yoshida, T. Matsuda, N. Osakabe, A. Tonomura, I. Matsui, and K. Kitazawa, Fine crystal lattice fringes observed using a transmission electron microscope with 1 MeV coherent electron waves, *Appl. Phys. Lett.* **76**(10), 1342-1444, (2000).

33. A. Tonomura, H. Kasai, O. Kamimura, T. Matsuda, K. Harada, Y. Nakayama, J. Shimoyama, K. Kishio, T. Hanaguri, K. Kitazawa, M. Sasase, and S. Okayasu, Observation of individual vortices trapped along columnar defects in high-temperature superconductors, *Nature* **412**(6847), 620-622, (2001).

34. A. Tonomura, H. Kasai, K. Kishio, T. Hanaguri, K. Kitazawa, M. Sasase, and s. Okayasu, Lorentz microscopy observation of vortices inside Bi-2212 thin film with columnar defects, *Physica C* **369**(1-4), 668-676, (2002)

35. T. Matsuda, O. Kamimura, H. Kasai, K. Harada, T. Yoshida, T. Asahi, A. Tonomura, Y. Nakayama, J. Shimoyama, K. Kishio, T. Hanaguri, and K. Kitazawa, Oscillating rows of vortices in superconductors, *Science* **294**(5551), 2136–2138, (2001).

36. A. Tonomura, H. Kasai, O. Kamimura, T. Matsuda, K. Harada, T. Yoshida, T. Akashi, J. Shimoyama, K. Kishio, T. Hanaguri, K. Kitazawa, T. Masui, S. Tajima, N. Koshizuka, P. L. Gammel, D. Bishop, M. Sasase and S. Okayasu, Observation of structures of chain vortices inside anisotropic high-*Tc* superconductors, *Phys. Rev. Lett.* **88**(23), 237001, (2002).

37. A. Tonomura, Direct observation of thitherto unobservable quantum phenomena by using electrons, *Proc. Natl. Acad. Sci. USA* **102**(42), 14952-14959, (2005).

38. A. Tonomura, Electron phase microscopy for observing superconductivity and magnetism, in *Proceedings of the 9th International Symposium on Foundations of Quantum Mechanics in the Light of New Technology—ISQM-Tokyo '08*, S. Ishioka and K. Fujikawa (Eds.), (World Scientific, Singapore, 2009), pp. 301–306.

39. See for example, D. Cyranoski, Japan rolls out elite science funds FIRST scheme targets large grants to world-leading researchers, *Nature* **464**(7291), 966-967, (2010).

Tonomura FIRST International Symposium on "Electron Microscopy and Gauge Fields"

Yoshimasa A. Ono

FIRST Tonomura Project, Japan Science and Technology Agency
c/o Hitachi Central Research Laboratory, Hatoyama, Saitama 350-0395, Japan
E-mail: yoshimasa.a.ono@gauge.jst.go.jp

1. Introduction

On May 9 and May 10, 2012. Tonomura FIRST International Symposium on "Electron Microscopy and Gauge Fields" was held in downtown Tokyo hosted by Japan Science and Technology Agency together with Hitachi, Ltd., RIKEN, Ministry of Education, Culture, Sports, Science and Technology, Okinawa Institute of Science and Technology School Corporation, The Asahi Shimbun Company, Nikkei Inc. and Nikkei Science Inc. Here FIRST stands for "Funding program for world-leading innovative research and development on science and technology." In 2010 as principal investigator Dr. Akira Tonomura was awarded this Japanese-government-funded project aiming at developing a 1.2 MV holography electron microscope capable of observing quantum phenomena in the microscopic world. The symposium was also intended to celebrate his 70th birthday, but on May 2, just a week before the opening of the symposium, Tonomura passed away during the course of a medical treatment on pancreatic cancer. Consequently, what had been planned to be a grand meeting with many of his scientific friends from all over the world, then turned as a memorial conference.

Fig. 1. Symposium participants gathered around a photograph of the late Akira Tonomura.

From all over the world 145 scientists attended the Symposium including 33 invited speakers (11 from abroad and 22 from Japan), 10 poster presenters, and 102 attendees from various fields. Invited speakers included distinguished physicists such as Chen Ning Yang (Nobel prize in physics 1957), Anthony J. Leggett (Nobel prize in physics 2003), Makoto Kobayashi (Nobel prize in physics 2008), and Yakir Aharonov (Wolf prize 1998).The photograph above shows the participants gathered around a photograph of Akira Tonomura.

2. Presentations on Day 1 (May 9, 2012)

The Symposium started with Tonomura's video letter recorded just before his death, where he said the following:

> "I am not certain whether I will be able to participate in the entire symposium, but I'm planning to attend the first part at the very least. I wish to show my true enthusiasm for the project. I hope to see you then. Good-bye."

In addition, in the Abstract booklet, Tonomura presented the following message in the "Greeting from Akira Tonomura" section:

> Welcome everyone. Thank you all for traveling here on such short notice to participate in the Tonomura FIRST International Symposium on "Electron Microscopy and Gauge Fields". I cannot express how much I appreciate your attendance here today.

> About a year ago, I was diagnosed with pancreatic cancer and I underwent a major operation that involved removing half my pancreas, nine-tenths of my spleen, and my entire stomach. Your warm, encouraging and supportive words have contributed immensely to my recovery from the operation. Today I am extremely delighted to be among many dear friends from all over the world whom I once thought I would never have the chance to see again.

> The FIRST Tonomura Project aims at developing a 1.2 MV holography electron microscope capable of observing quantum phenomena in the microscopic world. This project is a compilation of 48 years of my personal research on electron microscopy. It is my hope that you will continue to support the Tonomura Project and its success. I trust you'll enjoy the symposium. Thank you all again!

The symposium covered two important scientific and technological areas where Tonomura worked for four decades: electron microscopy and quantum physics. The first day (May 9) covered quantum physics related topics and the second day (May 10) covered electron microscopy related topics.

Prof. Yang of Tsinghua University and Chinese University of Hong Kong made the keynote address entitled "Topology and gauge transformation in physics." He discussed how topology, a mathematical concept, has progressed physics, in particular quantum physics, and emphasized the impact of the Aharonov-Bohm (AB) effect and its experimental verification by Tonomura.

> This experiment of Tonomura and the company showed, first, the topological meaning of the vector potential in electromagnetism. Second, it showed that fluxes in superconducting rings are quantized in

a unit of *ch/2e*, indicating the existence of Cooper pairs. This beautiful work is the first direct experimental proof of the importance of topology in physics.

Fig. 2. Professor Yang making presentation.

Prof. Aharonov of Tel-Aviv University and Chapman University discussed historical recollection of the development of the AB effect from his graduate student days in Bristol to the developments of its applications. Regarding Tonomura's experimental confirmation of the AB effect, he mentioned an interesting anecdote.

"I think his experiment to find my effect, i.e., the AB effect, was a marvelous one. I must say that when the first time Tonomura told me that he was going to try to do the AB effect with a completely closed flux inside the superconductor tube, I did not tell him that's not possible because I did not want to discourage him. I was sure that it couldn't be done because you needed to make a very, very small tube and to make sure that the beam of the electrons would still be coherent around it. But he was successful and it was an amazing experiment. I think it is one of the most beautiful and important experiments in modern physics."

Fig. 3. Professor Aharonov, Professor Leggett, and Professor Fujikawa.

Presentations on topology in physics include the following:
- "Majorana fermions and their possible occurrence of half-quantum vortices in helium-3 and strontium ruthenate" by Prof. Leggett of University of Illinois
- "Spin textures and gauge field in frustrated magnets" by Prof. Naoto Nagaosa of the University of Tokyo and RIKEN
- "Topological aspects in Bose-Einstein condensation" by Prof. Masahito Ueda of the University of Tokyo
- "Second quantization, hidden gauge symmetry, and geometric phases" by Dr. Kazuo Fujikawa of RIKEN

Presentations on quantum-physics related fields include the following:
- "Hanbury Brown-Twiss interferometry of charged particles: Coulomb vs. quantum statistics" by Prof. Gordon Baym of University of Illinois
- "A new approach to quantum interference" by Prof. Jeff Tollaksen of Chapman University
- Electron waves carrying orbital angular momentum" by Dr. Franco Nori of RIKEN

Presentations on the application of electron microscopy to physics include the following:
- Basic discoveries with the electron microscope" by Prof. Daisuke Shindo of Tohoku University
- "How the research to verify the Aharonov-Bohm effect was initiated at Hitachi" by Dr. Nobuyuki Osakabe of Hitachi, Ltd.

3. Presentations on Day 2 (May 10, 2012)

The second day (May 10) was devoted to discussion on electron microscopy.
Presentations on challenges in developing electron microscopes include the following:
- "Development status of the 1.2 MV holography electron microscope of the FIRST Tonomura Project" by Dr. Hiroyuki Shinada of Hitachi, Ltd. (in place of Tonomura)
- "Cs correction for the high-voltage electron microscope" by Dr. Maximillian Haider of CEOS.
- "Ultimate resolution in the electron microscopy" by Prof. David Smith of Arizona State University
- "Opportunities for chromatic aberration corrected high-resolution and Lorentz transmission electron microscopy" by Prof. Dr. Rafal Dunin-Borkowski of Peter Grünberg Institute Research Center Jülich

Presentations on application of electron microscope to physics include the following:
- "Carbon nanotube and electron microscopy" by Prof. Sumio Iijima of Meijo University and Dr. Kazutomo Suenaga of the National Institute of Advanced Industrial Science and Technology (AIST)

- "Visualizing magnetic monopoles, Dirac strings, and the AB effect" by Dr. Yimei Zhu of Brookhaven National Laboratory
- "Mapping atoms, electrons and fields with a tiny electron beam" by Dr. Christian Colliex of CNRS
- "Atomic-scale observation of inverse charge transfer at an oxide interface" by Prof. Cheng-Hsuan Chen of National Taiwan University.

4. Comments on Akira Tonomura and his research

During the Symposium we interviewed Professor C. N. Yang, Professor A. J. Leggett, and Professor Y. Aharonov on their impressions on Tonomura and his research. The following are the excerpts from these interviews.

Professor C. N. Yang:

In the early 1980s when I was visiting the University of Tokyo, Tonomura called me and he asked me whether I would like to visit his laboratory The next day I visited the Hitachi Central Research Laboratory, where he asked me whether it was useful and important to do experiment on the Aharonov-Bohm effect because he had some new ideas about how to do the that difficult experiment. So that started our long association of more than 30 years.

Tonomura is one of the most important, distinguished experimental physicists today. His contributions are in two directions. From the experimental side early in his career he produced the best coherent electron source, with that he became a leader in the science of electron holography and that of course continued to be his specialty and continued to be his main effort in sciences including the FIRST Tonomura Project, in which he has been engaged deeply in the last few years.

On the other hand, he also made first-class contributions to theoretical physics. That came about because of his experiment on the Aharonov-Bohm effect. The Aharonov-Bohm effect is important because it clarified some deep meaning of electromagnetism when quantum mechanics was discovered. Before quantum mechanics was discovered, electromagnetism was already understood through Maxwell's equations. But with the discovery of quantum mechanics, we now understand electromagnetism in a deeper way. This deeper way resides in two aspects, two symmetries of electromagnetism. The first symmetry is the Lorentz symmetry, which was the work of Einstein in special relativity. Before Einstein, Maxwell's equations already had that symmetry, but people did not realize it. And Einstein who pointed out that Maxwell's equations had that symmetry. The second symmetry was more subtle. It was pointed out mostly through the work of great mathematician Herman Wyle and is called gauge symmetry. With these two symmetries electromagnetism in quantum mechanics acquired certain characteristics which were not present in classical electromagnetism and one of the first important aspects of this characteristic was the Aharonov-Bohm experiment. The experiment was pointed out, or the fact was pointed out in 1969, but the experiment to do it was extremely difficult. This is because you need a very small specimen and you need to have very accurate measurements of magnetic fields. And so for many years starting from 1959 to the early 1980s there were only controversial, so called, verification of the Aharonov-Bohm effect.

Tonomura in the early 80s realized that with his electron holograms, he will be able to make fine experiments which would verify the effect. And that is how he got interested in the Aharonov-Bohm effect and he called me. So we have had lots of physics discussions after that. And I was very happy that in 1983 and then again in 1986 he had those beautiful and definitive experiments on the Aharonov-Bohm effect. In fact it's quantitative, not just qualitative, experiments. After that he organized a series of conferences in Hitachi called "(International Symposium on) Quantum Mechanics in the Light of New Technology". And I came to these conferences almost all of them.

After the discovery of high-temperature superconductivity in 1986, Tonomura became very much interested in magnetic flux movements in superconductors because that was deeply related to the existence of such high–temperature superconductors and he did beautiful experiments in that area.

I was very happy in recent years to hear that he has been chosen by the Japanese Government to lead this next big project. And everybody who had known him knew that in time after his project is finished, the world would have more powerful and more resolution clarification of atomic phenomenon. It was a great shock to Japan, to the science of physics, and to me a great personal loss when I learned that he became critically ill about a two months ago. I still hoped that I would be able to see him at this conference but unfortunately that was not to be. But I know that his project would very likely continue and I expect that the project will eventually be successful so as to fulfill his vision and his dream.

Professor A. J. Leggett:

I should say that I've mostly been an admirer from a certain distance from Tonomura's own work. I have always felt he was right at the very forefront field of electron holography and some of the advances he made are totally unique in the world. And in particular, I think both his observation of the Aharonov-Bohm effect, and essentially, what should I say, in foolproof circumstances. And that also his imaging of vortex, vortices, and vortex motions in superconductors has been totally unique and major contributions to the field.

I think FIRST Tonomura Project is a very imaginative and worthwhile project. Dr. Tonomura already showed that he could push the boundaries of what we could have done with the electron holography by orders of magnitude. Now, as I understand it, you would like to continue his work and take it to at least one order of magnitude further so that one will be able to image atoms and atomic motion on a genuinely sub-atomic scale. And this will again certainly be the world first, I think, and undoubtedly extremely worthwhile.

Professor Y. Aharonov:

First of all I must say that I was extremely sad that he was not even able to participate in this conference in his own, although I was hoping very much that he will be able to live it. In fact, I was devastated when I heard about his sickness some months ago. That's very sad because I had always admired Tonomura from the first time I met him nearly 30 years ago in the first conference I came here. And I always admired his passion for physics. He was so deep into trying to do things that other people cannot do. And I think he has marvelous achievement. I think his experiment to find my effect, the Aharonov-Bohm effect, was a marvelous experiment. In fact, when he told me first that he was going to do it, I did not want to discourage him that it couldn't be done, because it was such a complicated thing to do, to be able to really enclose the magnetic flux this way. But he was successful. I consider this to be one of the most important existing experiments in modern physics.

And I think that his achievements in building the best electron microscope in the world so far are marvelous and going to be certain.... I hope that his project will continue because it is very important. I think that if the project will be finished, we will have the best electron microscope in the world that could see things we hoped to see so far we cannot see.

Note: This article is a revised version of the article originally published in Asia Pacific Newsletter, Vol. 2, No. 1, pp. 9-11 (World Scientific, Singapore, 2013).

Tonomura FIRST
International Symposium on
Electron Microscopy
and Gauge Fields
May 9 Wed - 10 Thu, 2012
KEIO PLAZA HOTEL
TOKYO
C.N.Yang
1957 Nobel Laureate
A.J. Leggett
2003 Nobel Laureate
Y. Aharonov
1998 Wolf Prize
Register To Attend
KEIO PLAZA HOTEL
http://www.keioplaza.co.jp/
〒160-8330
東京都新宿区西新宿2-2-1
TEL 03-3344-0111
FAX 03-3345-8269
Sponsors
Japan Science and Technology Agency
Hitachi, Ltd.
Supported by
RIKEN
Ministry of Education, Culture, Sports, Science
and Technology
Okinawa Institute of Science and Technology
School Corporation
The Asahi Shimbun Company
Nikkei Inc.
Nikkei Science Inc.
Inquiry
tonomura-sympo@gauge.jst.go.jp
http://www.first-tonomura-pj.net/symposium/index.html

Greetings from Akira Tonomura

Welcome everyone. Thank you all for traveling here on such short notice to participate in the Tonomura FIRST international symposium on "Electron Microscopy and Gauge Fields". I cannot express how much I appreciate your attendance here today.

About a year ago, I was diagnosed with pancreatic cancer and I underwent a major operation that involved removing half my pancreas, nine-tenths of my spleen, and my entire stomach. Your warm, encouraging and supportive words have contributed immensely to my recovery from the operation. Today I am extremely delighted to be among many dear friends from all over the world whom I once thought I would never have the chance to see again.

The FIRST Tonomura Project aims at developing a 1.2 MV holography electron microscope capable of observing quantum phenomena in the microscopic world. This project is a compilation of 48 years of my personal research on electron microscopy. It is my hope that you will continue to support the Tonomura Project and its success. I trust you' ll enjoy the symposium. Thank you all again!

Akira Tonomura

外村 彰

(Picture: courtesy of RIKEN)

Thank You and Farewell to Tonomura-kun

Hidetoshi Fukuyama

Department of Applied Physics, Tokyo University of Science
1-3, Kagurazaka, Shinjuku-ku, Tokyo 162-8601, Japan
E-mail: fukuyama@rs.kagu.tus.ac.jp

"Science triggers technology; technology inspires science" is true, so is your case, when you proved partially[1] in 1982 and completely[2,3] in 1986 the Aharonov-Bohm(AB) effect, the principal manifestation of gauge field, based on the state-of-the-art electron microscope that you developed. You put "period," i.e., an end, on the controversy associated with the AB effect.

This success led to series of ISQM-Tokyo Symposia[4] (International Symposium on Foundations of Quantum Mechanics in the Light of New Technology). The first Symposium of this series was held in 1983 at Hitachi Central Research Laboratory (HCRL) in Kokubunji, Tokyo (See Fig. 1), and the latest (9th) Symposium in 2008 at Hitachi Advanced Research Laboratory (HARL), Hatoyama, Saitama.

Fig. 1. Akira Tonomura explaining his electron microscope to Chen Ning Yang in his lab. at Hitachi Central Research Laboratory during the First ISQM-Tokyo in August 1983. (Courtesy of Hitachi, Ltd.)

The scientific topics selected for this ISQM series were unique; they have touched every aspect of fundamental phenomena in materials ranging over several disciplines where consequences of quantum mechanics are disclosed thanks to the newly introduced experimental techniques. This has been the spirit of ISQM, which was created by Professor Sadao Nakajima, then the Director of the Institute for Solid State Physics, the University of Tokyo, and Dr. Yasutsugu Takeda, then General Manager of HCRL, for making bridges between basic science and technology. Up to the 5th Symposium in 1995, Professor Nakajima chaired the Organizing Committee, and from the 6th Symposium in 1998 I was in charge. This notion of importance of constructive and creative relationship between basic science and technology was not widely accepted in academia in the early 1980s, but is now fully appreciated in various sectors of science associated with materials.

After proving the AB effect, you demonstrated the wave-particle duality of electrons[5] through the double-slit experiment, which is so beautiful and convincing (See Fig. 2). Whenever I have to teach Quantum Mechanics in the class, I show the video of this experiment at the beginning of the lecture, one-minute version of which you kindly made for me for the public lecture I made in 2003 at the Yasuda Auditorium of the University of Tokyo. I remember that this was chosen as the best in the history of physics experiments. It is really beautiful and appealing to general audience with its clarity.

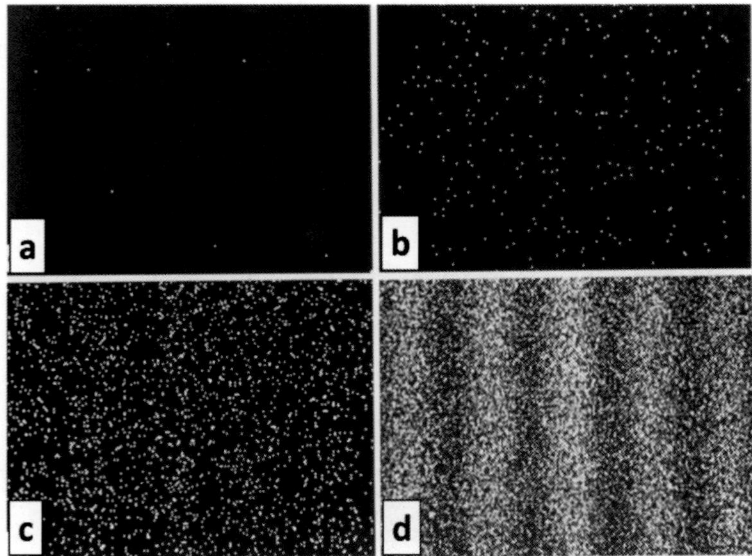

Fig. 2. Buildup of the electron interference pattern. (a) Number of electrons = 10; (b) Number of electrons = 100; (c) Number of electron = 3000; (d) Number of electrons = 70,000. (From Ref. 5) (courtesy of American Association of Physics Teachers)

You were hardworking as ever and jolly all the time. You lived close to the HARL, your working place and the site of the latter parts of ISQM series. In the premises of the HARL there is a new building where a 1.2 MeV holography electron microscope, your dream, is being built with which you planned to investigate further into the fine details of quantum mechanical processes in materials. You must have been very anxious to unveil many more mysteries through very sharp eyes of yours, as always. In this context, I wanted to discuss with you effects of gauge fields in solids, not in vacuum. In solids electron waves are characterized by Bloch energy bands. Once external magnetic field is present, motion of electrons is not confined within a particular Bloch band since the vector potential has finite matrix elements between different Bloch bands, resulting in "inter-band effects," which are the sources of all complications as well as of rich physics. This subject of "inter-band effects of magnetic field" was one of the most important scientific problems Professor Ryogo Kubo, our common mentor, was addressing in the 1960s. Possible important consequences of the inter-band effects of vector potential are expected for orbital magnetism and Hall effect[6]. To treat the effect theoretically, the Bloch representation does not work at all; instead, the representation introduced by Luttinger and Kohn[7] in 1955 is most suited since the way vector potential comes into play is clear and transparent and then gauge invariance is easily seen to be maintained.[8,9]

In 1969 I found that the anomalously large diamagnetism of bismuth (Bi), a long-standing mystery[10] since the early 1930s, was due to this inter-band effect of vector potential.[11] (Incidentally, electronic states of Bi near the Fermi energy are similar to 4×4 Dirac electrons.) Though this theory explains the experiment convincingly, the result is not trivial; diamagnetism takes the largest value when the chemical potential is located in the band gap, i.e., in the insulating state. Then, a natural question arises: What is happening in real space? Where is magnetic moment and where does the diamagnetic current flow?

I remember that one day soon after I became a postdoc at Harvard in 1971 I was invited to the office of Professor Clifford Shull at MIT who showed me a large sheet of graph papers pasted together on which many data points of neutron scattering are marked, and he said to me that he wanted to see where the diamagnetic current flows in the Bi crystal.

Forty years has passed since then, and this problem offers even more interesting questions now in the context of mutual relationship between orbital magnetism and Hall effects[12] and spin-Hall effects[13] in Bi. (It so turned out that in the case of massless Dirac electrons, such as graphenes[14] and molecular solids,[15] the interband effect plays crucial roles on Hall effect, especially when the chemical potential is located at the crossing point.[16]) I wanted to ask you the question which I am sure that you could answer with your fine techniques and sharp insight: where and how the diamagnetic currents flow in Bi? But it is too late.

We first met in the spring of 1963 at the age of 20 in the Department of Physics of the University of Tokyo. We were among 30 in the class, including Kazuo Fujikawa and Michiharu Nakamura. You were keen to perform experiments and were quite often absorbed in watching an oscilloscope as in the photo (Fig. 3) taken around that time. (You might remember that I was very unintentionally in charge of taking photos of my classmates for the record album of the class. But now this experience has turned out to be precious.)

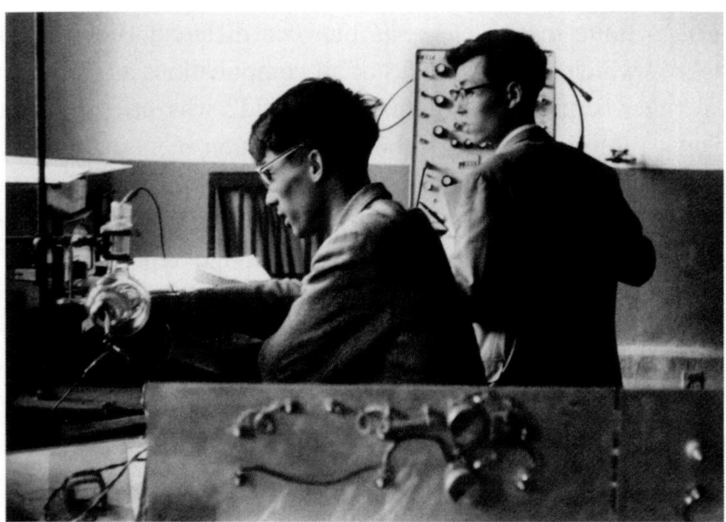

Fig. 3. Tonomura (left) in a student experimental laboratory in 1963. (Photo taken by H. Fukuyama)

At the stage when we had to decide which laboratory to do graduate study in the Department of Physics, everyone else was discussing such and such, so and so, but you were totally apart from these and were clearly determined not to go to the graduate school. Years later I learned that you had already determined at that time to study electron microscopy at Hitachi. And you did it and pursued it to the limit, and disclosed the beauty of nature. You worked very hard. Now it is time for you to take a peaceful rest.

I feel that I am very lucky to have met you in my life. I thank you and good-bye.

Acknowledgment

I acknowledge informative advice from Dr. Yoshimasa A. Ono for the preparation of the manuscript.

References

1. A. Tonomura, T. Matsuda, R. Suzuki, A. Fukuhara, N. Osakabe, H. Umezaki, J. Endo, K. Shinagawa, Y. Sugita, and H. Fujiwara, Observation of Aharonov-Bohm effect by electron holography, *Phys. Rev. Lett.* **48**(21), 1443-1446, (1982).
2. A. Tonomura, N. Osakabe, T. Matsuda, T. Kawasaki, J. Endo, S. Yano, and H. Yamada, Evidence for the Aharonov-Bohm effect with magnetic field completely shielded from electron wave, *Phys. Rev. Lett.* **56**(8), 792-795, (1986).
3. N. Osakabe, T. Matsuda, T. Kawasaki, J. Endo, A. Tonomura, S. Yano, and H. Yamada, Experimental confirmation of the Aharonov-Bohm effect using a toroidal magnetic field confined by a superconductor, *Phys. Rev. A* **34**(2), 815-822, (1986).
4. See for example, S. Nakajima, Y. Murayama, and A. Tonomura (Eds.), *Foundations of Quantum Mechanics in the Light of New Technology – Selected papers from the Proceedings of the First through Fourth International Symposia on Foundations of Quantum Mechanics* (World Scientific, Singapore, 1996).
5. A. Tonomura, J. Endo, T. Matsuda, T. Kawasaki, and H. Ezawa, Demonstration of single-electron buildup of interference pattern, *Amer. J. Phys.* **57**(2), 117-120, (1989).
6. R. Kubo and H. Fukuyama, Interband effects in orbital magnetism and Hall conductivity in *Proc. Intern. Conf. on the Physics of Semiconductors*, pp. 551-560, Cambridge, Mass., USA (1970).
7. J. M. Luttinger and W. Kohn, Motion of electrons and holes in perturbed periodic fields, *Phys. Rev.* **97**(4), 869-883, (1955).
8. H. Fukuyama, Theory of Hall effect II. Bloch electrons, *Prog. Theor. Phys.* **42**(6), 1284-1303, (1969).
9. H. Fukuyama, Theory of orbital magnetism of Bloch electrons. Coulomb interactions, *Prog. Theor. Phys.* **45**(3), 704-729, (1971)
10. D. Schoenberg and M. Z. Uddin, Magnetic properties of bismuth. I. The dependence of the susceptibility on temperature and addition of other elements, *Proc. Roy. Soc. (London)* **A156**, 687, (1936).
11. H. Fukuyama and R. Kubo, Interband effect on magnetic susceptibility II. Diamagnetism of bismuth, *J. Phys. Soc. Jpn.* **28**(3), 570-581, (1970).
12. Y. Fuseya, M. Ogata, and H. Fukuyama, Interband effects of magnetic field on Hall effects and orbital magnetism for Dirac electrons in bismuth, *Phys. Rev. Lett.* **102**(6), 066601, (2009).
13. Y. Fuseya, M. Ogata, and H. Fukuyama, Spin-Hall effect and diamagnetism of Dirac electrons, *J. Phys. Soc. Jpn.* **81**(9), 093704, (2012).
14. H. Fukuyama, Anomalous orbital magnetism and Hall effect of massless fermions in two dimension, *J. Phys. Soc. Jpn.* **76**(4), 043711, (2007).
15. A. Kobayashi, Y. Suzumura, and H. Fukuyama, Hall effect and orbital diamagnetism in zerogap state of molecular conductor α-(BEDT-TTF)$_2$I$_3$, *J. Phys. Soc. Jpn.* **77**(6), 064718, (2008).
16. For a review, H. Fukuyama, Y. Fuseya, M. Ogata, A. Kobayashi, and Y. Suzumura, Dirac electrons in solids, *Physica B: Condensed Matter* **407**(11), 1943-1947, (2012).

Remembering Akira Tonomura

Michael Berry

*H H Wills Physics Laboratory,
Tyndall Avenue, Bristol BS8 1TL, United Kingdom
E-mail: asymptotico@physics.bristol.ac.uk*

Three of Tonomura's fundamental quantum physics experiments are discussed from a personal perspective.

My memories of Akira Tonomura are of a gentle and quiet man, always courteous in his dealings with colleagues, a virtuoso experimenter who transformed the electron microscope into the Stradivarius of scientific instruments, on which he played beautiful physics music. As a theorist I cannot comment technically on his many contributions. Instead I will make brief remarks about three of them.

The first is his demonstration[1] of the Aharonov-Bohm (AB) effect, intended to settle controversies associated with the inevitable failure of the idealization, assumed in elementary presentations of the effect, that the electrons are completely isolated from the magnetic flux. To eliminate leakage from the ends of a conventional finite solenoid, he confined the flux within a toroidal magnet, and to eliminate almost all penetration by the electrons he coated the toroid with a superconductor. The principle had been proposed by Kuper[2] (with the inessential difference that the magnetic flux would be confined in a hollow torus rather than a solid one). Because the flux in a superconductor is quantized (in units of $h/2e$), the experiment did not test the general AB effect, for which the flux is arbitrary; but it did demonstrate the important special case where the AB phase shift is π – as well as providing direct evidence of the value of the flux quantum (if this had been h/e there would have been no effect).

Even with a superconductor there is always some penetration of the electrons, so the flux cannot "completely shielded" as claimed in the title of Tonomura's paper. This is important, because as had been proved by Roy,[3] if there is any penetration, however small, the AB phase shift can be interpreted in terms of fields rather than potentials. Nevertheless, the fact that the phase shift remains finite as the limit of zero penetration is approached supports the usual interpretation in terms of potentials. I commented on this[4] (in the same year – 1986 – as Tonomura's paper appeared) as an example of the need to be careful when considering idealizations in physics. Tonomura clearly appreciated the same point, commenting eloquently and wisely[1] "Since experimental realization of absolutely zero field is impossible, the continuity of physical phenomena in the transition from negligibly small field should be accepted instead of perpetual demands for the ideal; if a discontinuity is asserted, only a futile agnosticism results".

The second is his demonstration[5] in 1989 of electron two-slit interference, with the pattern developing gradually by the detection of individual electron impacts. This has been voted the most beautiful experiment in physics.[6] It illustrates convincingly and with the utmost simplicity the wave-particle duality that is fundamental to quantum physics. Its priority has been the subject of some controversy,[6] because the buildup of the pattern by individual electrons had already been observed in a pioneering experiment[7,8] by Merli *et al.* and published in 1976, together with an award-winning movie. However, as Tonomura points out,[6] his experiment improved on that of Merli *et al.* in several respects: (a) it had lower electron intensity (so the possibility of there being two or more electrons in the apparatus at any time is negligible), (b) it was sufficiently stable for the buildup to take place very slowly (during 20 minutes), and (c) it was sufficiently sensitive to detect the electrons with almost 100% efficiency. As with the AB torus experiment, Tonomura's demonstration was definitive.

The third is his creation of vortices (= phase singularities = nodal lines = wavefront dislocations) in an electron beam.[9] This was particularly gratifying to us in Bristol, where we have emphasized vortices[10] as generic singularities of waves of all types and have explored these topological features in detail theoretically[11-14] – including vortices generated by transmission through spiral phase plates,[15] exactly as employed in the experiment by Tonomura. His emphasis was on the orbital angular momentum carried by the vortex beam – an aspect much studied in recent years.[16] Almost all earlier experiments were carried out with classical light; the novelty of Tonomura's[9] was that it demonstrated vortices in the much more challenging quantum physics of electrons.

The Japanese government has agreed that the direction of research pursued by Tonomura will continue, and that is a welcome decision. Nevertheless the premature passing of this supremely talented experimenter leaves a sadness that is hard to overstate.

References

1. A. Tonomura, N. Okasabe, T. Matsude, T. Kawasaki, J. Endo, S. Yano, and H. Yamada, Evidence for the Aharonov-Bohm effect with Magnetic field completely shielded from electron wave, *Phys. Rev. Lett.* **56**(8), 792-795, (1986).
2. C. G. Kuper, Electromagnetic potentials in quantum mechanics: A proposed test of the Aharonov-Bohm effect, *Phys. Lett.* **79A**(5-6), 413-416, (1980).
3. S. M. Roy, Condition for nonexistence of Aharonov-Bohm effect, *Phys. Rev. Lett.* **44**(3), 111-114, (1980).

4. M. V. Berry, The Aharonov-Bohm effect is real physics not ideal physics, *in Fundamental aspects of quantum theory,* eds. V. Gorini and A. Frigerio, Plenum, NATO ASI series vol. 144, pp. 319-320 (1986).
5. A. Tonomura, J. Endo, T. Matsuda, T. Kawasaki, and H. Ezawa, Demonstration of single-electron buidup of an interference pattern, *Am. J. Phys.* **57**(2), 117-120, (1989).
6. R. Crease, The double-slit experiment, Physics World (online edition) September 2002. (http://physicsworld.com/cws/article/print/2002/sep/01/the-double-slit-experiment)
7. P. G. Merli, G. F. Missiroli, and G. Pozzi, On the statistical aspect of electron interference phenomena, *Am. J. Phys.* **44**(3), 306-307, (1976).
8. R. Rosa, The Merli-Missiroli-Pozzi two-slit electron-interference experiment, *Phys. Perpect.* **14**(2), 178-195, (2012).
9. M. Uchida and A. Tonomura, Generation of electron beams carrying orbital angular momentum, *Nature* **464**(8904), 737-739, (2010).
10. J. F. Nye and M. V. Berry, 1974, Dislocations in wave trains, *Proc. Roy. Soc. Lond. A* **336**, 165-90, (1974).
11. J. F. Nye, *Natural focusing and fine structure of light: Caustics and wave dislocations* (Institute of Physics Publishing, Bristol, 1999).
12. M. V. Berry, Much ado about nothing: optical dislocation lines (phase singularities, zeros, vortices...), in Singular optics, ed. M. S. Soskin, *SPIE* **3487**, 1-5, (1998).
13. M. V. Berry, Geometry of phase and polarization singularities, illustrated by edge diffraction and the tides, in Second International Conference on Singular Optics (Optical Vortices): Fundamentals and applications, *SPIE* **4403** (Bellingham, Washington), 1-12, (2001).
14. M. R. Dennis, K. O'Holleran, and M. J. Padgett, Singular optics: Optical vortices and polarization singularities, *Progress in Optics* **53**, 293-363, (2009).
15. M. V. Berry, Optical vortices evolving from helicoidal integer and fractional phase steps, *J. Optics. A* **6**, 259-268, (2004).
16. L. Allen, S. M. Barnett, and M. J. Padgett, *Optical Angular Momentum* (Institute of Physics Publishing, Bristol, 2003).

Akira Tonomura: An Experimental Visionary

Anton Zeilinger

University of Vienna & Austrian Academy of Sciences
A-1090 Vienna, Austria
E-mail: anton.zeilinger@univie.ac.at

It was one of the most memorable events in my scientific life when I got to know Akira Tonomura personally at the occasion of his visit to M.I.T., which must have been in the late 1970s or early 1980s. He visited our small neutron group which was under the direction of C. G. Shull at the Neutron Diffraction Laboratory. He gave a talk to a very small group of listeners, which mainly consisted of the graduate students, a few post-docs, and visitors interested in the foundations of quantum mechanics. Most importantly, his enthusiastic spirit was extremely contagious. I remember that we all discussed for a long time his beautiful work on electron interference and the possibilities it opened up for the future.

Another most significant event then was the "International Symposium on Foundations of Quantum Mechanics in the Light of New Technology" (ISQM-Tokyo) which was held in Tokyo in 1983. This conference - which actually turned out to be the first in a series - was triggered by Akira Tonomura's successful examples of electron holography. It is very remarkable that this symposium was held at the Central Research Laboratory of Hitachi, Ltd., a private company. What is also quite remarkable is the breadth of that conference: Not only did it cover the Aharonov-Bohm Effect and Gauge Fields, but virtually all areas of the foundations of quantum mechanics at the time, which turned out to attract lots of activities in subsequent years. It was the time of a new frontier for these foundations, because many experiments which had been gedanken experiments before made it possible to test ideas which had been only philosophical. The proceedings of the conference became certainly a classic in the field, and they contain many far-reaching ideas.

Among the most beautiful papers of the conference is the publication of the results of Tonomura and his colleagues on the Aharonov-Bohm Effect. In these experiments,[1,2] they had permalloy rings, where the magnetic flux was confined and the electron beam could penetrate both the rings and the area in between. In this way, by using electron holography they succeeded in clearly confirming the phase shift due to the Aharonov-Bohm Effect that is required by the necessity of the continuity of the phase lines.

This work was refined in 1986 when Tonomura and his colleagues published the decisive paper entitled "Evidence for Aharonov-Bohm effect with magnetic field completely shielded from electron wave".[3] This paper, to me, is one of the most significant classics of modern experimental physics. The challenge was to confirm experimentally that the Aharonov-Bohm Effect even arises when neither the interfering electron wave can penetrate into the region where the magnetic field is, nor the magnetic field can penetrate into the region where the interfering electron wave propagates. The question then is how to observe the Aharonov-Bohm Effect? I would have to admit that naïvely I did not expect at that time that this would be possible. The reason that it is possible is a very subtle one which arises when a superconductor is used for the shielding. This is founded on the fact that the charge carriers in superconductivity are Cooper pairs which contain two electron charges. Therefore, a flux which can be enclosed in a superconductor is half the flux which provides a 2π phase shift. As demonstrated beautifully by Tonomura *et al.*, the consequence is that the phase shifts on the electron beam can be either π or 2π. Thus, independent of how much flux is originally enclosed, only two patterns are possible, one where inside the rings the phase pattern is shifted by half a period and one where it is not shifted at all. All this was clearly seen in the experiment, indicating that this is the most beautiful experimental confirmation and visualization of Gauge Fields.

As is often the case in fundamental physics, important applications followed. Later on, Akira Tonomura expanded this work into extremely nice demonstrations of magnetic flux lines[4-8] and various applications of magnetic field imaging[9-10] using electron holography.

All these works show that Akira Tonomura was a visionary in two senses. Firstly, he was a visionary in inventing new, elegant experimental methods which clearly demonstrate a phenomenon. He was also a visionary in the sense that he provided the most convincing visual evidence for the phenomena he was investigating.

I consider it a privilege of my life to have met Akira Tonomura, and I remember with fondness our many discussions, some of which were accompanied by most exquisite Japanese food.

References

1. A. Tonomura, T. Matsuda, R. Suzuki, A. Fukuhara, N. Osakabe, H. Umezaki, J. Endo, K. Shinagawa, Y. Sugita, and H. Fujiwara, Observation of Aharonov-Bohm Effect by electron holography, *Phys. Rev. Lett.* **48**(21), 1443–1446, (1982).
2. A. Tonomura, H. Umezaki, T. Matsuda, N. Osakabe, J. Endo, and Y. Sugita, Electron holography, Aharonov-Bohm Effect and flux quantization. In *Proc. Int. Symp. On Foundations of Quantum*

Mechanics in the Light of New Technology (ISQM-Tokyo '83), Eds. S. Kamefuchi *et al.,* pp. 20-28, Tokyo, Japan (Aug., 1983) (Physical Society of Japan, Tokyo, 1984): also in *Selected Papers from the First through Fourth International Symposium on Foundations of Quantum Mechanics in the Light of New Technology,* Eds. S. Nakajima, Y. Murayama, and A Tonomura, pp. 18-26 (World Scientific, Singapore, 1996).

3. A. Tonomura, N. Osakabe, T. Matsuda, T. Kawasaki, J. Endo, S. Yano, and H. Yamada, Evidence for Aharonov-Bohm Effect with magnetic field completely shielded from electron wave, *Phys. Rev. Lett.* **56**(8), 792-795, (1986).

4. K. Harada, T. Matsuda, J. Bonevich, M. Igarashi, S. Kondo, G. Pozzi, U. Kawabe, and A. Tonomura, Real-time observation of vortex lattices in a superconductor by electron holography, *Nature* **360**, 51-53, (Nov., 1992).

5. T. Matsuda, K. Harada, H. Kasai, O. Kamimura, and A. Tonomura, Observation of dynamic interaction of vortices with pinning centers by Lorentz microscopy, *Science* **271**, 1393-1395, (Mar., 1996).

6. A. Tonomura, H. Kasai, O. Kamimura, T. Matsuda, K. Harada, Y. Nakayama, J. Shimoyama, K. Kishio, T. Hanaguri, K. Kitazawa, M. Sasase, and S. Okayasu, Observation of individual vortices trapped along columnar defects in high-temperature superconductors, *Nature* **412**, 620-622, (Aug., 2001).

7. T. Matsuda, O. Kamimura, H. Kasai, K. Harada, T. Yoshida, T. Akashi, A. Tonomura, Y. Nakayama, J. Shimoyama, K. Kishio, T. Hanaguri, and K. Kitazawa, Oscillating rows of vortices in superconductors, *Science* **294**, 2136-2138, (Dec., 2001).

8. A. Tonomura, Direct observation of thitherto unobservable quantum phenomena by using electron, *Proc. Nat. Acad. Sci. USA* **102**(42), 14952-14959, (2005).

9. J. J. Kim, K. Hirata, Y. Ishida, D. Shindo, M. Takahashi, and A. Tonomura, Magnetic domain observation in writer pole tip for perpendicular recording head by electron holography, *Appl. Phys. Lett.* **92**(16), 162501, (2008).

10. Y. Murakami, H. Kasai, J. J. Kim, S. Mamishin, D. Shindo, S. Mori, and A. Tonomura, Ferromagnetic domain nucleation and growth in colossal magnetoresistive manganite, *Nature Nanotechnology* **5**, 37-41, (2010).

Dr. Akira Tonomura: Master of Experimental Physics

Kazuo Fujikawa

RIKEN Nishina Center
Wako, Saitama 352-0198, Japan
E-mail: k-fujikawa@riken.jp

Dr. Akira Tonomura, Hitachi Fellow, passed away on May 2, 2012 at the age of 70. As a classmate at the University of Tokyo and his long-time friend, I would like to describe my personal memory of Tonomura and a brief review of his contributions to fundamental physics.

1. Goodbye Akira

There appeared a short column in the Asahi, a Japanese daily newspaper, by one of the members of the editorial board, Ms. Atsuko Tsuji, several days after Tonomura's funeral ceremony in Tokyo. This was titled "Goodbye Akira!" and about an address of Professor Chen Ning Yang at the ceremony. The appearance of this column tells how Tonomura was treated in the Japanese community and how his association with Prof. Yang was important for him.

Tonomura was probably the best known corporate physicist in Japan who made important contributions to fundamental physics that apparently has no direct connection with the business of the company such as Hitachi, Ltd.

The periods Tonomura made his major contributions, namely the 1980s and 1990s were the golden days of Japanese production industry. Many big Japanese companies established basic research laboratories. Hitachi, Ltd. also established a very impressive Advanced Research Laboratory (now a part of the Central Research Laboratory) in the suburb of Tokyo, where Tonomura built very efficient electron microscopes that produced his impressive list of scientific achievements. The above article by Ms. Tsuji tells us that Prof. Yang once told Tonomura, "It takes 10 years to grow trees but it takes 100 years to grow a good community of scientists." Tonomura's research results using the electron microscopes were also the results of accumulated know-how at the Hitachi laboratories that have a very strong tradition of producing excellent microscopes. I hear that Tonomura himself also made important contributions to the improvements of commercial electron microscopes.

The article by Ms. Tsuji closes by wishing that this tradition of the Japanese production industry will continue despite the present difficulties the Japanese industry faces.

2. Tonomura and electron microscope

After graduation from the University of Tokyo in 1965, Tonomura directly went to Hitachi, Ltd. and worked at the Central Research Laboratory to become an expert on electron microscopes. I myself went to the graduate school at the University of Tokyo for two years for Master's degree and then to the graduate school at Princeton University for my Ph.D. We thus had no contact with each other for several years.

When I was a postdoc at Enrico Fermi Institute of the University of Chicago in 1971, when Professor Yoichiro Nambu was a central figure, Tonomura happened to come to Chicago to see Professor Albert V. Crewe, an expert on electron microscopes. We had dinner together in Chicago downtown. Tonomura told me in some detail what he was doing, and apparently he was an expert on a *needle* of the electron gun that produces a nice coherent electron beam. This needle played an important role later in his experiments. I informally heard later that this electron gun also brought some profits to Hitachi, Ltd. As a typical courtesy of Tonomura, he sent me a nice picture postcard from Hawaii on his way back to Japan; at that time the airplane usually made a refueling stop at Hawaii for the trip from the West Coast of the USA to Japan.

After three years of training as a postdoc in Chicago and Cambridge, I came back to Institute for Nuclear Study, the University of Tokyo, as an assistant professor in September 1973. Tonomura of course continued to work at the Central Research Laboratory of Hitachi, Ltd., but I have no clear memory of what we talked to each other for several years after 1973. Sometime in 1981, Tonomura sent me a copy of his paper together with the copies of the referees' reports he received from Phys. Rev. Lett. on his first experiment on the Aharonov-Bohm effect.[1] He asked my opinion on these referees' reports. This was about 30 years ago and my memory is not so clear, but I at least clearly remember that one of the referees' reports stated something like "There is *no* Aharonov-Bohm effect as such; thus it is meaningless to test it by experiment." Apparently, Tonomura was in a difficult situation.

3. Prof. Yang and Tonomura

Before this occasion, I had a chance of spending the first six months in 1980 at Institute for Theoretical Physics, State University of New York at Stony Brook directed by Prof.

Yang. I was mainly occupied with the path integral formulation of quantum anomalies at that time, but I also read many past papers of Prof. Yang. Among those papers, I remembered that the famous paper by T. T. Wu and C. N. Yang[2] on the magnetic monopole briefly mentioned "possible non-Abelian generalization of the Aharonov-Bohm effect." I was thus confident that Prof. Yang believes in the Aharonov-Bohm effect. To rebut the above referees' reports, which indicated suspicion on the Aharonov-Bohm effect itself, I thought that the support of Prof. Yang is most effective. I thus suggested to Tonomura that he should get in touch with Prof. Yang and ask his opinion on this matter. Apparently, Tonomura did and he had a chance to talk to Prof. Yang quite soon later, when Prof. Yang visited Department of Physics at the University of Tokyo sometime in "1981". Since then Tonomura received advice from Prof. Yang on various aspects of fundamental physics related to gauge fields and phases in quantum physics in general. I was quite happy that I was helpful to Tonomura at the very beginning of his adventure into basic physics with his technology of electron microscopes. He was very successful indeed. My major contribution to Tonomura's research is that I suggested to him that he should talk to Prof. Yang!

4. Aharonov-Bohm effect and double-slit experiment

Tonomura made many important contributions to fundamental physics and also to the improvement of electron microscopes. Among his major contributions to fundamental physics, I count the following:

1. Confirmation[3-5] of the Aharonov-Bohm effect without any doubt. Very few people doubt the effect nowadays, in contrast to the situation in 1982.

2. Very beautiful (one of 10 most beautiful experiments in the history of physics) experiment[6] of the electron interference through a double slit.

3. Observation[7-10] of the movement of magnetic flux lines (Abrikosov vortices) in metal and high temperature superconductors by Lorentz microscopy.

Among these, the experiment on the Aharonov-Bohm effect is probably best known and also widely appreciated. In 1982 the first result of the experiment was published[3] but the experiment was not water-tight. There were several criticisms and doubts such as the leak of the magnetic field or the penetration of the electron into the region of magnetic flux. To avoid these criticisms, Prof. Yang made an ingenious suggestion to confine the magnetic field inside the superconductor. It took about four years for Tonomura and his team to perform this difficult experiment and the final result was published[4] in Phys. Rev. Lett. in 1986. This experiment is regarded as the perfect confirmation of the Aharonov-Bohm effect, and nobody nowadays doubts the existence of the effect. Tonomura and his team also mastered how to control a very small scale magnetic field and electron beam.

This technique was then fully utilized by Tonomura's team to demonstrate the electron interference pattern through a double slit. Professor Hiroshi Ezawa, a theoretical physicist and another mentor of Tonomura, joined Tonomura's team on this experiment. This is apparently a very difficult experiment: Feynman[11] once mentioned that the interference pattern through a double slit is theoretically very beautiful but it may be almost impossible to perform the actual experiment. Tonomura's team beautifully performed this experiment[6] and in fact a very impressive video of the experiment exist, which shows how the interference pattern is formed by an accumulation of events consisting of an individual electron. The textbook description of quantum mechanics is actually realized. This experiment was chosen as the most beautiful experiment performed in the history of physics[12] by *Physics World* in 2002: Among the top five of the list, you find Galileo's experiment of falling bodies, Millikan's oil drop experiment, Newton's decomposition of sunlight with a prism, and Young's light-interference experiment.

Sometime later, Tonomura and his team showed the actual movement of Abrikosov vortices in metal and high-temperature superconductors[7-10] by using Lorentz microscopy. One can now see explicitly the formation and movement of vortices.

All these experiments are very difficult ones, but in retrospect it appears that Tonomura and his team performed the experiments with great ease. My personal opinion is that Tonomura was a very brave physicist. He tried to perform very difficult experiments and produced the results without any doubt. This was partly due to the strong support of the administration of the Hitachi laboratories, but you need courage since mistakes or failures are not allowed. He often said that he does not understand difficult theories but he was a master of experimental physics, just like Faraday was a master of experimental physics.

5. ISQM and quantum physics

Finally, I would like to mention the benefits I received from his experiments and related activities, In particular, International Symposium on Foundations of Quantum Mechanics in the Light of New Technology (ISQM) supported by Hitachi, Ltd., which was initiated by Tonomura's experiment on the Aharonov-Bohm effect in 1982, influenced my research in theoretical physics. The organizers of ISQM included prominent theoretical physicists such as Prof. Sadao Nakajima, Prof. Hidetoshi Fukuyama, another classmate of ours at the University of Tokyo, and Prof. Naoto Nagaosa among others. About 20 years ago, I was also invited to be a member of the organizing committee of ISQM, which was a kind of salon of physicists interested in the fundamental aspects of quantum physics, and I learned a lot by organizing ISQM. The nice interaction among academics and industry on fundamental physics, which is quite different from my past experiences, is another side of my memory of Dr. Akira Tonomura at Hitachi, Ltd.

References

This article is a revised version of the article originally published in Asia Pacific Physics Newsletter, Vol. 2, No. 1, pp. 54-56 (World Scientific, Singapore, January 2013).

1. Y. Aharonov and D. Bohm, Significance of electromagnetic potentials in the quantum theory, *Phys. Rev.* **115**(3), 485-491, (1959).
2. T. T. Wu and C.N. Yang, Concept of nonintegrable phase factors and global formation of gauge fields, *Phys. Rev. D* **12**(12), 3845-3857, (1975).
3. A. Tonomura, T. Matsuda, R. Suzuki, A. Fukuhara, N. Osakabe, H. Umezaki, J. Endo, K. Shinagawa, Y. Sugita, and H. Fujiwara, Observation of Aharonov-Bohm effect by electron holography, *Phys. Rev. Lett.* **48**(21), 1443-1446, (1982).
4. A. Tonomura, N. Osakabe, T. Matsuda, T. Kawasaki, J. Endo, S. Yano, and H. Yamada, Evidence for Aharonov-Bohm effect with magnetic field completely shielded from electron wave, *Phys. Rev. Lett.* **56**(8), 792-795, (1986).
5. N. Osakabe, T. Matsuda, T. Kawasaki, J. Endo, A. Tonomura, S. Yano, and H. Yamada, Experimental confirmation of Aharonov-Bohm effect using a toroidal magnetic field confined by a superconductor, *Phys. Rev. A* **34**(2), 815–822, (1986).
6. A. Tonomura, J. Endo, T. Matsuda, T. Kawasaki, and H. Ezawa, Demonstration of single-electron buildup of interference pattern, *Amer. J. Phys.* **57**(2), 117–120 (1989).
7. K. Harada, T. Matsuda, J. Bonevich, M. Igarashi, S. Kondo, G. Pozzi, U. Kawabe, and A. Tonomura, Real-time observation of vortex lattices in a superconductor by electron microscopy, *Nature* **360**, 51–53, (1992).
8. T. Matsuda, K. Harada, H. Kasai, O. Kamimura, and A. Tonomura, Observation of dynamic interaction of vortices with pinning centers by Lorentz microscopy, *Science* **271**, 1393–1395, (1996).
9. A. Tonomura, H. Kasai, O. Kamimura, T. Matsuda, K. Harada, Y. Nakayama, J. Shimoyama, K. Kishio, T. Hanaguri, K. Kitazawa, M. Sasase, and S. Okayasu, Observation of individual vortices trapped along columnar defects in high-temperature superconductors, *Nature* **412**, 620–622, (2001).
10. A. Tonomura, Direct observation of thitherto unobservable quantum phenomena by using electrons, *Proc. Natl. Acad. Sci. USA* **102**(42), 14952–14959, (2005).
11. R.P. Feynman, R.E. Leighton and M. Sands, *Feynman Lectures on Physics*, Vol. III, Chap. 1, Sec. 1-4 (Addison-Wesley, Reading, Mass., 1965).
12. R. P. Crease, The most beautiful experiment, *Physics World,* September 2002.

Topology and Gauge Theory in Physics

Chen Ning Yang

Tsinghua University, Beijing, China
Chinese University of Hong Kong, Hong Kong, China
E-mail: castu02@tsinghua.edu.cn

This article is based on my speech delivered at Tonomura FIRST International Symposium on "Electron Microscopy and Gauge Fields" held in Tokyo on May 9-10, 2012.

I am deeply touched by the short message that Tonomura recorded for all of us today. It reminded me of my first visit to his laboratory in the early 1980s and the many conferences that I attended that he had organized in Japan. It also recalled for me the many discussions that he and I had, not only about the Aharonov-Bohm effect, but also about flux movement in superconductors, especially after the 1987 discovery of high-temperature superconductors. He and I had in these years many warm and fruitful meetings in Japan, in China, and in the United States. Last winter I was very happy to have received a photograph of him standing in a garden a few months after his operation. I thought he was on his way to full recovery. But that was not to be.

My title today is "Topology and Gauge Theory in Physics." In 1959, Aharonov and Bohm[1] proposed what is now known as the Aharonov-Bohm effect. If you want to appreciate the full meaning of this very important proposal, you have to look back into Maxwell's papers[2-5] of the 19th century. In his papers and his books, you will find that the vector potential **A** was *always* there: At the *very beginning* of his studies, he tried to express Faraday's amorphous concept of the *electrotonic state* in definite mathematical form.[6] He conceived of it as a vector, which later acquired a name: the vector potential **A**. So it was natural that it played an important role in all of his papers. It was after his death that Heaviside and Hertz eliminated the vector potential. Heaviside was a very enthusiastic and brilliant engineer. He was extremely happy that he had succeeded in eliminating the vector potential.[7] He said for example,

Heaviside was not exactly wrong in saying that the vector potential was a mathematical function of an *essentially indeterminate* nature. But in today's language we understand it is indeterminate only in the traditional sense. In fiber bundle language the vector potential is a "connection." The subtlety of this concept resides *precisely* in its *transformable* character.

After Heaviside there was a *dogma* that the field strengths **E** and **H** are observable, and the vector potential **A** is not observable. It is only of mathematical interest and can be eliminated. This was written into all textbooks when I was a graduate student. It was drilled into our minds that the important things are **E** and **H**, the vector potential being superfluous. Then Aharonov and Bohm proposed an experiment shown in Fig. 1. An electron beam is split at point A and recombines at point B. A long solenoid at the center contains magnetic flux perpendicular to the plane of the paper.

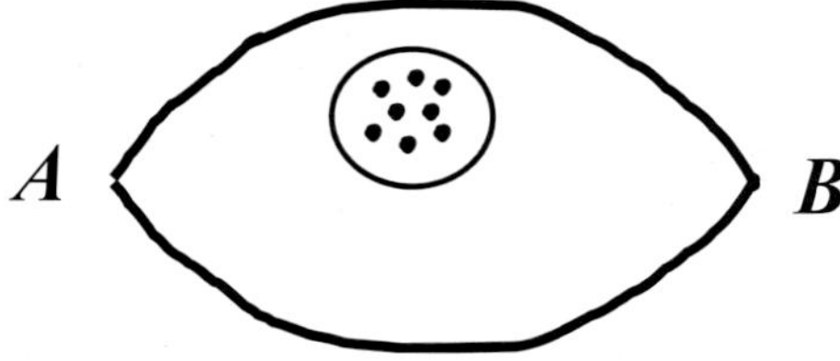

Fig. 1. Schematic experimental scheme proposed by Aharonov and Bohm.

If the flux is totally confined, then outside of the confinement, there is no **E**, no **H**. So the two paths of electron, starting from A to B, would not be subject to electromagnetic influences according to the traditional view. And so the image at B should not be dependent on the total flux inside of the solenoid. But Aharonov and Bohm in 1959 said the image at B *should be dependent* on the total flux.

The Aharonov-Bohm's proposal of 1959 was hotly debated. I remember I was in the Institute in Princeton at the time, and there were heated debates. One of the persons who absolutely refused to believe the Aharonov-Bohm experiment was Wigner. And those of us who knew Wigner knew it was absolutely impossible to change his opinion once he has formed a definite idea. That was, of course, part of his greatness as a physicist.

The A-B experiment was within a few years supposedly confirmed by Chambers.[8] But Chambers' experiment was not with a confined flux, but with a tapered needle. And there were endless discussions about what happened to the leaked flux around the tapered needle. It was finally the Tonomura experiments of 1983[9] and 1986[10] which settled the matter. With his beautiful electron holographic arrangements, he made the experiment which Dr. Zhu had just described to you.

This experiment of Tonomura and collaborators showed, first, the topological meaning of the vector potential in electromagnetism. Second, it showed that fluxes in superconducting rings are in quantized units of $ch/2e$. Notice the 2. That 2 is, of course, because of the *pairing* of fermions in a superconductor, which was the essence of the existence of superconductivity. Thus Tonomura's experiment quantitatively and explicitly showed, because of this factor of 2, the existence of Cooper pairs. This beautiful work is the first direct experimental proof of the importance of topology in physics.

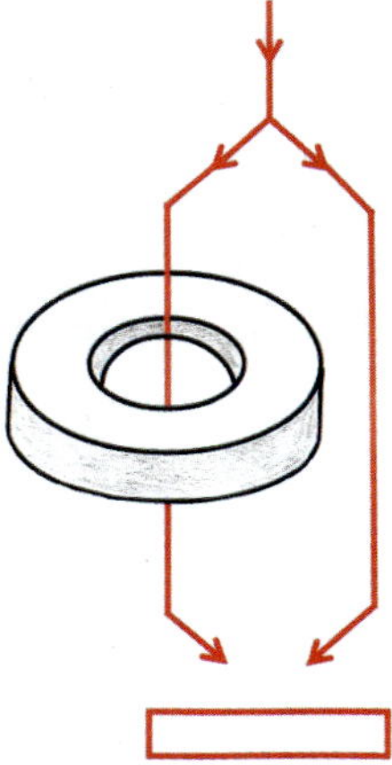

Fig. 2. Schematic diagram of Tonomura's experiment.

I shall now go over a number of other developments, experimental and theoretical, in modern physics, which are deeply related to topological concepts, and to gauge transformations. The first was the Van Hove singularity.[11] In 1953, Van Hove was a "member," as I was, at the Institute for Advanced Study. At that time, there were lots of computations with the then available computers of the vibrational frequency distribution in crystals. Papers by several people at the University of Maryland showed that there were ups and downs in such distributions. Figure 3 is reproduced from an early paper around that time. The abscissa is the frequency. The ordinate is the number of modes in each frequency range. (Such approximate computations were started, as you remember, already in 1905 and 1906 by Einstein, and later by Max Born. It was only in the 1950s, with computers, that people can make more accurate computations.) The resultant ups and downs caused quite a stir, as I vividly remember.

 C. N. Yang

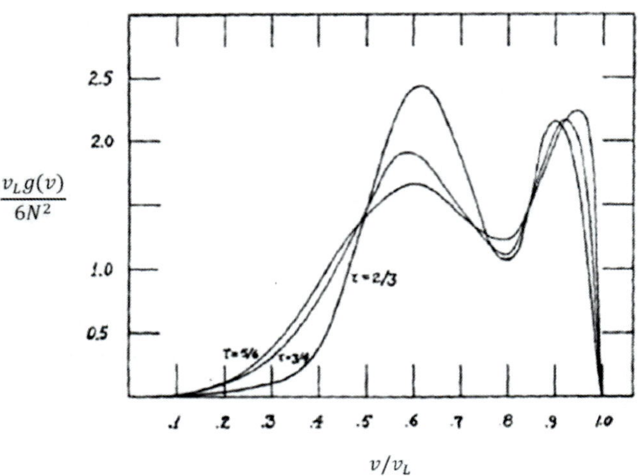

Fig. 3. Typical vibrational frequency distributions in crystals.

It was Van Hove, who had more mathematical training than most of us, who pointed out that for topological reasons, because of Morse theory, there should in fact be singularities in the spectrum, not just ups and downs. The result of today's computations is shown below (Fig. 4).

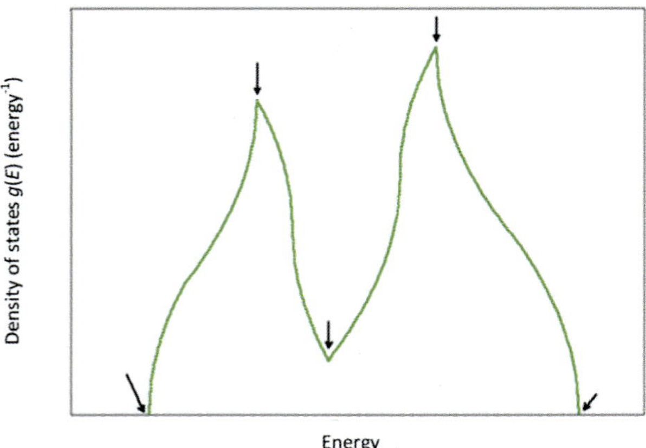

Fig. 4. A sketch of density of states g(E) versus energy E for a simulated three-dimensional solid.

Morse theory is a generalization to higher dimensions of a simple topological theorem: A continuous function $F(x)$ of one periodic variable x has at least one maximum and one minimum. In higher dimensions there must also be saddle points. Morse theory played a very important role in the early development of topology in the 20th century, and Van

Hove very elegantly pointed out that this is exactly what would explain those ups and downs in the frequency distribution in crystals.

Another important physical result related to gauge transformation and topology is the flux quantization experiment of 1961. This graph was from the original paper of Fairbank and his student[12] (Fig. 5).

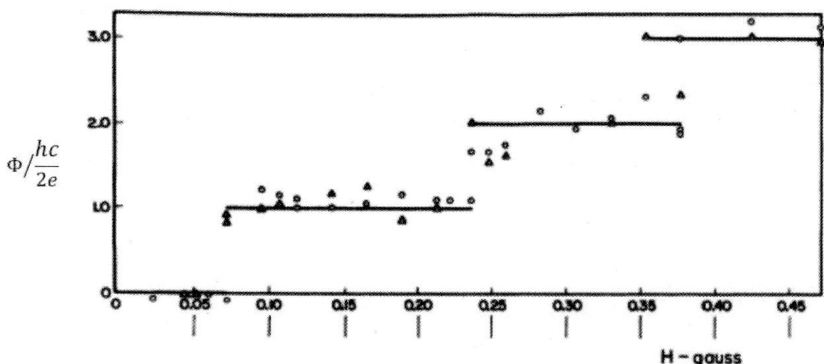

Fig. 5. Flux quantization in superconducting cylinders

The abscissa is the magnetic field. They had a superconducting ring and they were trying to measure the flux inside of it. This graph was in the first publication. If you look at today's publication, all those scattered points have coagulated into very good horizontal bars. The reason for this stepwise result is deeply related to gauge transformation and to topology.

Another very important idea which was finally understood in terms of topology, was Dirac's magnetic monopole of 1931.[13] I think everyone here knows this famous paper of 1931 of Dirac, who said that we have electrons with charge e; if there also exists a magnetic monopole with magnetic charge g, then

$$\frac{2eg}{\hbar c} = \text{integer}. \tag{1}$$

This was a paper which was very famous published in 1931. It was also famous for several other reasons. For example, in the introduction of that paper, Dirac said the progress of physics, of theoretical physics, depends on two avenues of attack: one based on experiments, and the other based on beautiful mathematics. As time goes on, he said, the avenue based on experiments is going to become more and more difficult. So we have to search for advances in physics by looking into beautiful mathematics. It is extremely interesting today to read this 80 year old philosophy of Dirac's.

Dirac's quantization rule (1) turns out to be an elegant topological result. To describe how I realized this please allow me to first tell you a story.

In 1968-1969 I taught a course on general relativity at Stony Brook. One day at the lecture I copied on the blackboard the famous curvature equation of Riemann:

$$R^{l}_{ijk} = \frac{\partial}{\partial x^{j}}\left\{{l \atop ik}\right\} - \frac{\partial}{\partial x^{k}}\left\{{l \atop ij}\right\} + \left\{{m \atop ik}\right\}\left\{{l \atop mj}\right\} - \left\{{m \atop ij}\right\}\left\{{l \atop mk}\right\} \tag{2}$$

As I did this it occurred to me that this equation looked very much like an equation in the 1954 papers of Yang-Mills:

$$F_{kj} = B_{k,j} - B_{j,k} + B_{k}B_{j} - B_{j}B_{k} \tag{3}$$

After the lecture I compared the two equations, and quickly realized that in fact if I define matrices β_k and R_{jk} by

$$\langle i|\beta_{k}|l\rangle = \left\{{l \atop ik}\right\}, \quad \langle i|R_{jk}|l\rangle = R^{l}_{ijk}, \tag{4}$$

then Equation (2) is the $\langle i| \, |l\rangle$ element of the matrix equation:

$$R_{jk} = \frac{\partial}{\partial x^{j}}\beta_{k} - \frac{\partial}{\partial x^{k}}\beta_{j} + \beta_{k}\beta_{j} - \beta_{j}\beta_{k}. \tag{5}$$

Equations (3) and (5) are thus *identical*. In excitement I went to consult Jim Simons, who was then the Chairman of Mathematics at Stony Brook. He told me both equations are *curvature* equations in fiber bundle theory. Later at my request Simons gave us physicists a serious of luncheon lectures about topology and fiber bundle theory. We thus understood *for the first time, that the mathematics of gauge theory, both Abelian and nonAbelian, had been independently developed by mathematicians in their beautiful fiber bundle theory*.

It is impossible to describe my *electrified* feeling upon realizing that electromagnetism, *a basic structure of the physical world*, is in fact based on subtle and beautiful mathematical constructs. In an article published later[14] I wrote:

"In 1975, impressed with the fact that gauge fields are connections on fiber bundles, I drove to the house of Shiing-Shen Chern in El Cerrito, near Berkeley. (I had taken courses with him in the early 1940's when he was a young professor and I was an undergraduate student at the National Southwest Associated University in Kunming, China. That was before fiber bundles had become important in differential geometry and before Chern had made history with his contributions to the generalized Gauss-Bonnet theorem and the Chern classes.) We had much to talk about: friends, relatives, China. When our conversation turned to fiber bundles, I told him that I had finally learned from Jim Simons the beauty of fiber-bundle theory and the profound Chern-Weil theorem. I said I found it amazing that

gauge fields are exactly connections on fiber bundles, which the mathematicians developed without reference to the physical world. I added "this is both thrilling and puzzling, since you mathematicians dreamed up these concepts out of nowhere." He immediately protested, "No, no. These concepts were not dreamed up. They were natural and real."

Also in 1975 T. T. Wu and I published a *dictionary*[15] of the terms used in physicists' gauge theory and the corresponding ones used in mathematicians' fiber bundle theory. This is shown in Table 1.

Table 1. Gauge field terminology versus bundle terminology

Gauge field terminology	Bundle terminology
Gauge (or global gauge)	Principal coordinate bundle
Gauge type	Principal fiber bundle
Gauge potential b_μ^k	Connection on a principal fiber bundle
S (Eq. (8))	Transition function
Phase factor Φ_{QP}	Parallel displacement
Field strength $f_{\mu\nu}^k$	Curvature
Source (electric) j_μ^k	?
Electromagnetism	Connection on a U_1 bundle
Isotopic spin gauge field	Connection on a SU_2 bundle
Dirac's monopole quantization	Classification of U_1 bundle according to first Chern class
Electromagnetism without monopole	Connection on a trivial U_1 bundle
Electromagnetism with monopole	Connection on a nontrivial U_1 bundle

One entry, Source J_μ^k, is the charge-current 4-vector, which is, of course, of *fundamental importance* to physicists. But Simons told us "we do not consider such quantities," and that explains the "?" in the dictionary.

Why did the mathematicians not consider J_μ^k? I think there may be a subtle reason: In 1944 fiber bundle theory became important in topology with Chern's generalization of the Gauss-Bonnet theorem to four dimensions. Chern considered an integral of

$$f_{\mu\nu}^k \widetilde{f}_{\mu\nu}^k, \tag{6}$$

where $\widetilde{f}_{\mu\nu}^k = \varepsilon_{\mu\nu\alpha\beta} f_{\alpha\beta}^k$, is the dual of $f_{\mu\nu}^k$. (It exchanges **E** and **H**.) Now conceptually J_μ results from a variation of the Lagrangian which is the integral of

$$f_{\mu\nu}^k f_{\mu\nu}^k \tag{7}$$

I think mathematicians were so fascinated with (6) that they neglected to consider expression (7) which is of *primary importance* to physicists.

Be that as it may, when I. M. Singer visited Stony Brook I showed him the Wu-Yang paper.[14] He brought it to Oxford. The result was a paper by Atiyah, Hinchin and Singer[16] which dealt with *sourceless* gauge fields, i.e. gauge fields for which

$$J^k_\mu = 0, \tag{8}$$

and for which the integral of expression (7) reaches an extremum. It is no exaggeration to say[17] that this three-man paper was one of the catalysts that ushered in the close collaboration between physicist and mathematicians in the last thirty odd years.

We now return to Dirac's quantization rule (1). Topologically it originated from the subtle structure of the vector potential around a magnetic monopole. When I showed it to Simons, he said "Dirac had anticipated Chern by more than a dozen years".

"Anomaly" was in a way first discovered in 1949 by the famous experimental physicist Steinberger when he calculated the rate of the process $\pi^0 \to \gamma + \gamma$. π^0 is a pseudoscalar and this process is related to expression (6) above. Much later it became clear that such anomalies are inherent in the topology of gauge fields.

More recently there are great excitements about topological insulators. One source of its remarkable properties is related again to expression (6) above. I think it is safe to predict that topology will continue to play new and subtle roles in physics.

Except for the Van Hove singularity, all of the examples mentioned above of the entry of topology into physics are basically related to the mathematical structure of gauge theory. It is *truly amazing*[18] that:

(1) Two hundred years ago, Michael Faraday (1791-1867), who knew little mathematics, was able to conceive of a vague and somewhat slippery concept of the *electrotonic state*.
(2) Maxwell (1831-1879) was able to formulate it into a mathematical quantity which was called the vector potential.
(3) The conceptual beauty of this quantity was independently discovered by the mathematicians *without reference to the physical world,* and was called the connection.

References

1. Y. Aharonov and D. Bohm, Significance of electromagnetic potentials in the quantum theory, *Phys. Rev.* **115**(3), 485-491, (1959).
2. J. C. Maxwell, On Faraday's lines of force, Trans. Camb. Phil. Soc. **10**, 27 (1856) [in *Scientific Papers of J. C. Maxwell*, ed. W. D. Niven, Vol. 1, p. 155 (Dover Publications, 1890)].
3. J. C. Maxwell, On physical lines of force, *Phil. Mag.* **21**, 281; 338, (1861).
4. J. C. Maxwell, *Phil. Mag.* **23**, 12, 85 (1862) [in *Scientific Papers of J. C. Maxwell*, ed. W. D. Niven, Vol. 1, p. 451 (Dover Publication, 1890)].

5. J. C. Maxwell, A dynamical theory of the electromagnetic field, *Phil. Trans. Roy. Soc. (London)* **155**, 459, (1865) [in *Scientific Papers of J. C. Maxwell*, ed. W. D. Niven, p. 526 (Dover Publication, 1890)].

6. A. C. T. Wu and C. N. Yang, Evolution of the concept of the vector potential in description of fundamental interactions, *Int. J. Mod. Phys. A* **21**(16), 3235-3277, (2006).

7. O. Heaviside, *Electromagnetic Theory* (1983) (The Electrician Printing and Publishing Company).

8. R. G. Chambers, Shift of an electron interference pattern by enclosed magnetic flux, *Phys. Rev. Lett.* **5**(1), 3-5, (1960).

9. A. Tonomura, T. Matsuda, R. Suzuki, A. Fukuhara, N. Osakabe, H. Umezaki, J. Endo, K. Shinagawa, Y. Sugita, and H. Fujiwara, Observation of Aharonov-Bohm effect by electron holography, *Phys. Rev. Lett.* **48**(21), 1443-1446, (1982).

10. A. Tonomura, N. Osakabe, T. Matsuda, T. Kawasaki, J. Endo, S. Yano and H. Yamada, Evidence for Aharonov-Bohm effect with magnetic field completely shielded from electron wave, *Phys. Rev. Lett.* **56**(8), 792-795, (1986).

11. L. Van Hove, The occurrence of singularities in the elastic frequency distributions of a crystal, *Phys. Rev.* **89**(6), 1189-1193, (1953).

12. B. S. Deaver, Jr. and W. M. Fairbank, Experimental evidence for quantized flux in superconducting cylinders, *Phys. Rev. Lett.* **7**(2), 43-46, (1961).

13. P. A. M. Dirac, Quantized singularities in the electromagnetic field, *Proc. Roy. Soc. London, Ser. A* **133**(821), 60-72, (1931).

14. C. N. Yang, Einstein's impact on theoretical physics, Lecture given at the Second Marcel Grossmann Meeting held in honor of the 100th anniversary of the birth of Albert Einstein, *Physics Today* **33**, 42,0 (June 1980).

15. T. T. Wu and C. N. Yang, Concept of nonintegrable phase factors and global formation of gauge fields, *Phys. Rev. D* **12**(12), 3845-3857, (1975).

16. M. F. Atiyah, N. J. Hinchin, and I. M. Singer, Self-duality in four-dimensional Riemann geometry, *Proc. Roy. Soc. London, Ser. A* **362**(1711), 425-461 (1978).

17. I. M. Singer, Some problems in the quantization of gauge theories and string theories, *Proc. Symposia in Pure Math.* **48**, 198-216, (1988).

18. I had commented on these amazing historical facts in a talk in Tokyo: C. N. Yang, Reflections on the development of theoretical physics, *Proc. 4th Int. Symp. Foundations of Quantum Mechanics, Tokyo, 1992*, Eds. M. Tsukada, S. Kobayashi, S. Kurihara, and S. Nomura, JJAP Series **9**, 3-9, (1993); Reprinted in *Foundations of Quantum Mechanics in the Light of New Technology—Selected Papers from the Proceedings of the First through Fourth International Symposia on Foundations of Quantum Mechanics*, Eds. S. Nakajima, Y. Murayama, and A. Tonomura, pp. 399-405 (World Scientific, Singapore, 1996).

On the Aharonov-Bohm Effect
and Why Heisenberg Captures Nonlocality Better Than Schrödinger

Yakir Aharonov

School of Physics and Astronomy, Tel-Aviv University,
Ramat-Aviv 69978, Israel,
Institute of Quantum Studies and Faculty of Physics, Chapman University,
1 University Drive, Orange, CA 92866, USA, and
Iyar, The Israeli Institute for Advanced Research,
Rehovot, Israel
E-mail: yakir@post.tau.il, aharonov@chapman.edu

This article is based on my lecture delivered at Tonomura FIRST International Symposium on "Electron Microscopy and Gauge Fields" held in Tokyo on May 9-10, 2012. I discuss in detail the history of the Aharonov-Bohm effect in Bristol and my encounters with Akira Tonomura later on. I then propose an idea that developed following the publication of the Aharonov-Bohm effect, namely the importance of modulo momentum and Heisenberg representation in dealing with non-local quantum phenomena.

1. Introduction

It is both a sad and a happy occasion for me. Sad because Professor Akira Tonomura is not with us anymore. He was a good friend of mine. We first met about 30 years ago in a conference that he organized in Hitachi. It was a very interesting conference and also an interesting meeting for me with him. He was very interested in my work on the Aharonov–Bohm (A–B) effect[1] and other questions of quantum mechanics. Since then we met many times and I always found him extremely enthusiastic about physics, with a lot of knowledge, not only about experimental physics, but also on theoretical questions. I must say that the first time he told me that he was going to try to do the A-B effect with a completely closed flux inside the superconductor tube, I did not want to tell him that it is possible, but I was sure that it would not work, because you needed to create a very, very small tube and to make sure that the beam of the electrons around it is still coherent. But he *was* successful and eventually performed an amazing experiment! This, then, is the happy side of this occasion. I think this experiment is one of the most beautiful ones in modern physics. I was hoping that he would be able to complete this very ambitious attempt to build the best electron microscope in the world. I now hope that his team will continue to work on it in his memory and complete his life goal.

2. Historical recollection of development of the Aharonov-Bohm effect

Let me begin by telling you a little about the history of the A-B effect, how I came to think about it, and how I collaborated with David Bohm, who was my thesis advisor at that time at Bristol University. It started, in effect, because I was a graduate student and a very ignorant one. Being ignorant, I knew nothing about gauge transformations. I did not know that you are not supposed to think about a potential, which is only a function of time, as something that should not have any effect at all. If I knew that, I wouldn't find the effect.

I was very excited when I learned about the Bloch functions. These are quantum functions. When you have a periodic potential in position and you look at the energy level, you find that there is a gap in energy because of the degeneracy for positive and negative momentum. If the periodic potential connects them, you get a gap between the two energies that are separated by this potential. So, as an ignorant student I thought the following: Why not look for an analogy? If I exchange position with time, and momentum with energy, then I'd have a periodic potential, which is a function of time. I will have gaps, not in energy, but in momentum. So, I tried to play around. I knew how to solve Schrödinger's equation. Then I added the periodic time-dependent potential to the equation. Lo and behold! The only thing I got, as everybody knows today, but *I* didn't know at that time, is just a phase. I kept on getting a phase, and a phase doesn't do anything, because you can't measure it! I was very frustrated. That night I could not sleep, and in the morning it occurred to me that maybe if I have, in one region of space, one potential which is only a function of time, I will get a phase, and in another region of space, where I have a different potential that is only a function of time, I will get a *different* phase. And then – maybe – I will see the relative phase as having a meaning.

So I thought about looking at two Faraday cages. In one cage, I have half of the electrons and the other half of the electrons in the other cage. I next considered the superposition of the electrons in both of them. In one Faraday cage I put the potential. So, inside it there will be nothing, just the potential. In the other Faraday cage, I will ground it so there will be no potential. And then when you solve the Schrödinger equation, you find indeed that in the first Faraday cage you do get the phase, which is the integral of the potential of the time proportional to it. In the other cage you get no potential. Now you remove the potential, so there is no effective field anywhere. Luckily, the phase still "remembers" not what the potential is now, but what the integral of the potential was. So now I open the two cages, send the two parts of the beam together. Lo and behold! You get an effect, which is a change in interference, even though the particles never touched the electric field.

So I came to my professor, David Bohm, and I showed it to him. And he said, "That's interesting. That's interesting. It's an interesting idea, but we have to think some more what we can do with it." And it lay there for a few months. And we thought about it only as a curiosity, nothing more. We did not connect it yet with the basic idea of electromagnetic potential.

Then Bohm sent me to a summer school that was at Oxford. At that time, it was very popular for people to think about physics in the language of analyticity. This was the time when people did not know yet how to renormalize field theories. I confess it was very boring.

Anyhow I was listening to some talks there, and there was one talk about what happens when you perform scattering with a vector potential. Suddenly it struck me that maybe there could also be an effect similar to the scalar potential with the vector potential. Excited I came back to Bristol and told Bohm, "Look what happens if, instead of just two different time-dependent scalar potentials, we have a solenoid with a vector potential around it. We should get an effect analogous to the one that we discussed before." Then *he* got also excited and said, "Okay, first of all, let's modify it. Instead of your two Faraday cages let's talk about two long cavities when you have different potentials. We send the two wave packets of the electron through them, and when the wave packets are inside the cavities we switch the potential in one of them and switch it off before the wave packet gets out." Then he said, "In order to really be able to publish it, I would like you to solve the problem of what happens if you have an infinitely thin solenoid and you do a scattering of electron beams when they go through this infinitely thin solenoid. Classically, they should not be affected at all, but quantum mechanically there should be some interesting scattering.

So I had to solve that problem. And it turned out that in order to solve it you have to sum up infinite series of Bessel functions, but unfortunately, with fractional order. So if you have to sum up Bessel functions, you only look at the Watson mathematical encyclopedia. I looked at the Watson, and there was nothing like that. I was walking around, didn't know what to do. Luckily, there was the chairman of the Physics Department at Bristol University, Price. And when Price looked at me and saw me walking like this, he asked, "What's the problem?" I told him, "I don't know how to sum these series." He said, "I think that there could be some interesting differential equation that could be connected to this. And if you could solve this, you will be able to continue." I took his suggestion, and indeed was able to do it, and solved the exact formula for the scattering. And, in fact, Bohm and I suggested to Price that he would also be a coauthor of this article. But he decided that he did not contribute enough to deserve that.

In Bristol, like in any other universities in England, there is a tradition of afternoon tea. Professors and graduate students get together, discussing different problems in physics. In one of these meetings, I told them that there is this interesting new possibility of seeing an effect on charges without field. But I said, "Unfortunately, I don't believe that it could be observed experimentally because we will need to have an extremely thin solenoid in order to have the possibility of seeing the effect." Among the professors attending that afternoon tea was Sir Frank, who was an expert on solid state, and he said, "Hey, hey, I believe it can be done with magnetic whiskers." There are magnetic whiskers which are very, very thin, line of magnetic flux inside crystals. And then Chambers was sitting next to him, and he said, "Okay, I'm going to do the experiment!" And that was how the first experiment was done, just a few doors away from my office. So I was very excited. Every morning I would come to the experiment to see how it is developing, and giving them ideas. One day Chambers took me aside and said, "Look, if you want this experiment ever to be finished, please continue with your theoretical work, and let us do our experiment without your intervention." I decided to accept his kind suggestion, and did not intervene any more. And indeed the experiment was finished,[2] but, as Professor Yang said, it was not an ideal experiment because the magnetic whisker, even though it was fairly straight, contained also lines of magnetic field that did go out. Therefore it could not be claimed that the electron beam is completely free of a magnetic field. Then, there was a better attempt[3] by Möllenstedt, who actually built a very, very thin solenoid, and did the experiment with it. But still, because it was not completely shielded, people did not accept this as a true experimental test.

This is where Tonomura enters the story. It was he who finally managed to beautifully do this experiment[4-6] with the magnetic field completely shielded by a superconductor surrounding it. And the electron beam did show the effect: the difference of one fluxon (half a fluxon from the point of the view of the electron) from the point of view of the two-charge carrier. And there was indeed a shift in interference fringes that showed the effect.

3. Wave or particle? Heisenberg's picture re-assessed

This is about the history. Now I want to tell you now some ideas connected with the effect which have developed since then. Let me start by saying that the way interference is introduced in books is misleading, because if we look at quantum interference, for example a two-slit quantum interference, we see a picture very similar to classical interference. Apparently in both classical and quantum interference, you see a wave: One part of the wave goes through one slit and the other part goes through the other, then the two beams meet and create an interference pattern. There seems to be no difference between the classical and the quantum picture, I now argue that there *is* a fundamental difference.

In classical theory, there is local information in the two slits that tells you what will happen when the two parts of the wave finally meet. But in quantum mechanics, what tells you where are maxima and minima is the relative *phase*, which goes through the one slit minus the phase in the other. But this phase cannot be observed, because quantum mechanics allows changing only the phase of the *whole* wave function by a constant, which therefore makes no difference. So the only thing that can be observed in quantum interference is the difference of the phases between the two slits. But no experiment could be done to say what the phase in each slit is individually. This is a truly *non-local* phenomenon, not obvious in Schrödinger's picture. Classically, when you speak about electromagnetic waves, water waves, or any classical wave giving rise to interference, there will be always a possibility of doing a local experiment at each slit that tells what the phase is. For example, if it's an electromagnetic field, the local experiments can tell what the electric field is at a given point. So the relative phase is really a difference between truly local properties in the first slit and truly local properties in the second one. But in quantum mechanics this is not true because quantum interference is truly non-local. And this non-local phenomenon remains hidden, when you think about it, in the Schrödinger representation.

So when I thought about it, I thought that perhaps one could gain a better understanding of what happens in quantum interference if one thinks about it in the Heisenberg picture instead of the Schrödinger picture. The question is as follows: what are the properties of interference that we see in the Heisenberg picture? In this picture the wave function is a constant thing: You only look at the time dependence of operators that are functions of position and momentum. How does this description catch these non-local features of quantum interference?

The insight that drove me to do it in the Heisenberg picture came from a solution of the following paradox, which I myself invented. Consider a grating with distance L between the slits (Fig. 1). I send through it a very weak electron beam, one electron after the other. What will happen? We all know that the beam can go in one out of some quantized directions, so that the difference between the lengths of neighboring rays is an integer times the wavelength:

$$P_y = sin\theta_a P = n\lambda P/L = nh/L,$$

hence only this direction will get constructive interference.

Now suppose I put solenoids behind each part of the grating, so that each solenoid, being half a fluxon, gives a relative phase of π between adjacent beams. This means that now, the new directions of constructive interference will be shifted, as shown in Fig. 1b:

$$P_y = sin\theta_b P = (n+1/2)h/L.$$

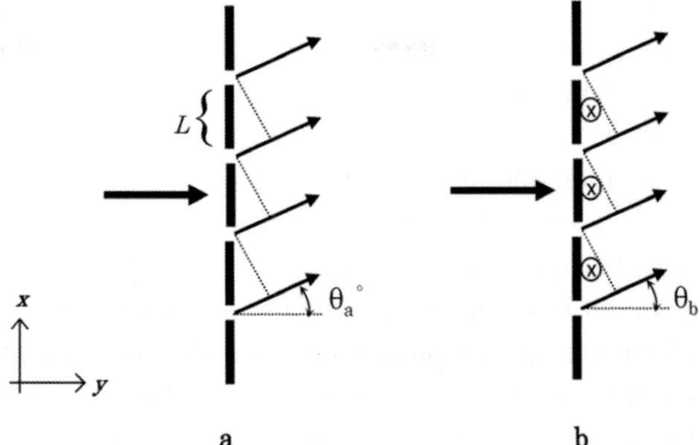

Fig. 1. Electron beam diffracting through a grating.

Now, you need an extra $L/2$ in order to get constructive interference. Let us see what this means in terms of momentum. If I call the vertical direction y, and the horizontal direction x; then without the solenoid, I know that the change in the electron's momentum was an integer times h/L. *With* the solenoids, this momentum change will be an integer plus $h/2L$. What happens to momentum conservation? I know that the sum of the momenta of the electron plus the grating plus the solenoid must be conserved. I know that the grating can give the electrons only an integer times h/L momentum, due to its periodicity. So it is the solenoids that must give the missing momentum, which is at least $\pm h/2L$. And that happens *without any force*. I can make sure that the electron does not touch the solenoid. I can even put a barrier to make sure that the electron never touches the magnetic field. So without forces at all, we know that there is a change of momentum between the array of solenoids and the electron which is at least $\pm h/2L$.

So far it's no big deal, because we know that the positions of the solenoids cannot be better defined than L because they have to be hidden from the electron. So the uncertainty in the momentum of the solenoid is h/L, which is bigger than the momentum exchange with the electron $h/2L$. But suppose now we send one electron after the other and ask what will happen after N electrons went by. We expect each electron to give at least $\pm h/2L$. After N electrons go through, the array of solenoid should get something like $h\sqrt{N}/2L$. Because there is no memory, each time is like a random walk. Each time the solenoid should get either $h/2L$ or $-h/2L$. So eventually, without any force, the array of solenoids should start to move. But that's against the classical intuition, because classically if the solenoids don't feel any force, they shouldn't move at all. So here was a paradox. How could you solve it? The solution led me to this new way of thinking about interference in the Heisenberg picture.

First I told myself the following. If instead of the random walk on a *straight line*, I will have a random walk on something like a *circle*, then the momentum will not keep on going but will stay bounded. And indeed once I thought about it like this, I understood the following: With no solenoid, I know that the change in the momentum is equal to some integer times h/L. With the solenoid, the change in the momentum is equal to an integer $\pm h/2L$. That means that the relevant quantity that is changing here is not the momentum itself, but the momentum *modulo h/2L*.

I realized that this is a very important quantity to describe this kind of *topological* effects. In topological effects, you don't change just the ordinary moments of the momentum, but instead you have an exchange between the solenoids and the electrons of this modulo variable which has no classical limit. The reason is that in the classical limit, when I keep L fixed and h goes to 0, the modular momentum becomes unobservable – it is oscillating infinitely fast. This is why I began thinking that the correct way to look at quantum interference is by using this kind of *modulo* variables. Let me show you how this works.

First, let us look at what happens when we have just two slits. Let's look at the wave function of the electron just after it passes the two slits. It is the superposition of two wave packets, ψ_1 and ψ_2 (Fig. 2). The distance between them is L. And for simplicity I assume that there is no overlap between them, just to make the mathematics simpler.

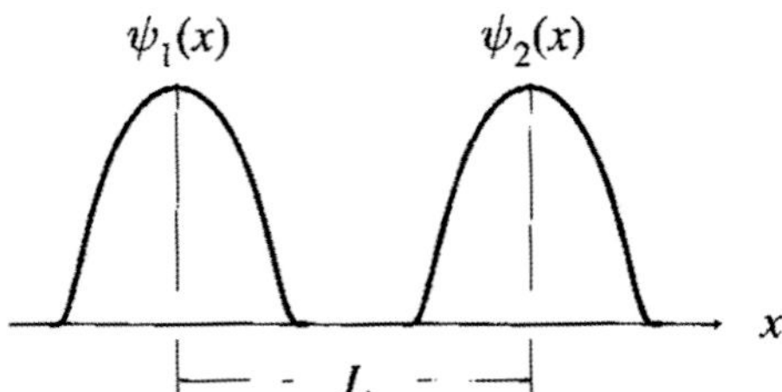

Fig. 2. Two non-overlapping wave functions on the x axis.

Now I ask: What kind of physical variable will be sensitive to the relative phase between the two? Let us define

$$\psi_\alpha \equiv \frac{1}{\sqrt{2}}(\psi_1 + e^{i\alpha}\psi_2),$$

where I assume that ψ_1 and ψ_2 are just shifted by L relative to each other. I now ask: Which operator, in the Heisenberg picture (being a function of position and momentum), will be sensitive to this relative phase α?

It turns out that if I look at any operator

$$f(x,p) = \sum_n \sum_m a_{mn} x^n p^m$$

symmetrized, it is insensitive to α. In other words, the average of this operator will be independent of α. Just to see why this is the case, look, for example, at the following integral:

$$\int (\psi_1{}^* + e^{-i\alpha}\psi_2{}^*)\, p^n\, (\psi_1 + e^{i\alpha}\psi_2)\, dx \,.$$

It will be the average of the *n-th* power of the momentum. You can see immediately, because the momentum acts like a derivative, that the only thing that will depend on α will be these mixtures. But because the *n-th* derivative of ψ_1 is only localized in one slit, its product with ψ_2 everywhere is 0. Therefore, there will be no dependence on α for the average of any power of the momentum. Certainly there will be no dependence on α if ever there is a power of position or the product of position and momentum.

So what is going on? Which operator is sensitive to it? Well it turns out that if I look at the operator $e^{ipL/\hbar}$, it acts like a translational operator. In order to prove that you don't necessarily have to use power expansion. You can use directly the Fourier transform, and as long as you have the function of x of the Fourier transform you can show that $e^{ipL/\hbar}$ is a translational operator. So when I take $e^{ipL/\hbar}$ and calculate each average, it will be equal to $e^{i\alpha/2}$. You can easily check that, because $e^{ipL/\hbar}$ now takes ψ_1 and brings it to ψ_2, or vice-versa. You can immediately say: "but you can expand $e^{ipL/\hbar}$ as an infinite power series of momentum!" Well, when you do expand it, it looks as if it is a Taylor expansion; that is what you take from the two slits. But because there is no overlap, this function *cannot* have a Taylor expansion. The Taylor expansion doesn't converge. And therefore, the average of the power series is not equal to the series of the averages. This is an example where the average $(dp/dt)e^{ipL/\hbar}$ does indeed depend on the electric phase. But if you look separately at each one of the components in the expansion, you see that this sum doesn't converge, because there is no Taylor expansion.

I argue that this variable, which is the modulo momentum, is the correct variable to think of when considering what happens in quantum interference. It has all the right properties. In the classical image, it disappears because it doesn't exist. In the quantum limit it is indeed sensitive to the relative phase.

So the idea of thinking about quantum interference in the Heisenberg picture is the following. When a particle comes through a pair of slits, with a solenoid behind them, the solenoid will not change any of the moduli, any of the powers of the momentum, because it changes only the *relative* phase. All the momentum powers, the average of all the moments, will be unaffected by it. The only things that will be affected in the A-B effect will be the modulo variable. And therefore we can think about the A-B effect as an exchange; a non-local exchange of this modulo momentum. The modulo momentum has been conserved too. Let's look at two interacting systems and write, for example, $\cos(p_1 L / \hbar)$ as the modulo variable of the first one and $\cos(p_2 L / \hbar)$ over the second. We know that p_1+p_2 is a conserved quantity. Then it is very easy to show that their conservation comes from the fact they move on an ellipse. That means that if I change the modulo variable of one system, there must be also a corresponding change of the other. So I can think about the interaction between the solenoid and the wave packets of the electron as an exchange, not of momentum, but an exchange of *modulo momentum.*

This exchange happens non-locally. That means that if I look at the Heisenberg equation of motion, it is fundamentally different from the classical equation of motion. Usually people think that the fundamental difference, or the most important difference, between classical and quantum mechanics is in the kinematics; namely, that the quantum particle is described by a wave, whereas a classical particle is described by point, position, velocity. But when you look at the equation of motion, you've changed the Poisson brackets to commutators. So it looks as if the difference between the classical and quantum accounts is just corrections of the order of h. It turns out that this is not true. The fundamental difference between the classical and quantum equation of motion emerges when you look at the variables that are relevant for interference. Assume we have a Hamiltonian for a particle which is the free part of the Hamiltonian plus, say, some potential $V(x)$. In classical physics, it is true that if you take any function of momentum $f(p)$, the time derivative of this function is equal simply to

$$\frac{df(p)}{dt} = \frac{\partial f}{\partial p}\frac{dp}{dt} = -\frac{\partial f}{\partial p}\frac{dV}{dx},$$

meaning that in classical physics, for the particle's momentum to change, it must be in a place where there is a force. But in quantum mechanics, we have seen that the relevant quantity is the modulo variable. If we look at the time derivative, for example, of $e^{ipL/\hbar}$, we find that it is equal to

$$\frac{d}{dt}e^{ipL/\hbar} = \frac{i}{\hbar}[V(x) - V(x+L)]e^{ipL/\hbar}.$$

So you suddenly realize that in quantum mechanics, when you look at the relevant variables for interference, they include not only the potential where the particle is, but also the potential distance L away from the particle.

Here is a new way to think about the mystery of the two-slit interference. Feynman[7] has once made a famous statement about the double-slit experiment that implies that nobody really understands what goes on in quantum interference. I beg to differ! There is an intuitive way to understand what is going on as described in the following: You send the particle and indeed, you do not know whether it goes this way or that way. But you would like to say that it goes through one of them, only you don't know which. The minute you say that, you have to ask: "Ah, but how does the particle that goes *there* knows that the potential *here*, whether this slit is open or not?" The answer I propose is that there is a non-local equation of motion of the relevant variable, which is the modulo momentum. It tells you that if one slit is closed, it affects the modulo momentum differently than if it is open. But then you say, "Hey, wait a minute. Doesn't it break causality? If the electron is now moving *here*, and it knows whether the other slit *there* is open or not, doesn't it violate causality? After all, opening or closing the slit is a last-minute decision!" Well, quantum mechanics saves us in a very interesting way. It shows that this modulo momentum, *p mod (h/L)*, has a very interesting property: if the particle is confined to a location limited by L, and if you look at the modulo momentum of any L equal to that of the latter, that is, *p mod (h/L)* becomes completely uncertain. What does it mean "completely uncertain"? You can think about it as a circle because it is limited. If you have an equal probability for any value, then of course if you rotate the circle you will not see any difference. It turns out that when the electron goes through one slit, its modulo momentum is completely uncertain. Therefore, although the modulo momentum has an effect, it is unobservable. But when the electron's position is in superposition of being in one slit and in the other, then this modulo variable is known. And then the non-local effect can be observed.

So, the basic idea is to point out the truly non-local side of quantum interference. Nonlocality, which will later emerge even more dramatically in the EPR experiment,[8] is presented by variables that have no classical analog. These variables can be associated with each individual particle, not like the Schrödinger wave, which, in essence, is a property of an *ensemble*. The beautiful thing about the Heisenberg representation is that variables can be observed on *individual* particles. This, by the way, is what we accomplish nowadays with weak measurements.[9] So we have to learn again to think not only in the Schrödinger representation, which is very simple mathematically though very confusing intuitively. We have to learn how to observe quantum phenomena from both the Schrödinger and Heisenberg points of view.

Finally please allow me to share with you an interesting historical anecdote. I met Heisenberg in the Max Planck Institute a few years before he died. That was in the early 1960s. I met him in his office and asked him, "Professor Heisenberg, how do you think about interference phenomena using your representation?" Lo and behold, he said, "I don't know how to do it. I've never thought about it." Really! Then I offered him this idea. He was so excited and so happy about it, and I want to propose the reason for that. There is something very interesting in the development of quantum mechanics. Schrödinger and Heisenberg came practically the same time, Heisenberg a little earlier. Everybody tried to solve everything using Heisenberg's matrix representation. It was very complicated. It could solve only very few problems in the beginning. Then came Schrödinger and immediately everybody jumped on his wagon because it was so much easier to work with. Heisenberg was naturally unhappy. He tried to fight this change but could not slow that bandwagon. So the idea I proposed on that day showed him that it *is* possible to understand quantum interference in his language. He was so happy about it that I heard indirectly from his assistant, Hans-Peter Dürr, that since that day every visitor who came to visit Heisenberg, the first thing he would show them is how to understand the interference pattern in the Heisenberg language. That made him extremely happy.

Closing the circle, I think this is also the correct way to think about the A-B effect. It is an exchange of these dynamical variables.

So I recommend to everybody to pay more attention, not only to Schrödinger's representation, but also to Heisenberg's. Because if you learn to do it you will look at the Schrödinger representation as only a mathematical simplification aiding the solution of complicated problems. But then things that appear have only to do with an ensemble, *not* with individual particles. You can't say any more that there is a wave function moving in space. There is nothing like that! But if you think about Heisenberg representation, you can think about variables that can really be associated with individual particles. When you look at functions of position and momentum – these are observables. You can look at individual particles and play around with the Heisenberg equation of motion. You get much more intuition. Much of my later work on weak measurements[9,10] presents this intuition's fruits.

4. Discussion (Questions and answers)

Q: I was wondering whether these ideas about interference can also shed some light about collapse, which is basically a loss of non-locality. Have you thought about this? I'm sure you have.

A: Yes and No. First of all, even in the Heisenberg representation, there is no indication for collapse. The collapse will have to come from someplace else. So far, there is no way that you can say otherwise, apart from some kinds of very ugly attempts to introduce non-localities into the Schrödinger equation. Up to now there is no solution of the collapse, although I have some ideas which are not connected with this. They are connected with what will be discussed by my colleague Jeff Tollaksen. But it is not finished yet so I'm not going to talk about it. The collapse is still an open issue. But at least out of collapse I am saying there is a way to develop more intuition about what happens in the quantum domain provided you learn to think about it in the classical language, and see the difference between the classical language when you change from the dynamical equation of motion by Poisson brackets to the dynamical equation of motion by commutators. That's where the basic new difference between quantum and classical mechanics is - The non-locality of the quantum equation of motion. That is the thing that was missed.

Q: How do I think about having detectors at both slits in this language?

A: The point is that once you put the detector, in any one of them, it makes the modulo momentum completely uncertain. And that is the beautiful thing. This complete uncertainty saves you from breaking causality,

I want to emphasize the difference between complete uncertainty and the usual Heisenberg uncertainty. Heisenberg has certain principles that are quantitative. The complete uncertainty is a qualitative one. There must be some reason for it. And my question is "Why does God play dice?" If God wants to have non-local equations of motion, and at the same time to save causality, you had better have uncertainties. But if you have uncertainties which are complete, you could not detect the violation of causality. And this complete uncertainty of the modular momentum.... It's a very strong one. From it you can deduce the usual Heisenberg uncertainty, but not vice versa. So this is the more fundamental way to think about the quantum uncertainties. There is a reason why there must be uncertainties if you want to combine non-locality and causality together.

References

1. Y. Aharonov and D. Bohm, Significance of electromagnetic potentials in the quantum theory, *Phys. Rev.* **115**(3), 485-491, (1959).

2. R. G. Chambers, Shift of an electron interference pattern by enclosed magnetic flux, *Phys. Rev. Lett.* **5**(1), 3-5, (1960).

3. G. Möllenstedt und W. Bayh, Messung der kontinuierlichen Phasenschiebung von Elektronenwellen im kraftfeldfreien Raum durch das magnetische Vektorpotential einer Luftspule, *Naturwiss.* **49**, 81, (1962).

4. A. Tonomura, T. Matsuda, R. Suzuki, A. Fukuhara, N. Osakabe, H. Umezaki, J. Endo, K. Shinagawa, Y. Sugita, and H. Fujiwara, Observation of Aharonov-Bohm effect by electron holography, *Phys. Rev. Lett.* **48**(21), 1443-1446, (1982).

5. A. Tonomura, N. Osakabe, T. Matsuda, T. Kawasaki, J. Endo, S. Yano, and H. Yamada, Evidence for Aharonov-Bohm effect with magnetic field completely shielded from electron wave, *Phys. Rev. Lett.* **56**(8), 792-795, (1986).

6. N. Osakabe, T. Matsuda, T. Kawasaki, J. Endo, A. Tonomura, S. Yano, and H. Yamada, Experimental confirmation of the Aharonov-Bohm effect using a toroidal magnetic field confined by a superconductor, *Phys. Rev. A* **34**(2), 815-822, (1986).

7. R. P. Feynman, *The Feynman Lectures on Physics*, eds. R. P. Feynman, R. B. Leighton, and M. Sands, (Addison-Wesley, Reading, Mass, 1964), Vol. 3, Chap. 1.

8. A. Einstein, B. Podolsky, and N. Rosen, Can quantum-mechanical description of physical reality be considered complete?, *Phys. Rev.* **47**(10), 777-780, (1935).

9. Y. Aharonov, E. Cohen, and A. C. Elitzur, Broadening the scope of weak quantum measurements I: A single particle accurately measured yet left superposed, (2013). http://xxx.lanl.gov/abs/1207.0667.

10. Y. Aharonov, E. Cohen, and S. Ben-Moshe, Unusual interaction of a pre-and-post-selected particle, (2013). http://xxx.lanl.gov/abs/1208.3203.

How the Test of Aharonov-Bohm Effect Was Initiated at Hitachi Laboratory

Nobuyuki Osakabe

Central Research Laboratory, Hitachi, Ltd.
Kokubunji, Tokyo 185-8601, Japan
E-mail: nobuyuki.osakabe.hu@hitachi.com

I joined the Tonomura's team in 1980. Since then, I have seen his enthusiasm and creativity in science as a member of his team and later as director of the laboratory. I will discuss in this article how the industrially driven technologies met science at Hitachi Central Research Laboratory in the case of verification of the Aharonov-Bohm effect and other scientific achievements by Akira Tonomura.

1. Introduction

In 1942, the development of electron microscopy began as one of the first research subjects at the newly founded Hitachi Central Research Laboratory (HCRL). The development was successful and the related business was launched. In 1956 this technology first met science when Dr. Hiroshi Watanabe of HCRL verified[1] the Bohm-Pines theory[2] of quantized electron plasma in a solid by means of electron energy loss spectroscopy, the derivative technique of electron microscopy.

Akira Tonomura, who had just graduated from the University of Tokyo, was quite fascinated by Watanabe's work and joined his research group at HCRL. At the suggestion of Watanabe, Tonomura started research on electron interferometry. He spent a year at Professor Gottfried Möllenstedt's laboratory in Tübingen. He was greatly inspired and perhaps it was through this experience that he reached the idea to verify the Aharonov-Bohm effect[3] and wave-particle duality.

Soon after returning to Japan, he started to develop a field-emission transmission electron microscope (FE-TEM), which is the key apparatus for electron interferometry. After years of unwavering efforts, in 1979 Tonomura successfully developed[4] an FE-TEM and began to pursue his ideas incubated at Tübingen.

The first attempt to verify the AB effect was a part of the project to improve the bit-densities of magnetic recording media through understanding of micromagnetics using electron interferometry for visualization of magnetization pattern.[5] Industrial necessity drove the project. The experiments were successfully conducted using a ferromagnetic toroid made by microfabrication technology, and this was followed by verification experiments[6-8] using toroidal ferromagnets completely shielded by superconducting niobium. The continuous encouragement and advice received during this study from many eminent scientists including Professor C. N. Yang was also a key factor in its success.

2. From Tonomura's birth to joining HCRL

Akira Tonomura was born in Nishinomiya, Hyogo in 1942. Interestingly, HCRL was also founded in the same year at Kokubunji, Tokyo. One of the first projects of HCRL was the development of electron microscope. It was also involved in a national project, which was composed of six universities as well as of industrial players, such as Hitachi, Toshiba, JEOL, and others. In this way the laboratory started its project in an open-innovation manner with strong contacts with academia. Thus began the development of electron microscope and grew into business, and the products were even exported to advanced countries.

Based on this successful business, the HCRL focused on the research and development of electron microscopes. In 1956 Dr. Hiroshi Watanabe[1] succeeded in demonstrating the presence of quantized plasma oscillation in metal solids by electron microscopy. Using a single picture, he clearly showed the momentum vs. energy dispersion relation of plasma oscillation, verifying the Bohm-Pines theory[2] predicted in 1953.

Five years after the Watanabe's experiment, Tonomura entered Department of Physics of the University of Tokyo. There, he was quite impressed by lectures on quantum mechanics given by Professor Hiroomi Umezawa, a theoretical physicist working on quantum theory of fields. Tonomura's graduation thesis advisor was Professor Koichi Shimoda, who worked on the foundation of laser with Professor Charles H. Towns at Harvard University. When he graduated from the University of Tokyo, he selected to get a job in the industry. At that time most of the graduates of Department of Physics went on to the post-graduate school. Tonomura used to say "My colleagues were so gifted that I had to select other career path." This was his modest way of expressing his choice of career because he was so deeply moved by the experiment of Watanabe, who showed a deep physics by the single picture of his experiment. Because of this, Tonomura decided to join Hitachi.

3. Tonomura's early research activities at HCRL

After joining HCRL, Tonomura started to work on electron holography using electron beams through the suggestion of Watanabe. Figure 1 shows the hologram and reconstructed images of a ZnO particle taken by Fraunhofer in-line holography[9] in 1968. This was the first successful in-line holography experiment.

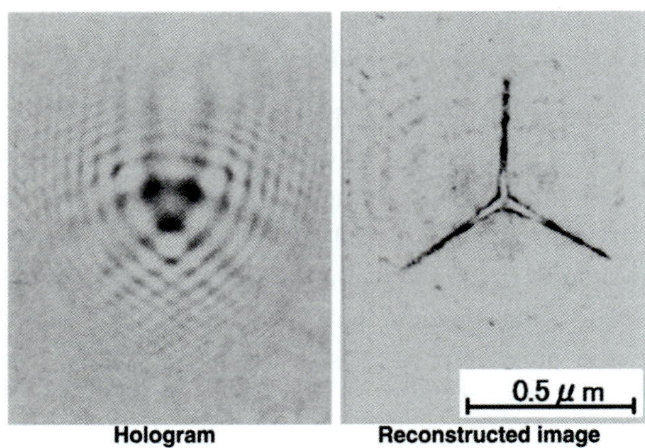

Fig. 1. First hologram and reconstructed images of a ZnO particle taken by Fraunhofer in-line holography.

Holography was invented by Dennis Gabor[10] in 1948. Tonomura left me a voice message about the exchanges of letters with Gabor; the message was recorded after Tonomura realized that his life was only one month left. In the following I pick out several passages from his voice mail message.

> Gabor invented holography for which he received the Nobel Prize in 1973.... He wanted very much, however, to attain a method to remove aberration in electron microscope. So in 1968, when I worked on in-line (Gabor-type) holography, he requested me to work on aberration correction, which led to many exchanges of letters.... At the time, he was quite persistent and so Komoda-san and I worked hard, but it was impossible to achieve without highly coherent electron beam such as a laser. So, Komoda-san and I decided to realize practical holography. This was the beginning of my lifework of the last 30 years to develop a coherent electron beam. Gabor passed away just before we achieved good results and so regrettably, I was never able to let him know, but that is how things happened.
>
> (red letters by Osakabe)

Akira Tonomura at Saitama Medical University International Medical Center, March 31, 2012.

Figure 2 shows a copy of the actual letter from Gabor, suggesting aberration correction using this technology. Tonomura started to develop a new microscope that had a good coherence. Field emission electron microscope is supposed have several orders of magnitude brighter electron beams. By using the first field-emission electron microscope designed by Tonomura and his colleagues, he was able to take interference patterns in 1970, but coherence of the electron beam was far less than that of his expectation because of mechanical vibrations and electro-magnetic induction.

1st April 1968

Dear Dr. Tonomura,

Many thanks for sending me your note on Optical Reconstruction from a Fraunhofer Electron Hologram. It is a very respectable result for a first effort.

I hope you will continue your work towards higher resolution. May I point out that the "twin" image does not seriously disturb the resolution if one opens up the aperture sufficiently for the spherical aberration to go up to, say, 16 fringes, and if one compensates this in the reconstruction, because the virtual image will have twice the spherical aberration, which means that it will be smeared out almost evenly.

I enclose a little paper of mine, with my compliments.

Yours sincerely,

Prof. D. Gabor, F.R.S.

Fig. 2. Copy of the letter from D. Gabor in 1968 (red letters by Osakabe).

In 1973 Tonomura had a chance to work with Professor Gottfried Möllenstedt at University of Tübingen. During the stay at Tübingen, he was stimulated very much by discussions with Professor Möllenstedt, because at that time Professor Möllenstedt did an earlier attempt to verify the Aharonov-Bohm effect[3] using his technology with an electron biprism.

After coming back to Japan in 1974, Tonomura started to develop a highly coherent electron beam with his colleagues. They succeeded in making the 125-kV field-emission electron microscope.[4] By using this microscope with a Möllenstedt-type electron biprism, they succeeded in taking three thousand interference fringes, indicating high coherence of the electron beam. Furthermore, they demonstrated magnetic domain structures inside the Co particle[11] as shown in the interference micrograph (Fig. 3), where magnetic lines of force in a fine cobalt particle are clearly demonstrated by using electron holography.

     N. Osakabe

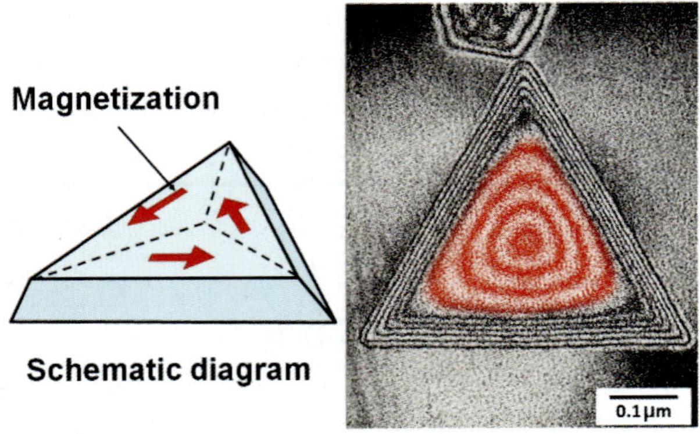

Fig. 3. Interference micrograph of Co particle.

4. Verification experiments of the Aharonov-Bohm effect

By this time Tonomura decided to start experiments to verify the Aharonov-Bohm (AB) effect. An important thing was that at that time the HCRL had a microfabrication technology for semiconductor, magnetic and superconducting devices. These were very important tools to verify the AB effect. The more important thing was Professor Chen Ning Yang's encouragement. Tonomura wrote a letter to Professor Yang about his plan to verify the AB effect through the introduction of Professor Kazuo Fujikawa, Tonomura's classmate at Department of Physics. In the summer of 1981, when Professor Yang was visiting Professor Miyazawa's laboratory at the University of Tokyo, he received a telephone call from Tonomura. The next day Professor Yang made a sudden visit to HCRL. I clearly remember the occasion when he came to encourage Tonomura. Everybody in the laboratory was so surprised. His visit gave a great power to Tonomura to start a new experiment to verify the AB effect.

Tonomura, however, had to negotiate with many scientists and engineers in other departments in the laboratory to help him make devices for the verification, because those people engaged in microfabrication were so busy in product-driven developments. With a passion of Tonomura they finally agreed to co-operate with him by supporting Tonomura's ideas. Actually, the project to verify the AB effect was a joint project to evaluate the magnetic recording capability[5] for higher recording densities at that time. Therefore, the project had dual purposes: one was to develop industrially useful technology by enhancing the recording density and the other was to study the scientifically important concept. By using toroidal ferromagnetic samples we were able show phase shifts passing through the center of the ring as shown in Fig. 4 and verified[6] the existence of the AB effect.

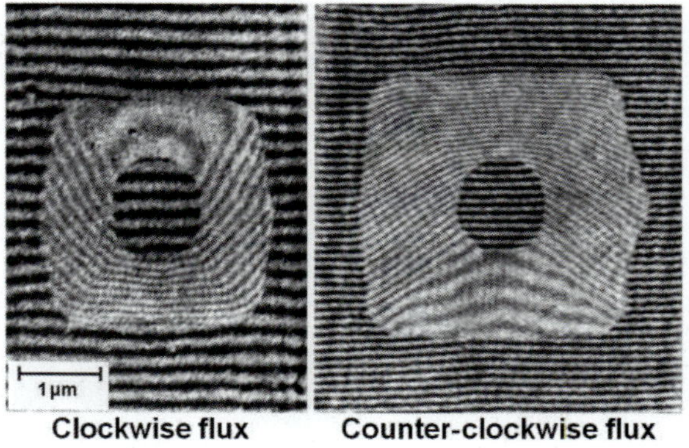

Fig. 4. Electron phase distribution of electrons passing through ring-shaped magnets.

When we succeeded in this experiment, General Manager of HCRL, Dr. Yasutsugu Takeda, decided to sponsor an international symposium on related issues with a strong support of Professor Yang. This is the International Symposium on Foundations of Quantum Mechanics (ISQM[12]) with the subtitle of "in the Light of New Technology," reflecting a nature of the conference.

At that time Tonomura's first experiment was denied by some physicists because of a possible leakage of magnetic field and also possible touching of electron beam to the magnetic field when passing through the ferromagnet. At the First ISQM in 1983, Professor Yang suggested that if the magnetic material was completely covered with a superconductor, an interesting thing might happen. Figure 5 shows the schematic diagram of such an experimental device.

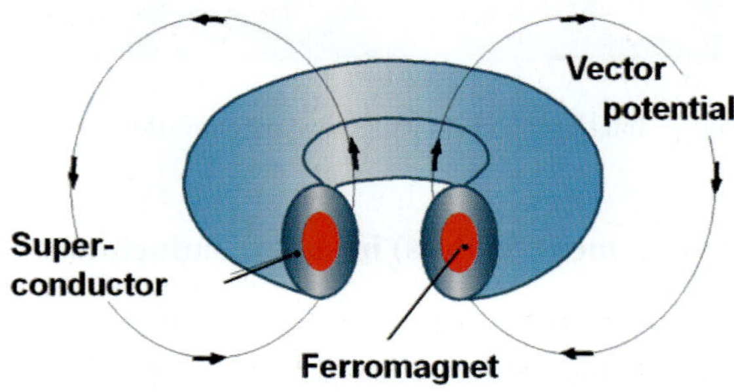

Fig. 5. Schematic diagram of the experimental device.

Figure 6 shows actual devices for this experiment in the form of toroidal ferromagnets covered with Nb superconducting thin films. They were connected to the stem to keep high thermal conductivity, because we had to cool down below the transition temperature of niobium, 9.2 K.

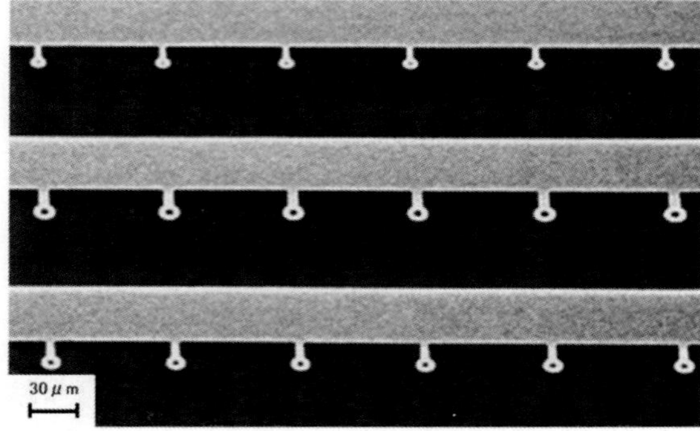

Fig. 6. Toroidal ferromagnet devices covered with Nb.

Using this device we obtained the interferogram shown in Fig. 7. This shows the magnetic flux quantization in units of $ch/2e$ and the electron beam phase shifts of half of the wavelength when the beam passed through the center of the ring. In this way, we completely confirmed the existence of the Aharonov-Bohm effect.[7,8]

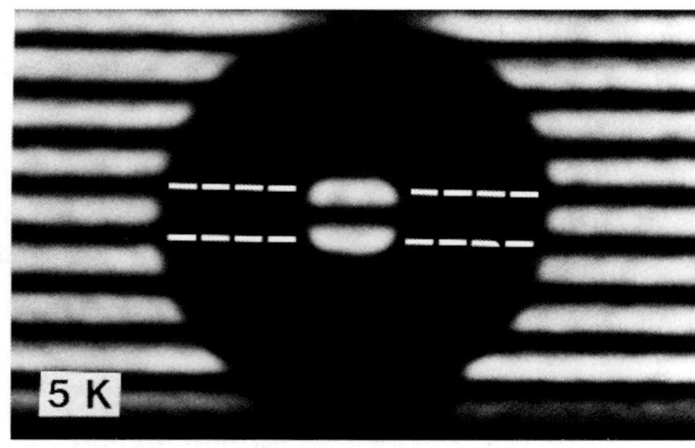

Fig. 7. Interferogram showing the Aharonov-Bohm effect.

5. Study on magnetic flux lines (vortices) in superconductors

For Tonomura it is natural to go on to the observation of magnetic flux lines (vortices) because at that time high-temperature (high-Tc) superconductors were found and the behavior of magnetic lux lines was of interest industrially as well as scientifically. Figure 8 shows results of the magnetic fields appearing from such quantized vortices in a superconducting Pb film.[13]

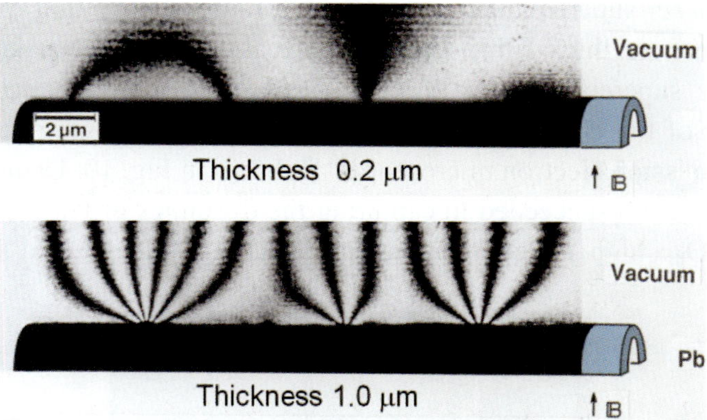

Fig. 8. Interference micrographs of magnetic flux lines penetrating superconducting Ph films.

The important thing for the physics and industrial applications, however, is to see the vortices inside the film. In order to perform that experiment he needed more penetrating power, i.e., higher accelerating voltage. For that purpose, 350-kV field emission electron microscope was developed, which was installed at the newly-founded Advanced Research Laboratory in Hatoyama in 1989. Using this new coherent beam Lorentz microscope Tonomura and his team succeeded in observing static lattice patterns of superconducting Nb and their dynamic behavior in real time.[14] Figure 9 shows the result: the specimen was field-cooled at 10 mT down to 4.5 K.

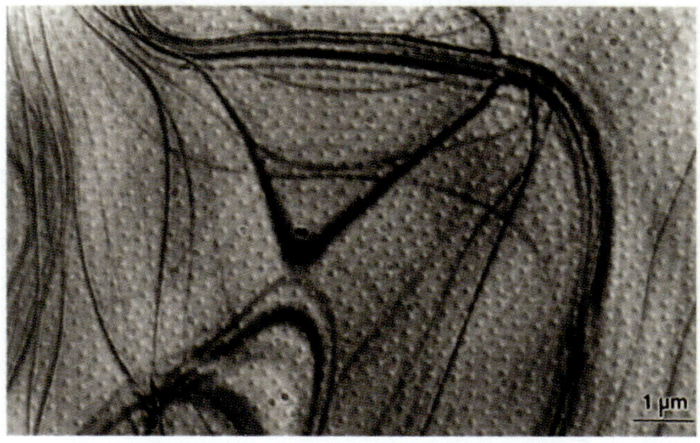

Fig. 9. Lorentz micrograph of Nb thin film.

The more important thing is high-Tc superconductor vortices because for industrial application pinning-down of vortices is extremely important to avoid energy dissipation when electric current passes through a superconductor under the magnetic field. In addition, there are a lot of fruitful physics in the vortex matter. In order to observe vortices in high-*Tc* superconductors, we need more penetrating power and more coherence to penetrate thick films to observe weak phase objects, i.e., magnetic flux lines, in high-*Tc* superconductors. For that purpose, we collaborated with Professor Koichi Kitazawa of the University of Tokyo under CREST program and we developed the 1-MV field emission electron microscope[15,16] shown in Fig. 10. Using this microscope Tonomura and his team succeeded in capturing the dynamics of vortices in $YBa_2Cu_3O_{7-\delta}$ and $Bi_2Sr_2CaCu_2O_{8+\delta}$ high-*Tc* superconductors.[17-19]

Fig. 10. 1-MV field-emission holography electron microscope.

6. FIRST Tonomura project to develop a 1.2 MV holography electron microscope

Tonomura's original motivation for developing holography was realization of high resolution. Toward this goal, he planned to start a new project to make a higher-resolution microscope. He tried for ten years to obtain government funding. In 2009, Japanese government announced an extremely huge-budgeted program called the FIRST program, i.e., Funding Program for World-Leading Innovative R&D on Science and Technology. He applied for it and luckily was selected as one of 30 distinguished researchers with six trillion-yen budget in five years. This FIRST Tonomura project is being conducted jointly by Hitachi, Ltd. and RIKEN to develop an atomic-resolution holography electron microscope. With this microscope, Tonomura wanted to realize his dream of reconstructing the wavefronts of electrons scattered by atoms first proposed by Gabor. Hitachi team is to develop the 1.2 MeV holography electron microscope and

RIKEN team is to prepare the applications. This team is preparing ideas for the experiments using the 1.2 MeV electron microscope to be completed in 2013.

In March 11, 2011, a strong earthquake and tsunami hit the development team. Many parts of the new electron microscope were damaged by this tsunami. For example, high-voltage cables were damaged, which were being fabricated at the firm facing the Pacific Ocean close to the epicenter of the earthquake. With the tremendous efforts of the Hitachi team, the microscopy development is now ongoing almost on time and actually the column of the electron microscope was almost completed in April 2012. But Tonomura passed away on May 2.

7. Quotation on R&D by Professor Ryoji Uyeda

Finally I quote words by Professor Ryoji Uyeda, who was a good advisor and mentor of Akira Tonomura. Tonomura told us that every time he came up with a result that he thought was impressive, he sent the report to Professor Uyeda, but after the jubilation subsided, Professor Uyeda always add the following words:

> You should not be satisfied with such a result. This may not be imitative, but it will not compete against the results achieved by scientists in western countries, and really doesn't go beyond the level of *basic practice*. It is only when your results become a great trunk supporting myriad boughs and branches, and then your work would be considered *basic research*. Cultivating the roots of robust science and technology traditions in Japan will take 60 years of concerted effort. You should maintain confidence in yourself, keep plugging away, and try not to get bogged down in pretty chores.
>
> Ryoji Uyeda

Tonomura always said these words of Professor Uyeda to us when he achieved something new. He was never satisfied with his achievement. But I think Tonomura worked a lot. Now, I think it's our turn to carry his baton, to complete the development of the microscope, and to use the microscope for creative and useful applications

Thank you very much, Tonomura-san. May you rest in peace.

Acknowledgments

The development of the 1 MV holography electron microscope and its application to the observation of vortices in high-Tc superconductors were financially supported by the Japan Science and Technology Agency through Core Research for Evolutional Science and Technology (CREST) and Solution-Oriented Research for Science and Technology (SORST) programs. The development of the 1.2 MV holography electron microscope was supported by a grant from the Japan Society for the Promotion of Science through the "Funding Program for World-Leading Innovative R&D on Science and Technology (FIRST Program)," initiated by the Council for Science and Technology Policy.

References

1. H. Watanabe, Experimental evidence for the collective nature of the characteristic energy loss of electrons in solids –Studies on the dispersion relation of plasma frequency–, *J. Phys. Soc. Japan* **11**(2), 112-119, (1956).
2. D. Bohm and D. Pines, A collective description of electron interactions: III Coulomb interactions in a degenerate electron gas, *Phys. Rev.* **92**(3), 609-625, (1953)
3. Y. Aharonov and D. Bohm, Significance of electromagnetic potentials in the quantum theory, *Phys. Rev.* **115**(3), 485-491, (1959).
4. A. Tonomura, T. Matsuda, J. Endo, H. Todokoro, and T. Komoda, Development of a field emission electron microscope, *J. Electron Microsc.* **28**(1), 1-11, (1979).
5. N. Osakabe, K. Yoshida, Y. Horiuchi, T. Matsuda, H. Tanabe, T. Okuwaki, J. Endo, H. Fujiwara, and A. Tonomura, Observation of recorded magnetization pattern by electron holography, *Appl. Phys. Lett.* **42**(8), 746-748, (1983).
6. A. Tonomura, T. Matsuda, R. Suzuki, A. Fukuhara, N. Osakabe, H. Umezaki, J. Endo, K. Shinagawa, Y. Sugita, and H. Fujiwara, Observation of Aharonov-Bohm effect by electron holography, *Phys. Rev. Lett.* **48**(21), 1443-1446, (1982).
7. A. Tonomura, N. Osakabe, T. Matsuda, T. Kawasaki, J. Endo, S. Yano and H. Yamada, Evidence for Aharonov-Bohm effect with magnetic field completely shielded from electron wave, *Phys. Rev. Lett.* **56**(8), 792-795, (1986).
8. N. Osakabe, T. Matsuda, T. Kawasaki, J. Endo, A. Tonomura, S. Yano, and H. Yamada, Experimental confirmation of the Aharonov-Bohm effect using a toroidal magnetic field confined by a superconductor, *Phys. Rev. A* **34**(2), 815-822, (1986).
9. A. Tonomura, A. Fukuhara, H. Watanabe, and T. Komoda, Optical reconstruction of image from Fraunhofer electron-hologram, *Jpn. J. Appl. Phys.* **7**(3), 295, (1968).
10. D. Gabor, A new microscopic principle, *Nature* **161**(4098), 777-778, (1948).
11. A. Tonomura, T. Matsuda, J. Endo, T. Arii, and K. Mihama, Direct observation of fine structure of magnetic domain walls by electron holography, *Phys. Rev. Lett.* **44** (21), 1430-1433, (1980).
12. *Foundations of Quantum Mechanics in the Light of New Technology—Selected papers from the Proceedings of the first through fourth International Symposia on Foundations of Quantum Mechanics.* Eds. S. Nakajima, Y. Murayama, and A. Tonomura, World Scientific, Singapore, 1996.
13. T. Matsuda, S. Hasegawa, M. Igarashi, T. Kobayashi, M. Naito, H. Kajiyama, J. Endo, N. Osakabe, A. Tonomura, and R. Aoki, Magnetic field observation of a single flux quantum by electron-holographic interferometry, *Phys. Rev. Lett.* **62**(21), 2519-2522, (1989).
14. K. Harada, T. Matsuda, J. Bonevich, M. Igarashi, S. Kondo, G. Pozzi, U. Kawabe, and A. Tonomura, Real-time observation of vortex lattices in a superconductor by electron microscopy, *Nature* **360**, 51-53, (1992).
15. T. Kawasaki, I. Matsui, T. Yoshida, T. Katsura, S. Hayashi, T. Onai, T. Furutsu, K. Myochinm, M. Numata, H. Mogaki, M. Gorai, T. Akashi, O. Kamimura, T. Matsuda, N. Osakabe, A. Tonomura, and K. Kitazawa, Development of a 1MV field-emission transmission electron microscope, *J. Electron Microsc.* **49**(6), 711-718, (2000).
16. K. Kawasaki, T. Yoshida, T. Matsuda, N. Osakabe, A. Tonomura, I. Matsui, and K. Kitazawa, Fine crystal lattice fringes observed using a transmission electron microscope with 1 MeV coherent electron waves, *Appl. Phys. Lett.* **76**(10), 1342-1444, (2000).

17. A. Tonomura, H. Kasai, O. Kamimura, T. Matsuda, K. Harada, Y. Nakayama, J. Shimoyama, K. Kishio, T. Hanaguri, K. Kitazawa, M. Sasase, and S. Okayasu, Observation of individual vortices trapped along columnar defects in high-temperature superconductors, *Nature* **412**, 620-622, (2001).

18. A. Tonomura, H. Kasai, O. Kamimura, T. Matsuda, K. Harada, Y. Nakayama, J. Shimoyama, K. Kishio, T. Hanaguri, K. Kitazawa, M. Sasase, and S. Okayasu, Lorentz microscopy observation of vortices inside Bi-2212 thin film with columnar defects, *Physica C* **369**(1-4), 68-76, (2002).

19. A. Tonomura, Direct observation of thitherto unobservable quantum phenomena by using electrons, *Proc. Natl. Acad. Sci. USA* **102**(42), 14952-14959, (2005).

Some Reflections Concerning Geometrical Phases

Anthony J. Leggett and Yiruo Lin

*Department of Physics, University of Illinois at Urbana-Champaign,
Urbana, IL 61801, U.S.A.
E-mail: aleggett@illinois.edu*

While the concepts of the Aharonov-Bohm and Berry phases are relatively clear for the case of a single particle, their application to many-body systems may be more problematic. We consider in particular the case of Bogoliubov and Majorana quasiparticles in a Fermi superfluid, with the conclusion that some rather basic questions appear at present not to have received a definitive resolution.

Foreword (by AJL)

While it is a matter of great sorrow that Dr. Akira Tonomura is no longer with us, I am very glad to have the opportunity to contribute to this volume, which I hope will honor and preserve his memory. I first met Dr. Tonomura in 1983 at the First International Symposium on the Foundations of Quantum Mechanics in the Light of New Technology (ISQM-Tokyo '83), a symposium that was motivated by his beautiful electron holographic experiments and that probed what in those days was largely untrodden ground. Since then we kept in touch for three decades, until his death last spring: I was particularly happy to see him at the 2010 meeting of the US National Academy of Sciences, to which he had been elected as a foreign member a few years earlier. I have always valued him both as a trail-blazing colleague and as a warm and hospitable personal friend. We will all miss him greatly.

1. Introduction

In this short essay, we would like to take up some questions which are associated directly or indirectly with Dr. Tonomura's work, and which we are inclined to believe are not quite as well understood theoretically as is generally believed. The peg on which we shall hang this discussion is the currently very active quest for a topological quantum computer, in particular based on so-called quasi-two-dimensional $(p + ip)$ Fermi superfluids such as

Sr_2RuO_4, which has generated a rich theoretical literature. The basic idea is that when a vortex of a certain kind is produced in such a (p + ip) Fermi superfluid, it may carry a peculiar type of excitation, a so-called "Majorana fermion," which in a sense to be explained is half of a real "Dirac-Bogoliubov (DB)" fermion; when a set of such vortices in a quasi-two-dimensional system (such as the individual RuO_2 planes in Sr_2RuO_4) is "braided," (i.e., the vortices are in effect carried around one another), the effect is to realize "non-Abelian" statistics and thus permit a form of quantum computing which is (partially) protected by topological considerations. Within the context of the standard approach based on the Bogoliubov-de Gennes (BdG) equations (see below), this idea is explained very elegantly in the original paper[1] by Ivanov, and amplified and/or reviewed in a number of subsequent papers, e.g., Refs. 2 and 3. In this brief and informal paper we will argue that there are a number of questions connected with this "standard" approach, some of them quite elementary in nature, which in our opinion (and to our initial surprise) have received insufficient discussion in the existing literature; until those questions are resolved (something we are unable to do here), we believe that the theoretical foundations for the effort to realize topological quantum computing (hereafter abbreviated TQC) in quasi-2D ($p + ip$) Fermi superfluids remain insufficiently secure.

The basic problem of interest relates to a system, either a single particle or a collection of particles, whose motion is effectively confined to a two-dimensional surface or plane; it concerns the behavior of the quantum state vector (wave function) of the system under various cyclic operations performed in this plane, e.g., encirclement of one vortex by another (The more commonly considered operation of interchange of the two vortices is "half" of this operation). So long as the motion is indeed strictly confined to the plane (but not otherwise!), it is possible to define the "sense" of such operations unambiguously.

2. Single massive particle with Aharonov-Bohm effect and Berry phase

Let us start with the case of a single massive particle, possibly endowed with charge and/or spin (the latter being 3-dimensional) and moving non-relativistically in the plane. We first briefly review two very well-known effects usually associated with the names of Aharonov and Bohm (AB) and Berry. The best-known form of the AB effect, of course, relates to the interference experiments for whose realization Dr. Tonomura is justly famous, in which a beam of electrons is directed around the two sides of a region of space containing a magnetic flux Φ (but with no magnetic field on the regions in which the beam actually propagates (see Fig. 1)). We will refer to this type of situation as "AB conditions." A bare-bones version of the original AB argument is to note that if in the absence of the flux, the time-dependent Schrödinger wave function has a solution corresponding to a quasiclassical wave packet $\psi(\mathbf{r}t)$, then in the presence of the vector potential $\mathbf{A}(\mathbf{r}t)$ associated with the flux, the equation is solved by the wave function and

$$\psi'(rt) \equiv \psi(rt)\exp - ie \int A(r)dr / \hbar,\qquad(1)$$

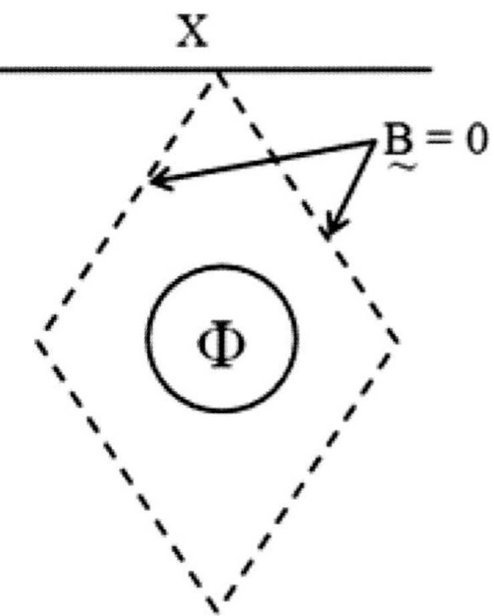

Fig. 1. Geometry for the AB effect.

hence two wave packets propagating around opposite sides of the flux will arrive at the point X (Fig. 1) with their relative phase shifted by the (gauge-invariant) quantity $(e/\hbar)\Phi$; this gives a corresponding effect on the interference fringes relative to those obtained in the absence of the flux, as seen in Dr. Tonomura's experiments. While this is the most directly observable version of the AB effect, it is useful to note that by inverting the argument noted below in the context of the Berry phase, it is possible to obtain the result that when a charged particle is transported, under "AB conditions," around a flux Φ, the extra (non-dynamical) phase acquired – call it φ_{AB} – is given by the expression

$$\varphi_{AB} = (e / \hbar)\Phi \equiv 2\pi(\Phi / \Phi_0),\qquad(2)$$

where we defined the (single-particle) flux quantum Φ_0 as h/e. Thus for example, if $\Phi = \frac{1}{2}\Phi_0$ as in the superconducting implementation used by Dr. Tonomura, $\varphi_{AB} = \pi$.

Let's next review the idea of a (scalar) Berry phase. To illustrate this notion, we consider the standard example of a particle of spin $\frac{1}{2}$, for the moment stationary in space, which is subjected to a Zeeman magnetic field $\mathbf{B}$ of constant magnitude whose direction varies cyclically in time, let us say rotating around the z-axis at an azimuthal angle θ. It is assumed that the rotation is sufficiently slow ($\omega \ll \gamma B$, where γ is the gyromagnetic ratio) that the spin direction follows that of the field adiabatically; then at any given time t (when the polar angle of the field is $\varphi(t)$) the spinor wave function, $\Psi(t)$, of the particle is of the form

$$\Psi(t) = \exp i\chi(t)\begin{pmatrix} \cos\dfrac{\theta}{2}\exp i\phi(t) \\[2ex] \sin\dfrac{\theta}{2} \end{pmatrix} \equiv \exp i\chi(t)\cdot\tilde{\Psi}(t), \tag{3}$$

where the overall phase $\chi(t)$ is for the moment undetermined. Note that the form of $\Psi(t)$, regarded as a function of φ, is single-valued. What Berry showed is that the overall phase $\chi(t)$ has not only the familiar "dynamical" contribution Et (where $E = -\gamma\hbar B/2$ is the Zeeman energy) but a second contribution of purely geometrical origin, which is given by the expression

$$\frac{\partial\chi_{geom(t)}}{\partial t} = \operatorname{Im}\tilde{\Psi}^*(t)\cdot\frac{\partial\tilde{\Psi}}{\partial t} \tag{4}$$

so that the value of $\chi_{geom}(t)$ after a complete cycle (the usually defined "geometrical" or "Berry" phase, φ_B) is (mod. 2π)

$$\varphi_B = \operatorname{Im}\oint\tilde{\Psi}^*(t)\frac{\partial\tilde{\Psi}}{\partial t}\,dt = 2\pi\cos^2\theta/2 \equiv 2\pi - S/2, \tag{5}$$

where S is the solid angle swept out by the field direction over the cycle (see Fig. 2).

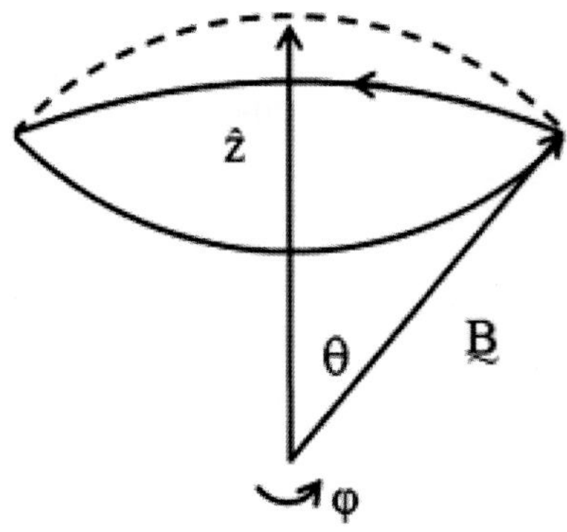

Fig. 2. Geometry for the Berry phase.

Note (a) that this expression is invariant under any choice of overall phase of $\Psi(t)$, which preserves its single-valuedness as a function of φ, and (b) that for $\theta = \pi/2$ (2π rotation in plane), we get $\varphi_B = \pi$, i.e., the standard minus sign of spinors under 2π rotation.

So far we assumed that the particle is stationary in space while the magnetic field varies in time. However, it is obvious that we can obtain the same result if we let the field be time-independent but spatially dependent, and allow the particle to follow a (classical) trajectory $r(t)$. Provided the adiabatic condition is maintained, we recover the result (5), with S the solid angle defined by the variation of the field direction along the path. Suppose now that we replace the classical trajectory by a corresponding quasi-classical wave packet, and consider a pair of such wave packets starting from O with the same overall phase and propagating around the two "arms" of the cycle to O' (see Fig. 3). A simple algebraic argument then shows that they arrive at O' with a relative overall phase which is exactly the Berry phase φ_B. As a result, the interference fringes should be shifted just as in the AB effect; this has been beautifully confirmed in neutron scattering experiments.[4,5]

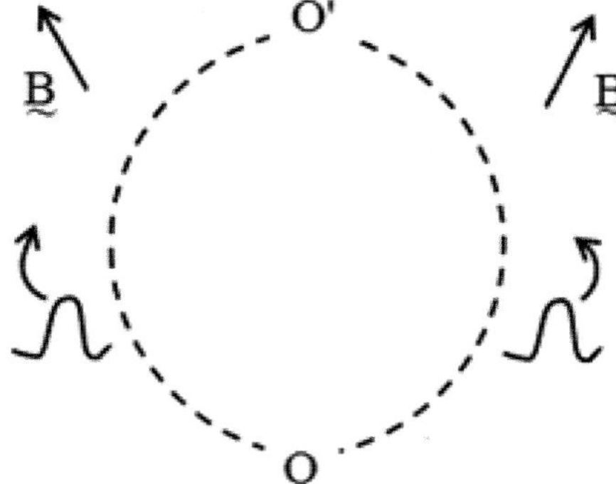

Fig. 3. Berry phase for interference of trajectories.

The question obviously arises; in what sense, if any, can the AB effect be regarded as a special case of the Berry-phase phenomenon? However, this question does not seem obviously relevant to the discussion below, so rather than address it, we just note one important generalization of the notion of Berry "phase," to a matrix quantity: in certain circumstances, if a system is carried adiabatically around a closed loop in the parameter space, the final state vector may not just have a non-trivial extra phase but may even have changed its <u>direction</u> in the Hilbert space. A simple example of this state of affairs is given by a spin-1 particle in a magnetic field with a Hamiltonian $(\mathbf{S}\cdot\mathbf{B})^2$, so that in the ground state $\mathbf{S}$ lies in the plane perpendicular to $\mathbf{B}$; if the direction of $\mathbf{B}$ varies cyclically around a nontrivial loop, then in general the final direction of $\mathbf{S}$ differs from its original one. (This is, of course, nothing but a Hilbert-space version of the phenomenon of "parallel transport" well-known in differential geometry and general relativity.)

The following application of the (scalar) Berry-phase idea may be relevant to our subsequent discussions: Consider a "heavy" particle which can tunnel between two sites, say O and R, so that its wave function when in isolation is of the form

$$\psi = a\psi_O + b\psi_R . \tag{6}$$

Moreover, consider a "light" particle which moves in the field of the heavy one so that it is always in a p-state with respect to it: then, ignoring irrelevant details like radial dependencies, the combined wave function of the two particles up to an overall phase $\chi(t)$ as in eqn. (3) is, of the (entangled) form

$$\tilde{\Psi} = a\psi_o \exp i\varphi_o + b\psi_R \exp i\varphi_R , \tag{7}$$

where φ_O, φ_R are the polar angles with respect to O and R respectively (see Fig. 4).

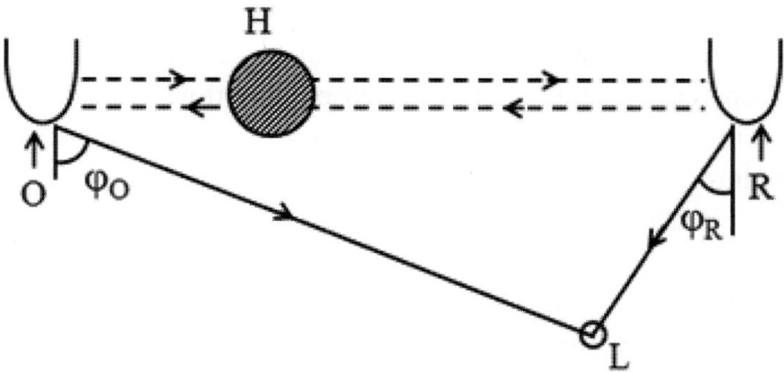

Fig. 4. A light particle constrained to be in a p-state relative to a heavy particle tunneling between A and R.

Needless to say, this wave function is single-valued as a function of (inter alia) the position of the light particle in the plane. Now imagine that we "pin" the light particle, e.g., by a strong additional external potential, and move it around *while continuing to insist that for every given position (i.e., any given pair of values φ_O, φ_R) the form of the wave function remains, up to an overall phase, as in eqn. (7).* It is then straightforward to evaluate the Berry phase using eqn. (4) with the result

$$\varphi_B = 2\pi\left\{|a|^2 n_o + |b|^2 n_R\right\} , \tag{8}$$

where n_O (n_R) is the number of times the path encircles O (R) (counterclockwise encirclements being, as always, counted with a negative sign). This result is intuitively plausible: the Berry phase is essentially a measure of how far the light particle circles the heavy one, and since when viewed semiclassically the fraction of its time the heavy one spends at O (R) is $|a|^2$ ($|b|^2$), this quantity is given by the expression in the brackets in eqn. (8). An application which is more complicated but similar in spirit is the Laughlin[6]-Halperin[7] argument for fractional statistics in the $v = 1/3$ quantum Hall state: since one electron encircling another gives a phase of 2π, and <u>on average</u> the extra number of electrons associated with a Laughlin quasiparticle is $1/3$, when one quasi-particle is encircled by another, the Berry phase is $2\pi/3$ (and hence the "exchange phase" is half of this, $\pi/3$ as usually quoted). So far, so good: at the single- (or two-) particle level things seem relatively clear.

3. Correlated many-particle systems—Fermi systems with Cooper pairing

Things get more interesting when we come to correlated many-particle systems, in particular Fermi systems with Cooper pairing. Let's start with the very simplest situation that is of interest in this context, a simple Abrikosov vortex in an extreme type-II superconductor. For such a vortex we find that at distances from the core large compared to the London penetration depth[a] λ there is no current flow, but a path encircling the vortex encloses a flux[b] $h/2e \equiv \frac{1}{2}\,\Phi_O$, while at distances much less than λ (but much larger than the core radius ξ) negligible flux is created but there is a current corresponding approximately to circulation $\kappa \equiv h/2m$ (where m is the electron mass). Note that the <u>canonical</u> angular momentum ℓ_O of a (paired) electron, which is $(mv_s/\hbar) + \Phi/\Phi_o$ (v_S = superfluid velocity) or $e/\hbar$ times the fluxoid, is independent of r and equal to $\frac{1}{2}$.

Let's consider the Berry phase picked up by a particle of charge e on encircling such a vortex (which by the argument used above for the spin-$\frac{1}{2}$ particles can be used to infer the behavior in the corresponding interference experiment). First, consider the relatively trivial case of a particle such as a muon which has no relation to the condensate. It is fairly clear that, ignoring possible subtle effects of the correlations induced by the Coulomb interaction with the condensate, such a particle will behave as in "free space," i.e., the Berry phase will be π for paths with radius $r \gg \lambda$, and will decrease slowly to become zero for $r \ll \lambda$. So far, no mystery.

---

[a]There is a slight technical problem here, since for a strictly 2-D superconductor λ is infinite; we shall implicitly assume that in practice there will be enough magnetic coupling between planes to make it finite.

[b]Note that throughout this paper Φ_O is the <u>single-particle</u> flux quantum, not the "superconducting" one $h/2e$.

The interesting case is that of a Bogoliubov quasiparticle. In considering the behavior of such a quasiparticle, it needs to be borne in mind that, contrary to a statement unfortunately made in many textbooks, it is <u>not</u> a quantum superposition of an electron and a hole; it is a superposition of (particle) and (hole-plus-Cooper-pair) so that the appropriate form of its creation operator α_i^+ is

$$\alpha_i^+ = \int dr \left\{ u_i(r)\psi^+(r) + v_i(r)\psi(r)\hat{C} \right\}, \tag{9}$$

when the operator $\hat{C}$ adds one extra pair to the condensate (cf. Ref. 6). Moreover, it is necessary to bear carefully in mind that the operator α_i^+ defined by (9) does not create a particle on the vacuum, but rather relates the many-body wave functions of the N-particle groundstate (which for simplicity we will consider to have N even) and the simplest (N+1)-particle (odd-parity) energy eigenstates.

To be sure that our question concerning adiabatic transport is meaningful, we need to consider a specific implementation. To this end let us imagine introducing a weak spatially dependent "pseudo-magnetic" field which couples only to the electron spins, not to their orbital motion[c], and configure this field so that it has a "dimple" such that while the Zeeman energy $\gamma\hbar B$ is small compared to the energy gap Δ (so that the effect on the condensate is at most of order B^2), there exists at least one bound state for a Bogoliubov particle of the appropriate spin. We then move the dimple around the vortex at an angular rate $\omega = \Delta/\hbar$, and ask what "geometric" phase (AB, Berry or both) is acquired by the (many-body) wave function.

In the "AB limit" ($r \gg \lambda$) the answer, as shown by one of us[9], appears to be unambiguously π. So there would seem to be two obvious alternative conjectures: (a) the geometric phase depends only on the canonical angular momentum ℓ_0 of the condensate, and hence is π for all r (or at least for all $r \gg \xi$). (b) the phase decreases smoothly, on decrease of r, from π to 0 (or something close to 0: cf. below). We should be able to discriminate between the conjectures (a) and (b) by studying a slightly simplified problem, namely that of a <u>neutral</u> Fermi system with s-wave pairing, which occupies a thin toroidal container in a state of superflow corresponding to circulation $h/2m$. Does a Bogoliubov quasiparticle transported in the way described above pick up a geometrical phase (a) π , (b) 0 , or (c) something else?

---

[c] For the most urgent case, $r \gg \lambda$, we could alternatively simply imagine turning off the electron charge.

                      A. J. Leggett and Y.-R. Lin

Surprisingly, we have been unable to find a definitive answer to this simply posed question in the existing literature[d]. At first sight it is tempting to regard the problem as completely analogous to that of the spin-½ particle discussed above, with the "particle" and "hole" components playing the role of the "up" and "down" spin components respectively. If this is correct, then the geometrical phase should be equal to 2π times the "weight" of the particle component (cf. eqn. (5), i.e. to the quantity $\int |u_i(\mathbf{r})|^2 \, d\mathbf{r}$. For a trapped quasiparticle, at least in the semiclassical limit where we can visualize it as being repeatedly Andreev-reflected from the walls of the potential, this quantity should be equal to ½, so we obtain the result (a). Unfortunately, we feel unable to regard this answer as completely convincing, inter alia because, so far at least, we have failed to reproduce it definitively within a proper particle conserving calculation.

It should be noted that should either (b) or (c) turn out to be the correct answer, this would constitute a severe prima facie difficulty for the whole idea of TQC in a $(p + ip)$ Fermi superfluid. In such a system the up- and down- spin populations form Cooper pairs independently, and the problem is that the "half-quantum" vortices which are believed to be a prerequisite for the TQC operations, unlike the simple Abrikosov vortices in an s-wave superconductor, have even at $r \gg \lambda$ a nonzero current for each spin population separately[10] (i.e., the situation is qualitatively similar to that obtaining for an s-wave superconductor for $r \sim \lambda$). Consequently, if the geometric phase is not simply a function of the (r-independent) quantity ℓ_O, it will not be π even in this limit. Worse, while for a state exactly symmetric between the up and down spin states it is (or at least might be) $\pi/2$, any slight asymmetry in the up- and down-spin superfluid densities[e] will push it away from this "nice" value (cf. Ref. 11). This would presumably be disastrous for the realization of TQC.

4. Majorana fermions

Let's now turn to the question of Majorana fermions (hereafter M.F.'s). In the context of Fermi systems with Cooper pairing, a M.F. is a localized solution of the Bogoliubov-de Gennes (BdG) equations with energy eigenvalue 0 and coefficients $u(r)$ and $v(r)$ which by a suitable choice of phase convention for the condensate can be made equal, (so that the "particle" so described is its own antiparticle, as in the original particle-physics context). As discussed e.g., in Ref. 12, such an excitation can be regarded as a quantum superposition of an "ordinary" DB fermion with zero energy and an operator which simply annihilates the many-body groundstate; by combining two M.F.'s with the

---

[d] At the Tonomura memorial conference in May 2012, one of us (AJL) posed the alternatives (a) and (b) to the audience; to his surprise he received a substantial number of votes in favor of (b) and none for (a).

[e] Such as might be due to the Zeeman effect of an <u>in-plane</u> magnetic field (which is not subject to Meissner screening).

appropriate relative phase one gets back a DB zero-energy fermion, and since M.F.'s are guaranteed always to occur in pairs it is probably simpler to think in terms of these DB fermions. However, since the positions of the M.F.'s in space may be arbitrarily distant, the DB fermions are spatially "split". In the case of a $(p + ip)$ Fermi superfluid, the BdG equations predict that M.F. solutions can occur on the possible "half-quantum" vortices (HQV's)[f] (that is, vortices in which for $r >> \lambda$ one spin state has circulation $h/2m$ and the other zero). Unfortunately, to the best of our knowledge no one has so far succeeded in writing down explicit many-body wave functions in the presence or absence of a pair of M.F.'s even for a state with just two HQV's. Indeed, the only model of a $(p + ip)$ Fermi system to date which to our knowledge allows such solutions is the celebrated 1D Kitaev quantum wire.[13] As discussed e.g., in Ref. 12, in this model the simplest independent fermionic excitations are associated not with the lattice sites but with the <u>links</u> between neighboring sites, and one way of obtaining a pair of M.F.'s is as follows (see Ref 12 for details): start with the sites arranged in a circle and associate a nonzero (but not necessarily equal) excitation energy E_i with each link i. Next, choose one open link, say $i = 0$, and turn E_O down to zero. Finally, break open the circle by cutting the link 0. The zero-energy DB fermion localized on this link is then "split" between the two ends of the chain, giving a single M.F. on each end.

In the context of the possible use of M.F.'s on HQV's in e.g., Sr_2RuO_4 for TQC, the crucial issue is the behavior of the states involved under "braiding" (adiabatic interchange) or, what is essentially equivalent, encirclement of one vortex by another (since encirclement can be regarded as two braiding processes sequentially performed with the same sense). Consider several possible thought-experiments involving the encirclement of one vortex by another. First, if the vortices are "empty" (no fermions of any kind associated with them) the Berry phase is presumably trivial (0 or, what is equivalent, $2n\pi$); a similar result follows if just one of the two vortices carries an "ordinary" DB fermion and the other is empty. Next, consider the case in which a single DB fermion commutes between the two vortices (regarded as effectively forming simple potential wells), giving rise to the standard even- and odd-parity energy eigenstates[g]. In this case one would expect that the Berry phase for encirclement is still $2n\pi$ in either energy eigenstate, but that for interchange (which now returns us to the same state) is more interesting: intuitively, it should be 0 for the even-parity groundstate but π for the excited state. However, it should be emphasized that this intuition, like the "BdG" argument used above for the annular problem, neglects the possible effect of reaction of the fermion on the condensate.

---

[f] In the case of an odd number of HQV's one M.F. is localized on the boundary of the system.

[g] In such a case the state localized on one vortex is a true DB fermion state, though not of course an energy eigenstate; there is no question of M.F.'s.

                            A. J. Leggett and Y.-R. Lin

We now come to the \$64K question: what is the interchange (or encirclement) phase when the pair of vortices is occupied by a pair of M.F.'s (or, equivalently, a DB fermion is occupying the <u>single</u> "split" $E = 0$ fermionic state)? According to Ivanov's argument (Ref. 1) based on the BdG equations, the exchange phase is $\pi/2$ (and the encirclement phase correspondingly π), and this result (and others which follow straightforwardly from it) is an essential ingredient in proposals for TQC in e.g., Sr_2RuO_4. However, as outlined above, our experience with the simpler "annular" problem suggests that conclusions based on the BdG equations may not always be reliable for this kind of problem; and while there is at least one very interesting recent attempt[14] to circumvent this consideration by solving a prima facie analogous problem for (a slight generalization of) the 1D Kitaev quantum wire, we would not regard the outcome as conclusive as regards the real 2D case. Thus we believe that an explicit solution of the Ivanov problem in terms of many-body wave functions should be a high priority.

Acknowledgments

This work was supported by the National Science Foundation under grant no. NSF-DMR-09-06921. We thank Alfred Goldhaber for helpful correspondence on the "annular" problem.

References

1. D. A. Ivanov, Non-Abelian statistics of half-quantum vortices in p-wave superconductors, *Phys. Rev. Lett.* **86**(2), 268-271, (2001).
2. A. Stern, F. von Oppen, and E. Mariani, Geometric phases and quantum entanglement as building blocks for non-Abelian quasiparticle statistics, *Phys. Rev. B* **70**(20), 205338, (2004).
3. C. Nayak, S. H. Simon, A. Stern, M. Friedman, and S. Das Sarma, Non-Abelian anyons and topological quantum computation, *Rev. Mod. Phys.* **80**(3), 1083-1159, (2008).
4. H. Rauch, A. Zeilinger, G. Badurek, A. Wilfing, W. Bauspiess, and U. Bonse, Verification of coherent spinor rotation of fermions, *Phys. Lett. A* **54**(6), 425-427, (1975).
5. S. A. Werner, R. Colella, A. W. Overhauser, and C. F. Eagen, Observation of the phase shift of a neutron due to precession in a magnetic field, *Phys. Rev. Lett.* **35**(16), 1053-1055, (1975).
6. R. B. Laughlin, Anomalous quantum Hall effect: An incompressible quantum fluid with fractionally charged excitations, *Phys. Rev. Lett.* **50**(18), 1395-1398, (1983).
7. B. I. Halperin, Statistics of quasiparticles and the hierarchy of fractional quantized Hall states, *Phys. Rev. Lett.* **52**(18), 1583-1586, (1984).
8. G. E. Blonder, M. Tinkham, and T. M. Klapwijk, Transition from metallic to tunneling regimes in superconducting microconstrictions: Excess current, charge imbalance, and supercurrent conversion, *Phys. Rev. B* **35**(7), 4515-4532, (1982).
9. Y.-R. Lin, Ph.D. thesis, University of Illinois at Urbana-Champaign (2013).
10. S. B. Chung, H. Bluhm, and E.-A. Kim, Stability of half-quantum vortices in $p_x + ip_y$ superconductors, *Phys. Rev. Lett.* **99**(19), 197002, (2007).

11. E. Babaev, Vortices with fractional flux in two-gap superconductors and in extended Faddeev model, *Phys. Rev. Lett.* **89**(6), 067001, (2992).

12. A. J. Leggett, in Doing Physics: a Festschrift for Tom Erber, pp. 173-183, (IIT Press, 2010).

13. A. Yu. Kitaev, Unpaired Majorana fermions in quantum wires, *Physics-Uspehki* **44**(Supplement), 131, (2001).

14. J. Alicea, Y. Oreg, G. Refael, F. von Oppen, and M. P. A. Fisher, Non-Abelian statistics and topological quantum information processing in 1D wire networks, *Nature Physics* **7**(5), 412-417, (2011).

Mesoscopic Aharonov-Bohm Interferometers:
Decoherence and Thermoelectric Transport

Ora Entin-Wohlman[1,2], Amnon Aharony[1,2], and Yoseph Imry[3]

[1]*Raymond and Beverly Sackler School of Physics and Astronomy, Tel Aviv University,
Tel Aviv 69978, Israel*
[2]*Department of Physics and the Ilse Katz Center for Meso- and Nano-Scale Science
and Technology, Ben Gurion University, Beer Sheva 84105, Israel*
[3]*Department of Condensed Matter Physics, Weizmann Institute of Science,
Rehovot 76100, Israel*

oraentin@bgu.ac.il, aaharony@bgu.ac.il, yoseph.imry@weizmann.ac.il

Two important features of mesoscopic Aharonov-Bohm (A-B) electronic interferometers are analyzed: decoherence due to coupling with other degrees of freedom and the coupled transport of charge and heat. We first review the principles of decoherence of electronic interference. We then analyze the thermoelectric transport in a ring threaded by an A-B flux, with a molecular bridge on one of its arms. The charge carriers may also interact inelastically with the molecular vibrations. This nano-system is connected to three terminals; two of them are electric and thermal, held at slightly different chemical potentials and temperatures, and the third is purely thermal, e.g., a phonon bath thermalizing the molecular vibrations. When this third terminal is held at a temperature different from those of the electronic reservoirs, both an electrical and a heat current are, in general, generated between the latter. Likewise, a voltage and/or temperature difference between the electronic terminals leads to thermal current between the thermal and electronic terminals. The transport coefficients governing these conversions (due to energy exchange between the electrons and the vibrations) and their dependences on the A-B flux are analyzed. Finally, the decoherence due to these inelastic events is discussed.

1. Introduction

Interference, resulting from the superposition of different amplitudes, is a basic attribute of Quantum Mechanics. Akira Tonomura Sensei has made decisive lasting contributions to the study of these effects for electron beams.[1] Here we review the analogous Physics in mesoscopic solid-state systems.[2] In those, small sizes and/or low temperatures are necessary in order to keep electrons coherent. We shall consider an Aharonov-Bohm (A-B) interference experiment on a ring-type structure.[3,4] This proves to be a convenient way to observe interference patterns in such samples, providing an experimentally straightforward way of shifting the interference pattern.

This experiment involves two electron wave packets, $\ell(x)$ and $r(x)$ (ℓ, r stand for left, right), crossing the ring along its two opposite sides and interacting with an environment, represented by a (usually thermal) bath. We assume that the two wave packets follow classical paths, $x_\ell(t), x_r(t)$ along the arms of the ring. The interference is examined after each of the two wave packets had traversed half of the ring's circumference. The combined wave function is

$$\psi = \ell(x) + r(x) \ . \tag{1}$$

The mixed terms, $2\mathrm{Re}[\ell(x) \times r^*(x)]$, constitute the interference contribution, clearly sensitive to a phase shift (e.g., A-B) introduced between the two partial waves $\ell(x)$ and $r(x)$. Interaction with the environment can reduce the strength of the interference. This process is called "decoherence" and it is of great relevance.

We will review the effects of interference, first on the purely electronic transport and then on coupled electric and thermal transport. When changes in the environment state, caused by the electrons, occur, decoherence may follow. These changes will be modeled by the effect of a local vibrational mode ("vibronic coupling"), which is in turn coupled to a reservoir that is restoring equilibrium to the local vibration. Thermoelectric effects in bulk conductors usually necessitate breaking of particle-hole symmetry, which can be substantial and controllable in mesoscopic structures. As a result, there is currently much interest in investigations of thermoelectric phenomena in nanoscale devices at low temperatures.

The coupling of the charge carriers to vibrational modes of the molecule should play a significant role in thermoelectric transport through molecular bridges. Indeed, the thermopower coefficient was proposed as a tool to monitor the excitation spectrum of a molecule forming the junction between two leads, and to determine both the location of Fermi level of the charge carriers and their charge.[5] Theoretically, when the coupling to the vibrational modes is ignored, the transport coefficients are determined by the usual energy-dependent transmission coefficient, replacing the conductivity.[6,7] Even when the corrections to the thermoelectric transport due to the coupling to the vibrational modes are small, their study is of interest because of fundamental questions related to the symmetries of the conventional transport coefficients, and since they give rise to additional coefficients connecting the heat transport in-between the electrons and the vibrational modes, as we have recently found.[8] We have considered the case where the molecule is (relatively) strongly coupled to a heat bath of its own, which maintains its own temperature, see Fig. 1, i.e., the relaxation time due to the coupling of the molecule to its own heat bath is short on the scale of the coupling of the molecular vibrations to the charge carriers. The phonon bath may be realized by an electronically insulating substrate or a piece of such material touching the junction, each held at a temperature T_{V}. Experimental realizations of three-terminal setups have been already discussed.[9]

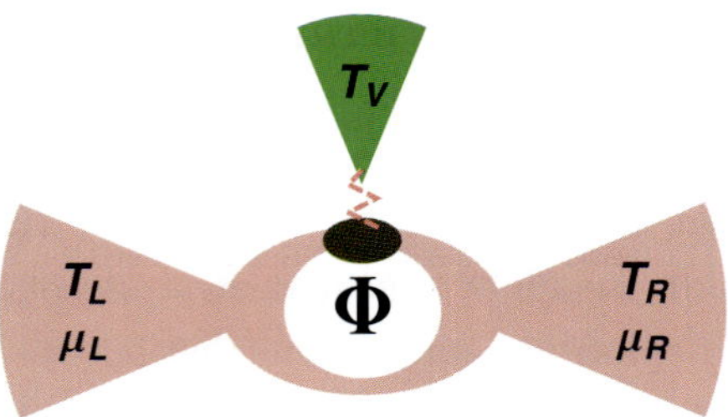

Fig. 1.   A mixed, thermal-electronic, three-terminal system, as explained in the text, threaded by a magnetic flux Φ.

Section 2 discusses the principles of decoherence. The simplest case of equilibrium persistent currents (PC) is discussed in Sec. 3. Our theoretical model is presented in Sec. 4 and the A-B oscillations of the transmission of the interferometer and their decoherence are discussed in Sec. 5. After a short review of thermoelectricity in Sec. 6, we give the main results of the analysis including the magnetic-flux dependence of the three-terminal transport coefficients. Section 7 presents these coefficients in the linear-response regime and verifies their Onsager symmetries. Section 8 is devoted to a discussion of the results in simple examples, and Sec. 9 summarizes the article.

2. Principles of decoherence

Interference effects are strongly affected by the coupling of the interfering particle to its environment, e.g., to a heat bath. The way such a coupling modifies quantum phenomena has been studied for a long time, both theoretically[10] and experimentally.[11] Some of the effect of the coupling to the environment may be described by the "phase-breaking" time, τ_ϕ, which is the characteristic time for the interfering particle to stay phase coherent as explained below.

Two points of view have been used by Stern *et al.*[12] to describe how the interaction of a quantum system with its environment might suppress quantum interference. The first regards the environment as measuring the path of the interfering particle. When the environment has all the information on that path, no interference is seen. The second description considers the phase uncertainty induced on the interfering particle by the interaction with the environment. The two descriptions were proven to be equivalent. Here we will review the first point of view, using the example of an A-B interference experiment on a ring. The electron (whose coordinate is x) and the environment [whose wave functions for the two paths are denoted by $\chi(\eta)$] interact during the traversal of the ring. Initially the combined wave function is $\left[\ell(x) + r(x)\right] \otimes \chi_0(\eta)$; At time τ_0 (measured with respect to the start of the experiment), when the interference is examined, the wave function is, in general,

$$\psi(\tau_0) = l(x,\tau_0) \otimes \chi_\ell(\eta,\tau_0) \; + \; r(x,\tau_0) \otimes \chi_r(\eta,\tau_0) \tag{2}$$

and the interference term is

$$2\,\mathrm{Re}\,\left[\ell^*(x,\tau_0) r(x,\tau_0) \int d\eta \chi_\ell^*(\eta,\tau_0) \chi_r(\eta,\tau_0) \right] . \tag{3}$$

Without the coupling to the environment, the interference term would have been just Eq. (1). So, the effect of the interaction is to multiply the interference term by the scalar product $\int d\eta \chi_\ell^*(\eta,\tau_0)\chi_r(\eta,\tau_0)$. The first way to understand the dephasing is clear from this expression, i.e. the reduction in the interference due to the response of the environment to the interfering waves. When the two states of the environment become orthogonal, the final state of the environment identifies the path the electron took. Quantum interference, which is the result of an uncertainty in this path, is then lost. The phase breaking time, τ_ϕ, is the time at which the two interfering partial waves shift the environment into states orthogonal to each other, i.e., when the environment has the full information on the electron path.

As seen from the above discussion, the phase uncertainty remains constant when the interfering wave does not interact with the environment. Thus, if a trace is left by a partial wave on its environment, this trace cannot be wiped out after the interaction is over. This statement can be proven also from the point of view of the change the interfering wave induces in its environment, and the proof follows simply from unitarity. The scalar product of two states evolving under the same Hamiltonian does not change in time. The only way to change it is by another interaction of the electron with the same environment, as discussed below. The interference will be retrieved only if the orthogonality is transferred from the environment wave function to the electronic wave functions which are not traced on in the experiment.

The above discussion was concerned with the phase $\phi = \phi_r$, accumulated by the right-hand path only. The left hand path accumulates similarly a phase ϕ_ℓ from the interaction with the environment. The interference pattern is governed by the *relative* phase $\phi_r - \phi_\ell$, and it is the uncertainty in *that* phase which determines the loss of quantum interference. This uncertainty is always smaller than, or equal to, the sum of uncertainties in the two partial waves' phases.

Often the same environment interacts with the two interfering waves. A typical example is the interaction of an interfering electron with the electromagnetic fluctuations in vacuum. In this case, if the two waves follow parallel paths with equal velocities, their dipole radiation, despite the energy it transfers to the field, does not dephase the interference. This radiation makes each of the partial waves' phases uncertain, but does not alter the relative phase. Another well-known example is that of "coherent inelastic neutron scattering" in crystals.[13] This process follows from the coherent addition of the amplitudes for the processes in which the neutron exchanges *the same* phonon with *all* scatterers in the crystal.

The last example also demonstrates that an exchange of energy is not a sufficient condition for dephasing. It is also not a necessary condition for dephasing. What is important is that the two partial waves flip the environment to *orthogonal* states. It does not matter in principle that these states are degenerate.[12] Thus, it must be emphasized that, for example, long-wave excitations (phonons, photons) cannot dephase the interference. But that is *not* because of their low energy but rather because they do not influence the *relative* phase of the paths.

We emphasize that dephasing may occur by coupling to a discrete or a continuous environment. In the former case the interfering particle is more likely to "reabsorb" the excitation and "reset" the phase. In the latter case, the excitation *may* move away to infinity and the loss of phase can then be broadly regarded as irreversible. The latter case is that of an effective "bath".[2]

We point out that in special cases it is possible, even in the continuum case, to have a finite probability to reabsorb the created excitation and thus retain coherence. This happens, for example, in a quantum-interference model due to Holstein[14] for the Hall effect in insulators. This model deals with phonon-induced hopping between two localized states with different energies, E_1 and E_2, which necessitates the (real) absorption of a phonon with energy $E_2 - E_1$ for $E_2 - E_1 > 0$. It focuses on transitions which occur via an intermediate localized state with yet another energy, E_3. We take, for definiteness $E_3 > E_1$. That intermediate state transition also involves the virtual absorption of a phonon with frequency ω_q, which is then re-emitted with the transition from 3 to 2. This higher-order process has a perturbation theory energy-denominator which at "resonance" ($\hbar\omega_q = E_3 - E_1$) produces a term $i\delta(\hbar\omega_q - E_3 + E_1)$. The integration over the continuous ω_q then yields a term with a $\pi/2$ phase shift in the transition amplitude between 1 and 2.

Note that these two interfering paths involve for $E_2 - E_1 > 0$ the absorption of *the same* phonon and they therefore stay coherent. A phase shift due to the A-B effect, $\phi = 2\pi\Phi/\Phi_0$, can be included between the direct and the compound transition paths ($\Phi_0 = hc/e$ is the flux quantum and Φ is the magnetic flux enclosed between the paths). Interestingly, it turns out that although this process creates a $\sin(\phi)$ term in some transition probabilities, it retains[15] the basic Onsager symmetry in Φ of the conductivity, $\sigma_{xx}(\phi) = \sigma_{xx}(-\phi)$, and also the "non diagonal" Hall conductivity, $\sigma_{xy}(\phi) = \sigma_{yx}(-\phi)$. Since a $\sin(\phi)$ dependence is allowed here by symmetry, there is no reason why it should not exist, and indeed it is the basic reason why the Holstein process furnishes an explanation for the Hall effect in insulators.

The adaptation of the Holstein idea to transitions between two leads with continuous energy spectra is actually simpler than the original model, in which an additional phonon is needed to conserve energy. Here an elastic, purely electronic, transition can be made from a full state on one side to an empty state on the other. This transition can again be accomplished via a superposition of the direct transition with a compound one through the molecule, involving again a "resonant" virtual phonon. Clearly, here this effect is of second order in the electron-phonon coupling.[16] There is no $\sin(\phi)$ term in either the electrical or the thermal conductivity. It exists however in other "non diagonal" transport coefficients,[16] as discussed in Sec. 8.

3. Interference in closed rings, mesoscopic persistent currents

It has been noticed rather early[17] that equilibrium properties, such as the average energy or magnetization [and therefore the circulating persistent current (PC)] of a small *free-electron* system with a simple, *ideal*, geometry, e.g., a perfect disc or ring (which is realized in aromatic benzene-type molecules) are very sensitive to a magnetic field. Later,[18,19] the connection with the A-B effect was made and oscillatory behavior was predicted for a ring (see Fig. 2) with an A-B flux Φ through its opening. The thermodynamic functions are periodic in Φ.[3] Very large fields, $\sim 10^5$ Tesla or more, are needed in order to observe the periodicities in molecules. Here we mainly have in mind man-made conducting rings, where several flux quanta are easily achievable.

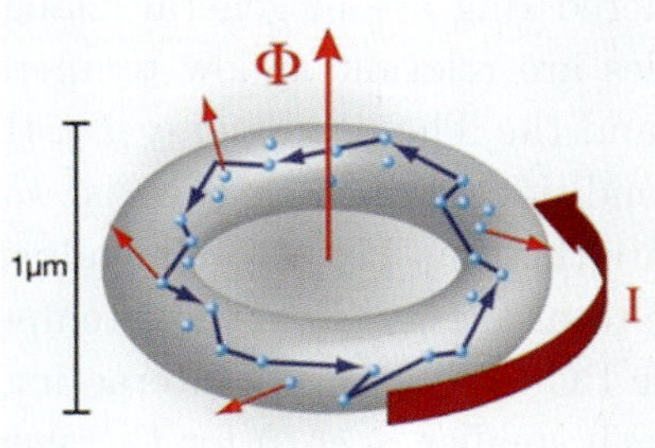

Fig. 2.   Schematic ring structure with impurities and threading flux.

At that time it was universally expected (and consistent with e-beam experiments) that any electron scattering will eliminate these effects in realistic systems. Impurities, defects and surface imperfections would limit the elastic mean free path, ℓ, at best by the ring arms' width and thickness, which are typically much smaller than, say, the ring's circumference L – which is the distance over which the electron's wave functions experience A-B type interference. Hence, these scatterings were expected to eliminate any interference effect and the PC would vanish in a realistic ring. (Nevertheless, Gunther and Imry[18] found persistent orbital currents in a system with a finite resistance.) However, for *static* defects in solids, this expectation is seriously wrong. *Elastic* scattering is due to *static* potential in which wave functions with well-defined phases exist. All physical properties of a general doubly-connected system with an A-B flux Φ through its opening,[3,4] are periodic in Φ with a period Φ_0.[20] The real issue is whether there is a sizeable sensitivity of, say, the energy levels to Φ.

A flux sensitivity of the ground state energy E_0 at zero temperature $T = 0$, or the free energy F at T$\neq$0, yields a circulating *equilibrium* current, given by[4]

$$I = -c\, \partial F/\partial \Phi \rightarrow -c\, \partial E_0/\partial \Phi \, . \tag{4}$$

These currents thus "never decay" as long as Φ is kept on, hence the name "persistent currents". It is, of course, well known that such currents exist in superconductors. There, they exist both in equilibrium and in metastable states (e.g., flux-quantized states of a superconducting ring or cylinder). Here, we consider only equilibrium persistent currents, at finite Φ, in a realistic nonsuperconducting ring. The existence of such currents has been greeted with some skepticism until the 90's, based on the misunderstanding of what destroys phase coherence, or on the insistence on taking the thermodynamic limit.

Büttiker *et al.*[21] were the first to understand the effect of elastic scattering, using the simple model of a one-dimensional ring with disorder. They showed clearly that all the levels of the disordered ring are in general changing with the flux. In that case, only two energy scales are relevant at low temperatures. The single-particle level spacing, $\delta \sim \hbar v_F/L$ and the Thouless energy E_c, the latter is defined via the sensitivity to boundary conditions, which in turn is $\sim$ the inverse of the time it takes an electron to circulate the ring. For a good conductor, $E_c >> \delta$. Estimates of the flux dependence of the total energy, E_0, at low temperatures[2] show that grosso modo, the low-temperature PC results for noninteracting electrons are $\sim eE_c/\hbar$ for a given sample and the much smaller $\sim e\delta/\hbar$ for the average over a large ensemble of similarly-prepared samples. The former is in agreement with experiment, while that latter is too small by two orders of magnitude. One order of magnitude can be gained by introducing electron-electron interactions,[22] the still lacking order of magnitude can be explained by a judicious manipulation of the renormalization of the interactions.[23] This is far beyond the scope of this article.

4. The model

In our analysis, the molecular bridge is represented by a single localized electronic level, standing for the lowest available orbital of the molecule; when an electron resides on the level, it interacts (linearly) with an Einstein oscillator. We do not include electronic interactions, but focus on the electron-vibron ones. Thus the Hamiltonian of the molecular bridge reads

$$\mathcal{H}_{\mathrm{M}} = \epsilon_0 c_0^\dagger c_0 + \omega_0 b^\dagger b + \gamma(b + b^\dagger)c_0^\dagger c_0 \,, \tag{5}$$

where ϵ_0 is the energy of the localized level, ω_0 is the frequency of the harmonic oscillator representing the vibrations, and γ is its coupling to the electrons. The Hamiltonian describing the tunneling between the molecule and the leads is

$$\mathcal{H}_{\mathrm{coup}} = \sum_k (V_k c_k^\dagger c_0 + \mathrm{H.c.}) + \sum_p (V_p c_p^\dagger c_0 + \mathrm{H.c.}) \tag{6}$$

[using $k(p)$ for the left (right) lead]. The leads' Hamiltonian is

$$\mathcal{H}_{\mathrm{lead}} = \sum_k \epsilon_k c_k^\dagger c_k + \sum_p \epsilon_p c_p^\dagger c_p + \left(\sum_{kp} V_{kp} e^{i\phi} c_k^\dagger c_p + \mathrm{H.c.} \right) . \tag{7}$$

The A-B magnetic flux is included in the phase factor $\exp[i\phi]$. Thus, our model Hamiltonian is $\mathcal{H} = \mathcal{H}_{\mathrm{lead}} + \mathcal{H}_{\mathrm{M}} + \mathcal{H}_{\mathrm{coup}}$ where the operators $c_0^\dagger$, $c_k^\dagger$, and $c_p^\dagger$ (c_0, c_k, and c_p) create (annihilate) an electron on the level, on the left lead, and on the right lead, respectively, while $b^\dagger$ (b) creates (annihilates) a vibronic excitation of frequency ω_0.

In the spirit of the Landauer approach,[2] the various reservoirs (which are assumed in the simplest case to be large enough to stay in equilibrium in spite of the small currents that flow from/to them) which supply charge and energy to the leads, are described by equilibrium distributions with given chemical potentials and temperatures. The electronic reservoirs on the left and on the right of the bridge are characterized by the electronic distributions f_{L} and f_{R},

$$f_{\mathrm{L(R)}}(\omega) = [1 + \exp[\beta_{\mathrm{L(R)}}(\omega - \mu_{\mathrm{L(R)}})]]^{-1} \,, \tag{8}$$

determined by the respective Fermi functions, with $\beta_{\mathrm{L(R)}} = 1/k_{\mathrm{B}} T_{\mathrm{L(R)}}$. The phonon reservoir (with $\beta_{\mathrm{V}} = 1/k_{\mathrm{B}} T_{\mathrm{V}}$) which determines the vibration population on the bridge, is characterized by the Bose-Einstein distribution, $N = [\exp(\beta_{\mathrm{V}} \omega_0) - 1]^{-1}$.

5. A-B oscillations of the interferometer and their decoherence

We first consider the interferometer having just a resonance level without vibrational coupling on one arm and a direct transmission on the other. The interacting with the vibrations is later included.

5.1. *A simple interferometer: no vibronic coupling*

The transport through the "direct" arm of the interferometer (the one not carrying the dot) is characterized by the dimensionless coupling

$$\lambda(\omega) = (2\pi)^2 \sum_{k,p} |V_{kp}|^2 \delta(\omega - \epsilon_k)\delta(\omega - \epsilon_p) , \tag{9}$$

and the bare transmission (reflection) amplitude, t_o (r_o), through that branch is

$$t_o^2(\omega) = 4\lambda(\omega)/[1 + \lambda(\omega)]^2 , \quad r_o^2 = 1 - t_o^2 . \tag{10}$$

The transport through the arm carrying the quantum dot is characterized by

$$\Gamma_{\rm L(R)}(\omega) = \pi \sum_{k(p)} |V_{k(p)}|^2 \delta(\omega - \epsilon_{k(p)}) , \tag{11}$$

and the total width of the level ϵ_0 is $\Gamma(\omega) = [\Gamma_{\rm L}(\omega) + \Gamma_{\rm R}(\omega)]/[1 + \lambda(\omega)]$. The current, I, and the (electronic) energy current, $I_{\rm E}$, across the interferometer are determined by the same transmission function,

$$I = -e \int \frac{d\omega}{2\pi}[f_{\rm R}(\omega) - f_{\rm L}(\omega)]\mathcal{T}(\omega) \quad I_{\rm E}(\omega) = \int \frac{\omega\, d\omega}{2\pi}[f_{\rm R}(\omega) - f_{\rm L}(\omega)]\mathcal{T}(\omega) , \tag{12}$$

where the Fermi functions are given in Eq. (8). The function $\mathcal{T}(\omega, \phi)$ (for a junction tunnel-coupled symmetrically to the leads) is[16]

$$\mathcal{T}(\omega, \phi) = t_o^2(1 - \Gamma {\rm Im} G^a + \Gamma^2 \sin^2 \phi |G^a|^2/4) + t_o r_o \Gamma \cos \phi {\rm Re} G^a + \Gamma^2 |G^a|^2/4 . \tag{13}$$

The first term in this expression, t_o^2, yields the conductance in the absence of the ring arm carrying the bridge while the last term yields the conductance of that arm alone. The other three terms in Eq. (13) result mainly from various interference processes (see below). The advanced Green function of the dot is

$$G^a(\omega) = [\omega - \widetilde{\epsilon}_0(\omega) - i\Gamma(\omega)/2]^{-1} , \tag{14}$$

with $\widetilde{\epsilon}(\omega) = \epsilon_0 + [\lambda(\omega)\Gamma_{\rm L}(\omega)\Gamma_{\rm R}(\omega)]^{1/2} \cos \phi/[1 + \lambda(\omega)]$. This result describes well many interferometer experiments. It satisfies the Onsager-related symmetry of the conductance, which follows from the property that $\mathcal{T}(\omega)$ is even in the flux Φ.

We now turn to the explanation of the three terms in Eq. (13) which are not direct transmissions through one of the arms. The second term in the brackets is just a correction to the direct transmission t_0 due to attempting to go through the bridge (and instead being reflected) either before or after the direct transition. The third term is the "weak localization" correction to the transmission, due to the increase of the total reflection via the two time-reversed paths which encircle the whole ring clockwise and anti-clockwise. The term before the last, having the product of t_0 and (real part of) the transmission amplitude through the bridge, $\Gamma {\rm Re} G^a$, resembles the mixed term in the good old two-wave interference. There is however a very important small difference: the latter should have the real part of $\Gamma G^a e^{i\phi}$, while our expression has $\cos\phi \mathrm{Re}G^a$. It thus has a pure $\cos\phi$ dependence on the flux as it should. This is so, because we consider a *closed* interferometer which *conserves particles*. This differs from the ordinary two-wave interference, in which particles are lost and therefore, a $\sin(\phi)$ dependence is also generated in the transmission.[24] Technically, this is due to the fact that here the Green's function of the dot, which is determined by paths starting at and returning to it, is dressed by paths (or "diagrams") in which the electron experiences the whole interferometer. Physically, the point is that the fact that electrons can enter and exit the interferometer from/to the two leads, does *not* break particle conservation for the scattering matrix of the interferometer. Therefore, the Onsager symmetry, for which particle conservation, or "unitarity", is essential,[24,25] holds.

5.2. *Interferometer with vibronic coupling*

The technical calculations of the transport coefficients with the electron-vibron coupling are rather complicated, and so far it has only been feasible to do them to order γ^2, which is sufficient to see the effects of vibronic excitation/deexcitation.[16] Suffice it to say that, for the electrical conductance, indeed the occurrence of a vibronic change of state does eliminate, in the term resulting from the interaction with vibrations, the $\cos\phi$ term in the conductance, as expected due to decoherence. It was however found that a $\cos^2\phi$ (or $\cos 2\phi$) dependence survives. This dependence resembles the weak-localization-type flux dependence, as in the third term of Eq. (13). This in fact results from a novel generalization of the usual weak-localization process, in which the two time-reversed paths change the vibrational state *in the same fashion*.[12] Thus, there is no decoherence of this special process! The situation with the thermal and thermoelectric coefficients is much more complicated and interesting. It will be reviewed in the rest of this paper.

6. Generalities on thermoelectric transport

The electronic part of thermoelectric linear-transport problem is fully characterized for the two-terminal situation by

$$\begin{pmatrix} I_e \\ I_Q^e \end{pmatrix} = \begin{pmatrix} G & L_1 \\ L_1 & K_e^0 \end{pmatrix} \begin{pmatrix} \delta\mu/e \\ \delta T/T \end{pmatrix}, \tag{15}$$

where I_e is the charge current and I_Q the heat current, $\delta T = T_\mathrm{L} - T_\mathrm{R}$ and $\delta\mu/e = (\mu_\mathrm{L} - \mu_\mathrm{R})/e \equiv V$ is the voltage between the left and right terminals. The 2×2 matrix contains the regular conductance G, the bare electronic thermal conductance, K_e^0

and the (off-diagonal) thermoelectric coefficients L_1. That the latter two are equal is the celebrated Onsager relation (valid for time-reversal symmetric systems; with a magnetic field, B, if one of them is taken at B, it is equal to the other one at $-B$). We remind the reader that since the (Seebeck) thermopower is the voltage developed due to a temperature difference at zero electrical current, $S = L_1/G$. For noninteracting electrons, all currents and transport coefficients in Eq. (15) are given in terms of integrals involving the energy-dependent elastic transmission, $\mathcal{T}(E)$ and the Fermi function, $f(E)$, at the common chemical potential μ and temperature T, $G = I_0; L_1 = I_1; K_e^0 = I_2$, where

$$I_n = \frac{2e}{h} \int_{-\infty}^{\infty} dE\, \mathcal{T}(E) f(E)[1 - f(E)](E - \mu)^n/(k_{\mathrm{B}}T). \tag{16}$$

We recall that since the heat conductance is defined as the heat current due to a unit temperature difference, *at vanishing electrical current*, the electronic heat conductance is given by

$$K_e = K_e^0 - GS^2 = I_2 - I_1^2/I_0 \ . \tag{17}$$

In a thermoelectric energy conversion device, the performance is governed by the well-known dimensionless "figure of merit", $zT = (L^2/KG) = (GS^2)/K$, where $K = K_e + K_{ph}$ is the full thermal conductance. Here K_{ph} is the phonon (+ any other neutral mobile excitation of the solid) thermal conductance. The efficiency is a fraction, $g(zT)$, of the, maximal allowed, Carnot efficiency. Large zT's yield better performance, with $g(zT)$ rising monotonically from 0 for $zT \to 0$ to 1 for $zT \to \infty$. An important remark here is that the determinant of the transport matrix in Eq. (15) is nonnegative. This implies that were the figure of merit based on the bare conductance $K^0 = K_e^0 + K_{ph}$, zT would be limited by unity. It is the "renormalization" of K as in Eq. (17), which allows larger zT values and opens the way to thermoelectric applications!

   Based on the above, Mahan and Sofo,[7] suggested that regarding $\mathcal{T}(E)f(E)[1 - f(E)]/G$ as a (positive, normalized) weight function, then $S = <E - \mu>$ and $K_e =< (E - \mu)^2 > - <E - \mu>^2$. One can then make K_e very small by having a very narrow transmission band away from the Fermi level. This is needed in order to have a finite $<E - \mu>$, not relying only on the, usually small, asymmetry of $\mathcal{T}(E)$ near E_{F}, to break electron-hole symmetry. This implies that zT would be limited only by K_{ph}.

   As an example we take an impenetrable barrier of height $W >> k_{\mathrm{B}}T$, whose transmission we take as $\mathcal{T}(E) \simeq \Theta(E - W)$. (Energies are measured from the common chemical potential μ.) Under these conditions, $f(E)$ approaches the Boltzmann distribution, leading to $f(E)[1 - f(E)] \simeq \exp[-E/(k_{\mathrm{B}}T)]$. Hence,

$$<E> = eL_1/G = W + k_{\mathrm{B}}T \ , \quad <E^2> - <E>^2 = (k_{\mathrm{B}}T)^2 \ , \tag{18}$$

with the charge and heat electron current given by Eqs. (12). To reiterate, the formulae above say that S is the average energy transferred by an electron, divided by e, while K_e^0/G is the average of the square of that energy, divided by e^2. Therefore K_e/G is proportional to the variance of that energy. The latter obviously vanishes for a very narrow transmission band. In the high barrier case that band is a few $k_{\mathrm{B}}T$ above W. Thus, it is not surprising that K_e is of the order of $(k_{\mathrm{B}}T/W)^2$.

7. Interference effects and their decoherence in thermoelectric transport

The linear-transport full set of thermoelectric coefficients[16] is given by

$$\begin{bmatrix} I \\ I_{\mathrm{Q}} \\ -\dot{E}_{\mathrm{V}} \end{bmatrix} = \mathcal{M} \begin{bmatrix} \delta\mu/e \\ \delta T/T \\ \Delta T_{\mathrm{V}}/T \end{bmatrix} , \quad \mathcal{M} = \begin{bmatrix} \mathrm{G}(\Phi) & \mathrm{K}(\Phi) & \mathrm{X}^{\mathrm{V}}(\Phi) \\ \mathrm{K}(-\Phi) & \mathrm{K}_2(\Phi) & \widetilde{\mathrm{X}}^{\mathrm{V}}(\Phi) \\ \mathrm{X}^{\mathrm{V}}(-\Phi) & \widetilde{\mathrm{X}}^{\mathrm{V}}(-\Phi) & \mathrm{C}^{\mathrm{V}}(\Phi) \end{bmatrix} \tag{19}$$

(with $\Delta T_{\mathrm{V}} = T_V - T$). This matrix satisfies the Onsager-Casimir relations: all three diagonal entries are even functions of the flux Φ, and the off-diagonal entries obey[16] $\mathcal{M}_{ij}(\Phi) = \mathcal{M}_{ji}(-\Phi)$. Explicitly,[16] there are three types of flux dependencies hiding in those six coefficients. First, there is the one caused by interference. The interference processes modify the self energy of the Green functions pertaining to the A-B ring, in particular the broadening of the electronic resonance level due to the coupling with the leads. The interference leads to terms involving $\cos(\phi)$. Secondly, there is the flux dependence which appears in the form of $\cos(2\phi)$ (or alternatively, $\sin^2(\phi)$). This reflects the contributions of time-reversed paths. These even-in-the-flux dependencies yield the full flux dependence of the diagonal entries of the matrix $\mathcal{M}$, and the even (in the flux) parts of the off-diagonal elements. Finally, there is the *odd* dependence in the flux, that appears as $\sin(\phi)$. This dependence characterizes the odd parts of the off-diagonal entries of $\mathcal{M}$. These terms *necessitate* the coupling of the electrons to the vibrational modes. They arise from Holstein-type processes, briefly discussed toward the end of Sec. 2.

8. Examples and discussion

The interesting effect induced exclusively by the coupling of the electrons to the vibrational modes is the possibility to create an electric current, or an electronic heat current, by applying a temperature difference ΔT_{V} on the phonon bath thermalizing this mode. These new thermoelectric phenomena are specified by the two coefficients X^{V} and $\widetilde{\mathrm{X}}^{\mathrm{V}}$. To make a closer connection with possible experiments, we introduce the (dimensionless) coefficients

$$\mathrm{S}^{\mathrm{V}}(\Phi) = e\beta \mathrm{X}^{\mathrm{V}}(\Phi)/\mathrm{G}(\Phi) \ , \quad \widetilde{S}^{\mathrm{V}}(\Phi) = \widetilde{\mathrm{X}}^{\mathrm{V}}(\Phi)/\mathrm{K}_2(\Phi) \ . \tag{20}$$

The first, S^{V}, gives the potential drop across the molecular bridge created by ΔT_{V} when the temperature drop there, ΔT, vanishes, and the second, $\widetilde{S}^{\mathrm{V}}$, yields the temperature difference created by ΔT_{V} when $\Delta\mu = 0$ [or the inverse processes, see Eqs. (19)]. For both the conductance, G, and the thermal conductance, K_2, we use below their leading terms, resulting from the coupling to the leads alone.

As mentioned above, the transport coefficients of our three-terminal junction obey the Onsager-Casimir relations. They do it though in a somewhat unique way: the "off-diagonal" elements are related to one another by the reversal of the magnetic field. However, they are not a purely odd functions of it. A special situation arises when the molecular junction is completely symmetric. Then, the two coefficients are odd functions of the flux [see Eqs. (9), (10), and (13)]

$$\mathrm{X}^{\mathrm{V}}(\Phi) = e\omega_0 \sin\phi \int \frac{d\omega}{\pi} \mathcal{T}^{\mathrm{V}}(\omega,\phi)\Big(t_o(\omega_-) - t_o(\omega_+)\Big) \ ,$$

$$\widetilde{\mathrm{X}}^{\mathrm{V}}(\Phi) = \omega_0 \sin\phi \int \frac{d\omega}{\pi} \mathcal{T}^{\mathrm{V}}(\omega,\phi)\Big(t_o(\omega_-)(\omega_- - \mu) - t_o(\omega_+)(\omega_+ - \mu)\Big) \ , \tag{21}$$

where $\mathcal{T}^{\mathrm{V}}$ is the inelastic transmission,[16] and $\omega_\pm = \omega \pm \omega_0/2$ which implies that the thermoelectric processes described by $\mathrm{X}^{\mathrm{V}}(\Phi)$ and $\widetilde{\mathrm{X}}^{\mathrm{V}}(\Phi)$ require a certain symmetry-breaking. when $\Phi = 0$, that is supplied by the spatial asymmetry of the junction; in the presence of a flux, those processes appear also for a junction symmetrically coupled to the leads, provided that the couplings to the leads depend on the energy.

For a spatially-symmetric molecular bridge the transmission amplitude of the direct bond between the two leads, t_o [Eq. (10], is an even function of ω. The transmission function $\mathcal{T}^{\mathrm{V}}(\omega,\Phi)$, Eq. (13), is not entirely even or odd in ω, and therefore *a priori* the integrals which give $\mathrm{X}^{\mathrm{V}}(\Phi)$ and $\widetilde{\mathrm{X}}^{\mathrm{V}}(\Phi)$ do not vanish. However, the asymmetry in the $\omega-$dependence of the integrand (which results from the $\omega-$dependence of the Green function) is not significant. As a result, S^{V} is extremely small, while $\widetilde{S}^{\mathrm{V}}$ is not (because of the extra ω factor in the integrand), see Fig. 3. These plots are computed using $\Gamma(\omega) = \Gamma^0 \sqrt{1 - (\omega/W)^2}$, and $\lambda(\omega) = \lambda^0[1 - (\omega/W)^2]$, where W is half the bandwidth, and all energies are measured in units of $1/\beta = k_{\mathrm{B}}T$ (we have set $\Gamma^0 = \lambda^0 = 1$ and $W = 50$).

The relative magnitudes of S^{V} and $\widetilde{S}^{\mathrm{V}}$ are significantly changed when the molecule is coupled *asymmetrically* to the leads, in the sense that the densities of states in the two leads have different energy dependencies. Let us assume that the left reservoir is represented by an electron band, with $\Gamma_{\mathrm{L}}(\omega) = \Gamma_{\mathrm{L}}^0 \sqrt{(\omega - \omega_c)/(\omega_v - \omega_c)}$, while the right reservoir is modeled by a hole band, with $\Gamma_{\mathrm{R}}(\omega) = \Gamma_{\mathrm{R}}^0 \sqrt{(\omega_v - \omega)/(\omega_v - \omega_c)}$. Here, ω_c is the bottom of the conductance band

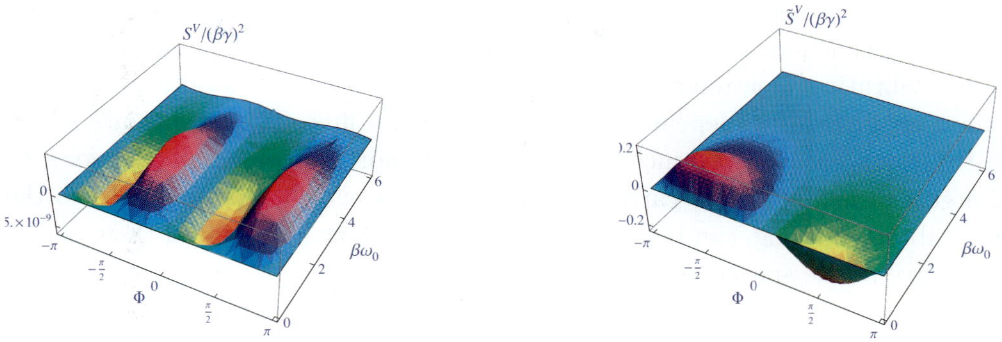

Fig. 3.   The transport coefficients S^V and $\widetilde{S}^V$ as functions of the flux (measured in units of the flux quantum) and $\beta\omega_0$, for a symmetric bridge.

(on the left side of the junction), while ω_v is the top of the hole band (on the right one). The energy integration determining the various transport coefficients is therefore limited to the region $\omega_c \leq \omega \leq \omega_v$. (For convenience, we normalize the Γ's by the full band width, $\omega_v - \omega_c$.) Note that the density of states is increasing (decreasing) with energy in the electron (hole) lead. Exemplifying results in such a case are shown in Fig. 4, computed with $\Gamma_{\mathrm{L}}^0 = \Gamma_{\mathrm{R}}^0 = \lambda^0 = 1$ and $\omega_c = -\omega_v = 100$, all in units of $k_B T$.

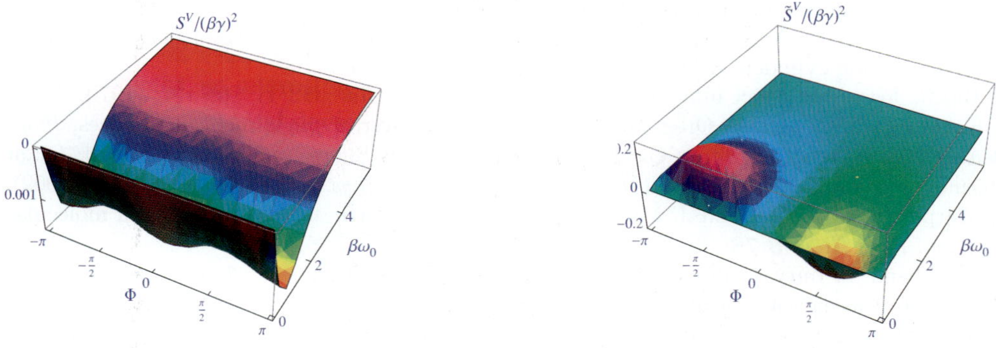

Fig. 4.   Same as Fig. 4, for an asymmetric bridge.

9. Conclusions

We reviewed the interference effects of mesoscopic-scale solid-state A-B interferometers, with special emphasis on how they are reduced/eliminated by decoherence. Coupling to local vibrations, to lowest significant order, as well as to a heat bath, was introduced. The latter makes the problem much richer. We first discussed just the electronic transport, with and without the vibronic coupling. Then, we reviewed the electric and thermal transport in an interferometer with the above-mentioned couplings. In particular, we have found that the thermoelectric transport coefficients through a vibrating molecular junction, placed on an Aharonov-Bohm interferometer, have an interesting dependence on the magnetic flux. The coefficients which relate the temperature difference between the phonon and electron reservoirs to the charge and heat currents carried by the electrons, which exist only due to the electron-vibron coupling, can be enhanced by the magnetic flux. One of them is also very sensitive to the spatial asymmetry.

Acknowledgments

This work was supported by the US-Israel Binational Science Foundation (BSF), by the Israel Science Foundation (ISF) and by its Converging Technologies Program. OEW and AA are indebted to the Albert Einstein Minerva Center for Theoretical Physics and to the Goldschleger Center for Nanophysics at the Weizmann Institute, for partial support.

References

1. M. Peshkin and A. Tonomura, *The Aharonov-Bohm Effect*. (Springer-Verlag, Berlin, 1989).
2. Y. Imry, *Introduction to Mesoscopic Physics*. (Oxford, 2nd edition, Oxford, 2002).
3. N. Byers and C. N. Yang, Theoretical considerations concerning quantized magnetic flux in superconducting cylinders, *Phys. Rev. Lett.* **7**(2), 46-49, (1961).
4. F. Bloch, Josephson effect in a superconduting ring, *Phys. Rev.* **B2**(1), 109-121, (1970).
5. J. Koch, F. von Oppen, Y. Oreg, and E. Sela, Thermopower of single-molecule devices, *Phys. Rev. B* **70**(19), 195107, (2004); C. M. Finch, V. M. García-Suárez, and C. J. Lambert, Giant thermopower and figure of merit in single-molecule devices, *Phys. Rev. B* **79**(3), 033405, (2009); P. Murphy, S. Mukerjee, and J. Moore, Optimal thermoelectric figure of merit of a molecular junction, *Phys. Rev. B* **78**(16), 161406(R), (2008).
6. U. Sivan and Y. Imry, Multichannel Landauer formula for thermoelectric transport with application to thermopower near the mobility edge, *Phys. Rev. B* **33**(1), 551-558, (1986).
7. G. D. Mahan and J. O. Sofo, The best thermoelectirc, *Proc. Natl. Acad. Sci. USA* **93**(15), 7436-7439, (1996).
8. O. Entin-Wohlman, Y. Imry, and A. Aharony, Three-terminal thermoelectric transport through a small junction, *Phys. Rev. B* **82**(11), 115314, (2010).
9. H. L. Edwards, Q. Niu, and A. L. de Lozanne, A quantum-dot refrigerator, *Appl. Phys. Lett.* **63**(13), 1815-1817, (1993); H. L. Edwards, Q. Niu, G. A. Georgakis, and A. L. de Lozanne, Cryogenic cooling using tunneling structures with sharp energy features, *Phys. Rev. B* **52**(8), 5714-5736, (1995); V. S. Khrapai, S. Ludwig, J. P. Kotthaus, H. P. Tranitz, and W. Wegscheider, Double-dot quantum ratchet driven by an independently biased quantum-point contact, *Phys. Rev. Lett.* **97**(17), 176803, (2006); J. R. Prance, C. G. Smith, J. P. Griffiths, S. J. Chorley, D. Anderson, G. A. C. Jones, I. Farrer, and D. A. Ritchie, Electronic refrigeration of a two-dimensional electron gas, *Phys. Rev. Lett.* **102**(14), 146602, (2009).
10. R. P. Feynman and F. L. Vernon, The theory of a general quantum system interacting with a linear dissipative system, *Ann. Phys. (N.Y.)* **24**(1), 118-173, (1963).

11. E. Buks, R. Schuster, M. Heiblum, D. Mahalu, and V. Umansky, Dephasing in electron interference by a 'which-path' detector, *Nature (London)* **391**(6670), 871-874, (1998).

12. A. Stern, Y. Aharonov, and Y. Imry, Phase uncertainty and loss of interference: A general picture, *Phys. Rev. A* **41**(7), 3436-3448, (1990).

13. See e.g., C. Kittel, *Quantum Theory of Solids*. (John Wiley and Sons, Hoboken, N.J., 1963).

14. T. Holstein, Hall effect in impurity conduction, *Phys. Rev.* **124**(5), 1329-1347, (1961); T. Holstein, Studies of polaron motion: Part II. The "small" polaron, *Ann. Phys. (N.Y.)* **8**(3), 343-389, (1959).

15. O. Entin-Wohlman, A. G. Aronov, Y. Levinson, and Y. Imry, Hall resistance in the hopping regime: A "Hall insulator"?, *Phys. Rev. Lett.* **75**(22), 4094-4097, (1995).

16. O. Entin-Wohlman and A. Aharony, Three-terminal thermoelectric transport under broken time-reversal symmetry, *Phys. Rev. B* **85**(8), 085401, (2012).

17. L. Pauling, The diamagnetic anisotropy of aromatic molecules, *J. Chem. Phys.* **4**(10), 673-677, (1936).

18. L. Gunther and Y. Imry, Flux quantization without off-diagonal long range order in a thin hollow cylinder, *Solid State Commun.* **7**(18), 1391-1394, (1969).

19. I. O. Kulik, Magnetic flux quantization in the normal state, *Sov. Phys. JETP* **31**(6), 1172-1174, (1970) and references therein.

20. The proof consists of a gauge transformation which establishes an exact equivalence between Φ and a phase change of the transformed full wave function by $\phi = 2\pi\Phi/\Phi_0$ when one electronic coordinate is rotated once around the ring. Thus the fluxes Φ and $\Phi + n\Phi_0$ are *indistinguishable*.

21. M. Büttiker, Y. Imry, and R. Landauer, Josephson behavior in small normal one-dimensional rings, *Phys. Lett.* **A 96**(7), 365-367, (1983).

22. V. Ambegaokar and U. Eckern, Coherence and persistent currents in mesoscopic rings, *Phys. Rev. Lett.* **65**(3), 381-384, (1990).

23. H. Bary-Soroker, O. Entin-Wohlman, and Y. Imry, Pair-breaking effect on mesoscopic persistent currents, *Phys. Rev. B* **80**(2), 024509, (2009).

24. O. Entin-Wohlman, A. Aharony, Y. Imry, Y. Levinson, and A. Schiller, Broken unitarity and phase measurements in Aharonov-Bohm interferometer, *Phys. Rev. Lett.* **88**(16), 166801, (2002).

25. L. D. Landau and E. M. Lifshitz, *Statistical Physics*, part 1 §120, (Butterworth-Heinemann, Oxford, 1996).

Spin Textures and Gauge Fields in Frustrated Magnets

Naoto Nagaosa and Yoshinori Tokura

RIKEN Center for Emergent Matter Science, Wako, Saitama 351-0198, Japan
Department of Applied Physics, The University of Tokyo,
7-3-1, Hongo, Bunkyo-ku, Tokyo 113-8656, Japan
nagaosa@riken.jp, tokura@riken.jp

Electrons in magnetic materials are subject to the internal gauge field, so called emergent electromagnetic field, originating from nontrivial spin configurations. Here we discuss its basic principle and its application to skyrmion spin textures in helical magnets. It is characterized by the solid angle that corresponds to the emergent magnetic field. This leads to the topological Hall effect and emergent electromagnetic induction.

1. Gauge field in magnets

The Aharonov-Bohm (AB) effect[1] demonstrates the vital role of vector potential $\mathbf{A}$ of electromagnetic field in quantum mechanics, where the quantum amplitude of the propagation is modified by the phase factor given by the contour integral of $\mathbf{A}$ over the trajectory. This vector potential $\mathbf{A}$ has geometrical meaning of "connection" and magnetic field $\mathbf{B} = \nabla \times \mathbf{A}$ as "curvature". This seminal theoretical discovery has been beautifully demonstrated experimentally by Dr. Akira Tonomura by using electron holography.[2] Intuitively, one can imagine that the space is "curved" for the quantum mechanical particles in the presence of the electromagnetic field, and the quantum mechanical amplitude for each trajectory is modified accordingly in the path integral quantization scheme.

On the other hand, M. V. Berry published in 1984[3] a seminal paper concerning the quantal phase associated with the adiabatic change of the system. This paper has revealed an essential role of geometry in quantum mechanics, and gives an extension of the AB effect in generalized spaces such as the momentum space and parameter space. During the adiabatic process, the system does not jump between different quantum states with different energy eigenvalues, which means that the wavefunction is confined within a subspace of the Hilbert space. This leads to nontrivial geometry of the Hilbert space, and the concept of "connection" and "curvature" arises, which is formulated by the vector potential a_μ and the field strength $f_{\mu\nu} = \varepsilon_{\mu\nu}\partial_\mu a_\nu$, respectively. An example of Berry phase is demonstrated in the spin 1/2 system, which is described by the 2×2 Hamiltonian

$$H = -\mathbf{B} \cdot \sigma, \tag{1}$$

where σ is the vector made of three Pauli matrices. The Berry phase of this spin 1/2 system can be easily obtained for an adiabatic cyclic change in the direction of $\mathbf{B}$ as $\Omega/2$ with Ω being the solid angle subtended by the loop on the unit sphere $\mathbf{B}/|\mathbf{B}|$.

One can consider a situation that this adiabatic change occurs for the electron spin when it moves on the background of the slowly varying magnetic structure; namely, when the conduction electron spin is coupled to the background spin $\mathbf{S}_i$ by the Hund's coupling J_H as

$$H = -\sum_{ij,\alpha} t_{ij} c_{i\alpha}^\dagger c_{j\alpha} - J_H \sum_{i,\alpha\beta} \mathbf{S}_i \cdot c_{i\alpha}^\dagger \sigma_{\alpha\beta} c_{i\beta}, \tag{2}$$

the solid angle made by $\mathbf{S}_i$, called scalar spin chirality, acts as the emergent magnetic field for the conduction electrons (See Fig. 1).

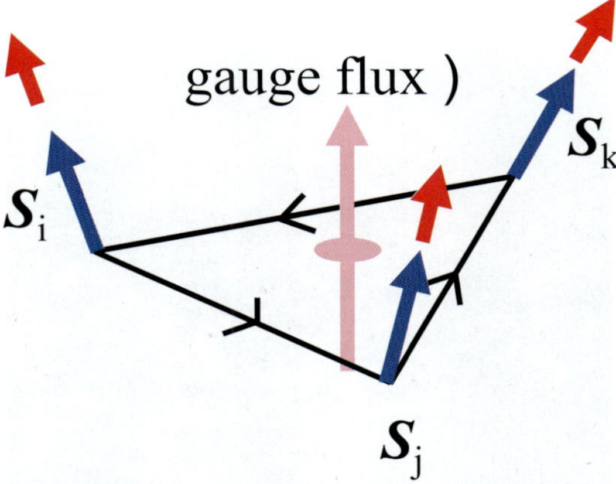

Fig. 1. When the three spins from a non-coplanar configuration, the conduction electron coupled to them feels the effective magnetic field and its flux Φ is given by the half of the solid angle subtended by the three spins.

This fact is the basics of the gauge theory of strongly correlated electronic systems.[4] In the case of spin liquids, the gauge field a_μ describing the scalar spin chirality is the dynamical degrees of freedom. This gauge field has several phases such as the confining phase, deconfining phase, and Higgs phase, representing the topological order of the system.[4] In the magnetically ordered systems, a_μ is the static background, but still offers many intriguing topological phenomena. In particular, real space observation of the spin structures is possible by electron microscopy in this case, and the combined analysis of neutron scattering, transport, and theoretical methods functions well.

2. Skyrmion in helical magnets

Skyrmion is a topological spin texture in which spins point in all directions to wrap a sphere; namely, one can define the mapping from $(x.y)$-plane to the unit sphere $\mathbf{n}(x, y)$. When the spins point to the common direction at $\sqrt{x^2 + y^2} \to \infty$, this mapping is characterized by the homotopy group $\pi_2(S^2)$ with the topological invariant given by the integral of the solid angle over the space as

$$N_{sk} = \frac{1}{4\pi} \int dx dy \, \mathbf{n} \cdot \left(\frac{\partial \mathbf{n}}{\partial x} \times \frac{\partial \mathbf{n}}{\partial y} \right). \tag{3}$$

This number N_{sk} is an integer counting how many times the unit sphere is wrapped by the mapping and is called a skyrmion number. The schematic picture of the skyrmion is shown in Fig. 2 with $N_{sk} = 1$. Originally, skyrmion has been proposed as a model for hadron in nuclear physics,[5] but is now discovered in condensed matter systems, such as quantum Hall system[6] and magnetic systems as discussed below.

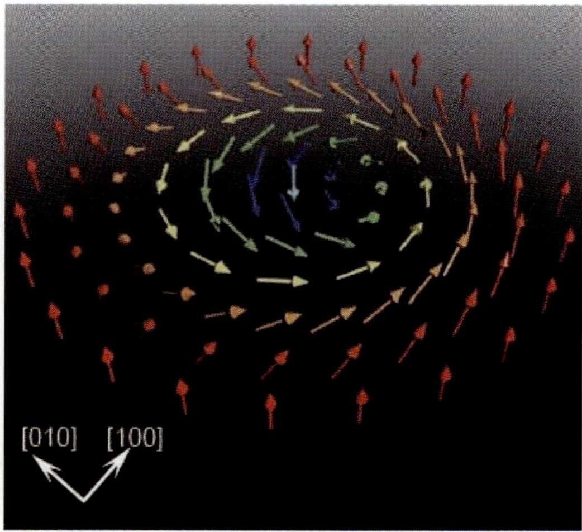

Fig. 2. Schematic view of the skyrmion configuration. One can define the mapping from (x, y) to the unit sphere $\mathbf{n}(x, y)$, and an associated skyrmion number which counts how many times this mapping wraps the unit sphere.

Target materials are the noncentrosymmetric magnets with the B20 structure such as MnSi and (Fe,Co)Si.[7,9] In these systems, the Dzyaloshinskii-Moriya (DM) spin-orbit interaction is allowed, which is described by the following Hamiltonian in the continuum approximation.

$$H = \int d\mathbf{r}\left[\frac{J}{2}(\nabla\mathbf{M}(\mathbf{r}))^2 + \alpha\mathbf{M}(\mathbf{r})\cdot(\nabla\times\mathbf{M}(\mathbf{r})) - \mathbf{B}\cdot\mathbf{M}(\mathbf{r})\right], \qquad (4)$$

where $\mathbf{M}(\mathbf{r})$ is the magnetization, $\mathbf{B}$ the magnetic field, J the ferromagnetic interaction, and α the DM interaction constant. The lattice constant is put to be unity. The DM interaction prefers the spiral spin configuration with the spins rotating in the plane perpendicular to the wavevector $\mathbf{q}$. The length of $q = |\mathbf{q}|$ is determined to be the ratio α/J, i.e., the wavelength is given by J/α and is much larger than the lattice constant when $\alpha << J$. In the Hamiltonian given by eq.(4), the degeneracy occurs with respect to the direction of $\boldsymbol{q}$, and usually this degeneracy is lifted by the weak spin anisotropies not included in eq.(4). The external magnetic field $\mathbf{B} \parallel +z$ prefers the magnetization along its direction, but the frustration occurs due to the DM interaction. These two competing interactions result in the skyrmion configuration.

The phase diagram of MnSi contains a puzzling phase, so called A-phase, in a small region in (T, B)-plane (T: temperature, $B = |\mathbf{B}|$: strength of the magnetic field). Recently, a neutron scattering experiment identified this A-phase as a skyrmion crystal state.[7] This skyrmon crystal is stabilized by a subtle effect of thermal fluctuation, while the conical spin structure is the most stable state under the magnetic field in the major part of the phase diagram in 3D crystal. On the other hand, a Monte Carlo simulation of the 2D magnet with the DM interaction revealed that the skyrmion crystal state is stable even at zero temperature under magnetic field.[8] Following this theoretical prediction, a real-space observation of the skyrmions has been done in thin films of (Fe,Co)Si employing Lorentz transmission electron microscope (TEM).[9] Figure 3 shows the phase diagram and Lorentz TEM images of the skyrmions and skyrmion crystals. The left panel shows the experimental results, while the right panel shows the results by Monte Carlo simulations, showing excellent agreement. Later, the skyrmions are observed also in FeGe[10] and MnSi[11] using Lorentz TEM.

Another source of the skyrmion structure is the long range magnetic dipolar interaction in ferromagnetic thin films combined with the magnetic anisotropy. Garel and Doniach[12] showed in 1982 that the phase diagram in the (T, B)-plane contains three phases, i.e., the striped phase, the triangular bubble crystal phase, and the ferromagnetic phase. The important feature of this skyrmion system is that the helicity is not fixed by the dipolar interaction and acts as a dynamical degree of freedom in sharp contrast to the DM magnet; namely, the helicity, which is the direction of the spin rotation in the plane perpendicular to the wavevector $\mathbf{q}$, is fixed by the sign of α in DM magnets, while it is not fixed for dipolar magnets. This

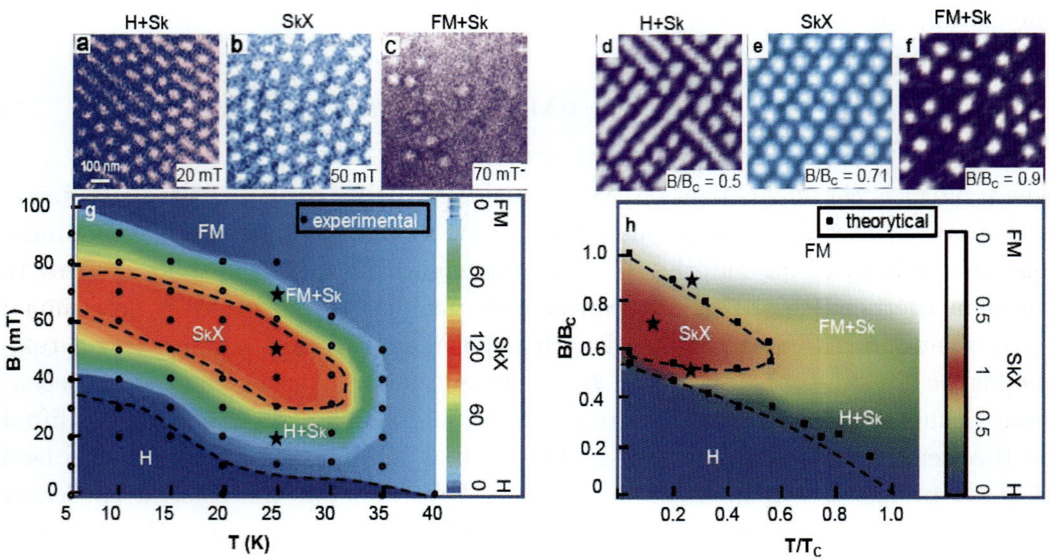

Fig. 3.　The real-space observation of the spin structures by Lorentz TEM and obtained phase diagram in a thin film of (Fe,Co)Si (left panel) and corresponding theoretical spin structures by Monte Carlo simulation (right panel). Reproduced from Ref. 13.

leads to many nontrivial spin textures in the latter case. The readers are referred to Ref. 13 for more details.

3. Dynamics of skyrmions and skyrmion crystal

The next important issue is the dynamics of the skyrmions and electrons coupled to them. As described in Section 1, the Berry phase and the vector potential a_μ play an essential role here. We have recently formulated[14] a theoretical framework for the coupled dynamics of skyrmions and conduction electrons. There appears a term in the Lagrangian describing the interaction between the conduction electron current $\mathbf{j}$ to the gauge field $\mathbf{a}$ as

$$L_{\text{int}} = -\frac{1}{c} \int dr \mathbf{j} \cdot \mathbf{a}. \tag{5}$$

From this term, an effect of the current on the dynamics of the spin direction $\mathbf{n} = \mathbf{M}/|\mathbf{M}|$ is derived and given by the Landau-Lifshitz-Gilbert equation

$$\dot{\mathbf{n}} = \frac{\hbar\gamma}{2e}[\mathbf{j} \cdot \nabla]\mathbf{n} - \gamma \left[\mathbf{n} \times \frac{\delta \mathbf{H_s}}{\delta \mathbf{n}}\right] + \alpha \left[\dot{\mathbf{n}} \times \mathbf{n}\right], \tag{6}$$

where α is the Gilbert damping coefficient. On the other hand, conduction electrons are subject to the effective electromagnetic field as

$$e_i = -\partial_i a_0 - \tfrac{1}{c}\dot{a}_i = \tfrac{\hbar}{2e}\left(\mathbf{n}\cdot\partial_i\mathbf{n}\times\dot{\mathbf{n}}\right),$$

$$b_i = [\nabla\times\mathbf{a}]_i = \tfrac{\hbar c}{2e}\delta_{iz}\left(\mathbf{n}\cdot\partial_x\mathbf{n}\times\partial_y\mathbf{n}\right). \tag{7}$$

In the skyrmion crystal phase, there exists a periodic array of the emergent magnetic field b_z of the same sign. This static distribution of b_z acts as an effective magnetic field and induces the Hall effect replacing the external B_z. We can estimate b_z to be approximately $4000\ T$ when the size of skyrmion ξ is 1 nm, and inversely proportional to ξ^2. This field induces the Lorentz force; from a value of the normal Hall effect, one can estimate the magnitude of this topological Hall effect. Recently the Hall measurement on MnGe has been performed and the contribution from this topological Hall effect has been analyzed.[15] The Hall resistivity $\Delta\rho_{yx}\cong 200$ nΩcm is due to this topological contribution, which corresponds to the effective magnetic field $b\cong 1100\ T$ for $\xi = 3$ nm. These values are compared with those of MnSi: $\Delta\rho_{yx}\cong 5$ nΩcm and $b\cong 28\ T$ for $\xi = 18$ nm. From these studies, it is concluded that the topological Hall conductivity is indeed inversely proportional to the size of the skyrmions.

When skyrmions move, even more interesting phenomena are expected to occur. In general, motions of the magnetic textures are driven by the spin transfer torque mechanism as described by the term containing $\mathbf{j}$ in eq. (6). Recently, it has been revealed that a current density as low as $j_c\cong 10^2$ A/cm^2 can drive the motion of skyrmion crystals in MnSi.[16] This value is orders of magnitude smaller than that is needed to drive the motion of a domain wall in metallic ferromagnets,[17] and hence skyrmions are promising candidate for applications to memory devices. Also in thin films of FeGe, similar very low critical current density is observed for the skyrmion crystal phase, but the single spiral does not move.[13] This sharp contrast between the single spiral and skyrmion phases is an important issue need to be scrutinized.

When skyrmions move, the emergent magnetic field distribution also moves resulting in finite $d\mathbf{b}/dt$ and hence finite $\mathbf{e}$ due to the emergent electromagnetic induction. This $\mathbf{e}$ field leads to an additional Hall effect. We predicted the reduction of the Hall signal as the skyrmions begin to move when the external current density becomes larger than the critical value[14] due to this electromagnetic induction. Remarkably, a recent experiment confirmed this prediction by an accurate measurement of the current dependence of the Hall signal in the A-phase of MnSi.[18]

The phenomena derived from the coupled dynamics of skyrmions and electrons are not limited to what we have discussed above, and eqs. (6) and (7) should provide the bases of applications to the future "skyrmionics", i.e., the utilization of skyrmions as the carriers of information.

Acknowledgments

The authors thank M. Mochizuki, H. J. Han, C. Jia, K. Nomura, M. Mostovoy for collaborations, and A. Rosch and C. Pfleiderer for useful discussions. This work was supported by Priority Area Grants, Grant-in-Aids under the Grant number 24244054 from the Ministry of Education, Culture, Sports, Science, and Technology, Japan, by Strategic International Cooperative Program (Joint Research Type) from Japan Science and Technology Agency, and by a grant from the Japan Society for the Promotion of Science through the "Funding Program for World-Leading Innovative R and D on Science and Technology (FIRST Program)", initiated by the Council for Science and Technology Policy.

References

1. Y. Aharonov and D. Bohm, Significance of electromagnetic potentials in the quantum theory, *Phys. Rev.* **115**(3), 485–491, (Aug, 1959).
2. A. Tonomura, Applications of electron holography, *Rev. Mod. Phys.* **59**(3), 639–669, (Jul, 1987).
3. M. Berry, Quantal phase factors accompanying adiabatic changes, *Proc. Roy. Soc. London A* **392**(1802), 45–57, (1984).
4. P. A. Lee, N. Nagaosa, and X.-G. Wen, Doping a Mott insulator: Physics of high-temperature superconductivity, *Rev. Mod. Phys.* **78**(1), 17–85, (Jan, 2006).
5. T. Skyrme, A unified field theory of mesons and baryons, *Nuclear Physics.* **31**, 556–569, (1962).
6. S. L. Sondhi, A. Karlhede, S. A. Kivelson, and E. H. Rezayi, Skyrmions and the crossover from the integer to fractional quantum Hall effect at small Zeeman energies, *Phys. Rev. B* **47**(24), 16419–16426, (Jun, 1993).
7. S. Mühlbauer, B. Binz, F. Jonietz, C. Pfleiderer, A. Rosch, A. Neubauer, R. Georgii, and P. Böni, Skyrmion lattice in a chiral magnet, *Science.* **323**(5916), 915–919, (2009).
8. S. D. Yi, S. Onoda, N. Nagaosa, and J. H. Han, Skyrmions and anomalous Hall effect in a Dzyaloshinskii-Moriya spiral magnet, *Phys. Rev. B* **80**(5), 054416, (Aug, 2009).
9. X. Z. Yu, Y. Onose, N. Kanazawa, J. H. Park, J. H. Han, Y. Matsui, N. Nagaosa, and Y. Tokura, Real-space observation of a two-dimensional skyrmion crystal, *Nature.* **465**(7300), 901–904, (2010).
10. X. Z. Yu, N. Kanazawa, Y. Onose, K. Kimoto, W. Z. Zhang, S. Ishiwata, Y. Matsui, and Y. Tokura, Near room-temperature formation of a skyrmion crystal in thin-films of the helimagnet FeGe, *Nature Materials.* **10**(2), 106–109, (2010).
11. A. Tonomura, X. Z. Yu, K. Yanagisawa, T. Matsuda, Y. Onose, N. Kanazawa, H. S. Park, and Y. Tokura, Real-space observation of skyrmion lattice in helimagnet MnSi thin samples, *Nano Letters.* **12**(3), 1673–1677, (2012).
12. T. Garel and S. Doniach, Phase transitions with spontaneous modulation-the dipolar Ising ferromagnet, *Phys. Rev. B* **26**(1), 325–329, (Jul, 1982).
13. X. Z. Yu, N. Kanazawa, W. Zhang, T. Nagai, T. Hara, K. Kimoto, Y. Matsui, Y. Onose, and Y. Tokura, Skyrmion flow near room temperature in an ultralow current density, *Nature Communications* **3**, 988, (2012).
14. J. Zang, M. Mostovoy, J. H. Han, and N. Nagaosa, Dynamics of skyrmion crystals in metallic thin films, *Phys. Rev. Lett.* **107**(13), 136804, (Sep, 2011).

15. N. Kanazawa, Y. Onose, T. Arima, D. Okuyama, K. Ohoyama, S. Wakimoto, K. Kakurai, S. Ishiwata, and Y. Tokura, Large topological Hall effect in a short-period helimagnet MnGe, *Phys. Rev. Lett.* **106**(15), 156603, (Apr, 2011).
16. F. Jonietz, S. Mühlbauer, C. Pfleiderer, A. Neubauer, W. Münzer, A. Bauer, T. Adams, R. Georgii, P. Böni, R. A. Duine, K. Everschor, M. Garst, and A. Rosch, Spin transfer torques in MnSi at ultralow current densities, *Science* **330**(6011), 1648–1651, (2010).
17. A. Yamaguchi, T. Ono, S. Nasu, K. Miyake, K. Mibu, and T. Shinjo, Real-space observation of current-driven domain wall motion in submicron magnetic wires, *Phys. Rev. Lett.* **92**(7), 077205, (Feb, 2004).
18. T. Schulz, R. Ritz, A. Bauer, M. Halder, M. Wagner, C. Franz, C. Pfleiderer, K. Everschor, M. Garst, and A. Rosch, Emergent electrodynamics of skyrmions in a chiral magnet, *Nature Physics.* **8**, 301–304, (Apr, 2012).

Gauge Theory and Artificial Spin Ices: Imaging Emergent Monopoles with Electron Microscopy

Shawn D. Pollard and Yimei Zhu

Department of Condensed Matter Physics, Brookhaven National Laboratory,
Upton, NY, 11973 USA
Department of Physics and Astronomy, Stony Brook University,
Stony Brook, NY, 11794, USA
E-mail: Zhu@bnl.gov

Since it was proposed, the Aharonov-Bohm (AB) effect has sparked profound debate on the nature of gauge fields in physics and offered the possibility for experimental proof for gauge theories. Proof of the AB effect, essential evidence for the theory of gauge fields, was given by Tonomura in his seminal work on the subject. In this article, we discuss its relation to the Dirac monopole and string. Furthermore, we utilize it to experimentally study emergent monopoles and flux channels in the condensed matter setting of artificial spin ices. We use phase imaging based on the AB effect to study ordering processes in the spin ice lattice during magnetic reversal and the effect that emergent monopoles and flux channels have on this reversal. We observe that monopole interactions govern defect densities within the spin lattice, and exploit these interactions to develop at a new method to achieve high degrees of ground state ordering in a frustrated system.

We dedicate this article to Dr. Akira Tonomura in memory of his lifelong contributions to the Aharonov-Bohm effect and magnetic phase imaging.

1. Introduction

Since its introduction by Maxwell in 1856, the magnetic vector potential was an enigma in physics. It was temporarily discarded by Heaviside and Hertz as lacking in physical meaning,[1] until the 20th century, when vector potentials came in vogue again. Vector potentials were essential to Yang and Mills' work towards unifying the nuclear and electromagnetic forces,[2] and were even used by Weyl in failed attempts to unify gravity and electromagnetism. In these contexts, the vector potentials are known as gauge fields, fundamental towards our understanding of high energy and particle physics, and are related to the phase factor of the electron wave function by

$$\exp\left[-\frac{ie}{\hbar}\left(\oint \vec{A}d\vec{s} - \oint Vdt\right)\right],$$

where e is the electron charge, $\hbar$ is the reduced Planck constant, $\vec{A}$ is the vector potential, and V is the scalar potential. The Aharonov-Bohm (AB) effect, proved experimentally by Tonomura, described in the next section, would offer the first and most complete experimental evidence of the theory of gauge fields.

2. The Aharonov-Bohm effect and magnetic monopoles

The ability to directly measure and image the magnetic and electronic structure and its ordering of materials in real space has proven to be invaluable in understanding the fundamental properties of a variety of systems, ranging from ferromagnets and ferroelectrics to superconductors.[3] In electron microscopy, techniques range from the qualitative Fresnel mode[4] to more quantitative methods such as electron holography and the transport-of-intensity method.[5] The physical mechanism behind these techniques is the AB effect, which describes the phase shift acquired by an electron wave which passes through a sample.[6,7] The phase shift φ is described mathematically by

$$\varphi = C_e Vt - \frac{e}{\hbar}\int \vec{A}d\vec{s},$$

where C_e is a constant determined by the accelerating voltage of the electron beam, V is the average potential across the specimen thickness, t is the thickness, and $\vec{A}$ is the vector potential.[3] Of note, this is the same as the phase factor described in the introduction, and is also the same used in electron interference experiments such as electron holography. Commonly, the second term is written in regards to the magnetic field, B, as

$$\varphi = C_e Vt - \frac{e}{\hbar}\int \vec{B}d\vec{S}.$$

However, this form of the equation disregards a fact realized by the individuals for which the effect is named. They realized that even in space where $\vec{B} = 0$, $\vec{A}$ may still have a nonzero value and result in a phase shift of an electron, which could then be measured with an interference experiment. The typical description of such an experiment in which this effect can be observed consists of an electron source, a detector, and in between a solenoidal field that is shielded from the electron wave-front. In theory, the solenoid should be infinite to eliminate the leakage field; however, this is difficult to achieve in practice.[8] Contrary to intuition, the electrons experience a phase shift despite

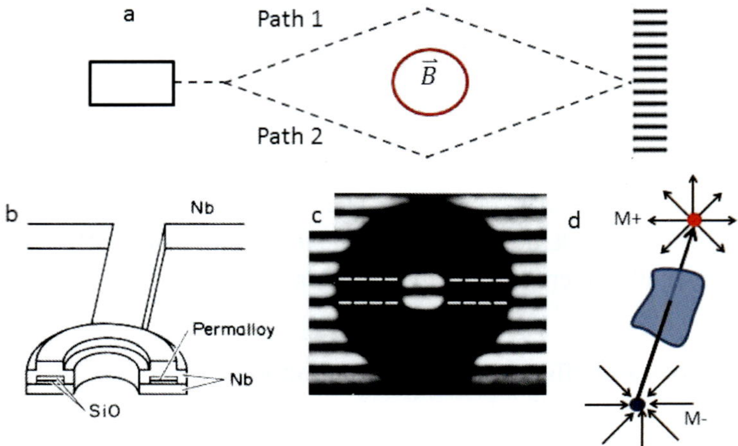

Fig. 1. (a) A sketch of a classic experiment demonstrating the AB effect. The magnetic field is zero outside of the cylinder shown. However, electrons travelling outside of the field are still phase shifted, a result of a non-zero potential. The beam will then exhibit a typical interference pattern, which may be measured. (b) Sketch of the toroidal magnet used by Tonomura *et al.* to experimentally prove the AB effect.. The outer layer is a superconducting Nb film, which prevents stray magnetic field from leaking into vacuum, as well as being thick enough to prevent electron penetration into the permalloy magnetic layer. (c) Hologram taken of the toroidal magnet showing a π phase shift. Figures (b) and (c) are taken from Ref. 1. (d) Sketch of a pair of monopoles and Dirac string connecting them. The monopoles appear as sources (red) and sinks (blue) of magnetic field. Classical electromagnetism is preserved in the presence of monopoles by the tensionless Dirac string, acting as a solenoid carrying magnetic flux.

experiencing no magnetic field, and the detector should record an interference pattern (Fig. 1a). Whether this effect was purely mathematical or if it was corresponding to physically observable phenomena heavily debated for many years.[9-11] The AB effect was finally proven by Tonomura *et al.*[12] using a toroidal permalloy magnet grown on SiO substrate coated by a niobium superconducting film (Fig. 1b).

This layer of superconductor not only shielded the toroidal magnet from the incident electron beam, but the Meissner effect precluded leakage of stray magnetic flux to the surrounding vacuum. Use of a toroidal magnet also avoided the inherent difficulties in working with an infinite solenoid, which is experimentally unattainable. While the toroidal magnet itself was not electron transparent, the beam still passed through vacuum external the toroid and internally in the circular gap, and the relative phase shift in the gap and external to it could be measured. In this set-up, the phase shift should be quantized in multiples of π. Electron interference measured with electron holography showed a clear phase shift of exactly 0 or π (Fig. 1c). Not only did this solve a problem in physics that had stood for over 30 years, it also provided evidence for the theory of gauge fields by demonstrating that the gauge itself, not the magnetic field, was responsible for the observed phase shift.

Of further importance is the AB effect's role in the understanding of magnetic monopoles. Since their prediction by Dirac[13,14] in 1931, the Dirac monopole and Dirac string have been sought after by physicists in a variety of strings, from the Relativistic Heavy Ion Collider at Brookhaven National Laboratory to the Large Hadron Collider, to detectors meant for monopoles produced not by colliders, but as relics from the early universe.[15,16] This interest is primarily a result of their relation to the quantization of the electric charge. In one formulation, the monopole field was described by a pair gauge equivalent potentials with singularities along the + or − z-axis (the Dirac strings),

$$\vec{A}_{\pm} = \mp \frac{m(1\pm cos\theta)}{r\,sin\theta}\,\hat{\phi},$$

with m being the magnetic charge, and produces a field which is the magnetic equivalent to a Coulomb charge. The line of singularity resulting from this potential is known as the Dirac string, and to reconcile quantum mechanics with electromagnetism, the Dirac string must carry a flux $4\pi m$ opposite the monopole flux, preserving $\nabla\cdot\vec{B} = 0$ (Fig. 1d). Furthermore, by considering that the wave function of an electron subjected to the corresponding vector potentials must be single valued in the space where the potentials overlap, i.e.,

$$\psi_{+} = \exp\left(-\frac{2iem\phi}{\hbar c}\right)\psi_{-}.$$

This condition requires that m is quantized in units of $m = \hbar c /2|e|$. This implies that if a Dirac monopole were to be found, it would imply the quantization of the electric charge. As we know, electric charges are indeed quantized. However, despite the intense interest, to this date all studies related to high energy or particle physics have given a null result with no evidence for a fundamental magnetic charge.

At first glance, the AB effect seems like a useful technique to observe monopole charges and Dirac strings. While the monopole would be observable to the AB effect and give the correct flux distribution, the Dirac string itself is invisible. Note that

$$\varphi = \frac{e}{\hbar c}\oint \vec{A}d\vec{s} = \frac{e}{\hbar c}(4\pi m) = 2\pi n,$$

with n being an integer, after replacing m using the quantization condition. Phase differences of multiples of 2π are invisible to the AB effect, and hence unobservable.

3. Magnetic monopoles and spin ices

Recently, intense interest has been focused on the condensed matter setting of pyrochlore oxides ($A_2B_2O_7$, where A = Dy, Ho, B = Ti) in which quasiparticles similar to the Dirac monopole and string can be experimentally accessed and studied with ready experimental techniques.[17-21] In the pyrochlore spin ices, magnetic moments within the pyrochlore lattice point along their Ising axis. This is similar to the dipole moments in water ice, and hence the name, spin ice. The low energy state at each vertex exists when four moments are arranged such that two moments point into, and two out of, the vertex (Fig. 2).

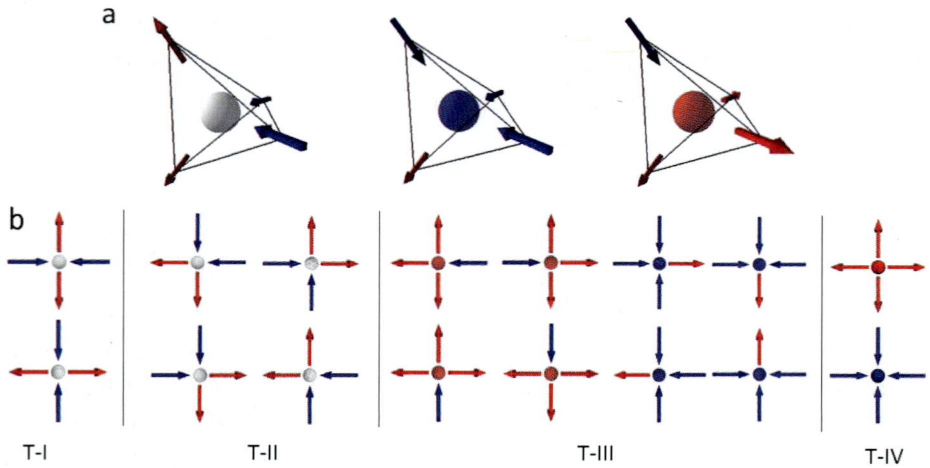

Fig. 2. (a) Schematic of possible magnetic moment configurations of a tetrahedron in the pyrochlore spin ice. Where the ice rules are satisfied (top left), there is no net moment (grey). Where the ice rules are violated, a net moment exists, either negative (blue) when 3 moments point into a vertex (middle), or positive (red) where 3 moments point out of a vertex. These vertices behave as magnetic monopoles. (b) In artificial spin ices, there are analogously 4 vertex types with different energies, two satisfying the 2-in/2-out ice rules and two violating them, labeled in terms of increasing energy. Shown here are the 16 possible vertex configurations (T-I to T-IV).

In such a situation the ice rules are satisfied (2in/2out). Defects arise when a moment is switched, resulting in 3 moments into one vertex, and 3 out of another, violating the ice rules (Fig. 2). These charges are then separated by switching subsequent adjacent magnetic moments. It was realized by Castelnova *et al.*[19] that these interact via a magnetic Coulomb force and could be separated without any tension between them, behaving as free charges, behaving similarly to the Dirac string and monopole. These predictions were confirmed experimentally in neutron scattering, muon spin relaxation, and heat capacity measurements in $Dy_2Ti_2O_7$ and $Ho_2Ti_2O_7$.[18,21] Despite the similarities, there are key differences between these emergent monopoles and flux channels and the Dirac monopole and strings proposed by Dirac. Spin ice monopoles are not quantized in such a way that they are unobservable to the AB effect, provided that the initial and final magnetic configurations of the lattice are known before and after application of some stimulus (i.e., reversal by a magnetic field).[22] Despite the divergence between these emergent monopoles and the fundamental monopole, their existence in condensed matter systems has opened up a new field in which magnetic charges with Coulomb interactions can be studied.

In the pyrochlore spin ices, these studies can be limited to a lack of suitable direct imaging techniques capable of atomic scale phase imaging. A new approach based on simulating these atomic scale magnetic moments using micron-scale magnetic islands that act as analogs to the atomic scale spins of true spin ices has recently been implemented.[23] These micro-scaled islands, when placed in either a square or kagome lattice,[24] form the basis for artificial spin ices. Shape anisotropy leads to the magnetization lying along their long axis, effectively behaving as Ising spins. All islands point into a vertex, with three elements forming the vertex in the kagome lattice, and four in the square lattice (Fig. 2b). The relative ease of the fabrication process has opened up a wide variety of parameter space, where coupling at vertices can be varied, disorder controlled, or even the ice rules can be modified.[24-28]

Great interest has been devoted to the ordering of these systems and how the presence of these quasiparticles affects the ordering process.[28-35] Real space imaging techniques allow for the location and path that of these defects tracked. It is observed that the monopole defects are connected by chains of switched islands, which act as magnetic flux tubes. These flux tubes are similar in appearance to the solenoid field within a Dirac string, but macroscopic, observable, and unquantized. This makes electron microscopy an ideal tool to study these systems, due to both the scale of the islands and the change in phase of the electron with the AB effect.

4. Tracking monopoles and ordering during magnetic reversal in spin ice

To directly image monopoles, Dirac strings, and related ordering during magnetic reversal we have performed experiments on artificial square spin ice (Fig. 3a) that are analogous to those performed on the pyrochlore spin ice in the work by Morris *et al.*,[21] in which a saturating field was applied along the [001] axis leading to a state that fully obeys the ice rules. Removing the field from saturation results in the appearance of monopole defects in order to minimize the system energy. These monopoles are connected by "Dirac strings," chains of switched magnetic moments.[21] We study a $14 \times$ 14 vertex array of 100 nm wide, 300 nm long, and 30 nm thick permalloy islands with a 450 nm center-to-center spacing for elements along the same axis, arranged in a "square" array by saturating the 2d lattice along the [11] axis, then reverse the field, switching the magnetization to lie along the [11] direction. At sufficiently high fields, in which magnetic reversal occurs, we track the creation of monopole defects and the flux channels connecting them. While similar to the reversal in the pyrochlores, the asymmetric coupling between neighboring islands in our system leads to a non-zero string tension. Various methods exist to remove this tension, such as a height off-set between sublattices, but this is extremely challenging from an experimental standpoint and to date has not been achieved.[36]

To image local moment configurations, we utilize phase retrieval electron microscopy – both Lorentz imaging and differential transport of intensity (D-TIE) (Fig. 3b). As monopole type defects and the connecting strings are defined only by their excitations above the initial ground state, D-TIE represents a unique technique in which these excitations may be imaged directly, without convolution of extraneous background fields. Real space imaging techniques further allowed for vertex fractions, remnant magnetization, and correlations between individual islands to be quantified throughout the reversal cycle.

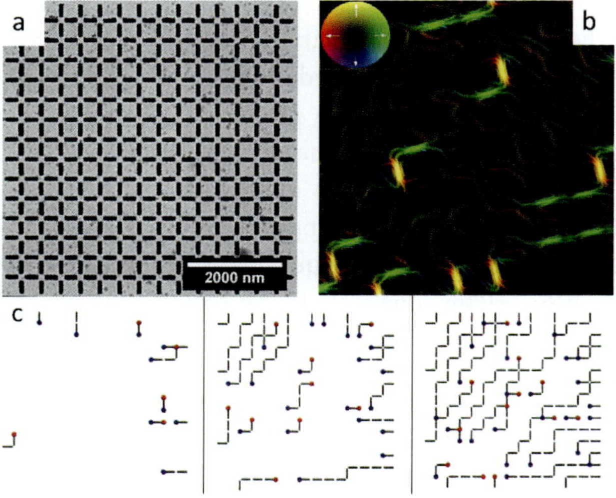

Fig. 3. (a) TEM image a typical square spin ice array studied. (b) Reconstructed differential phase image of a selected portion of the array after partial switching, showing only flux associated with the switched elements. The colored legend marks the magnetization direction. The ends of the chains corresponding to the locations of ice rule breaking vertices, with the chains acting as flux channels between the defects. (c) Schematic of the reversal, taken from real space images. Lines represent chains of switched elements, and red (blue) circles represent defects of charge +q (-q). Images were taken at 34, 38, and 42 mT, respectively.

We further link our experimental measurements to numerical results. In the artificial system, various methods have been used to simulate the system under different situations, most commonly a simple point dipole model, although a variety of models factoring in the energetics of different vertex types have also been used.[26,37]

In the point dipole model, each element of the array is modeled as a dipole with a critical switching field. We vary the critical switching field for each element using a random Gaussian distribution centered at the critical field determined from experiment.

Switching occurs when the critical switching field ($H_{c,i}$) of the *i-th element* is exceeded by the sum of the external field plus the dipole fields of all other elements with the array, i.e., $H_{ext} + \Sigma\ H_{ij} > H_{c,i}$, where H_{ext} is the external magnetic field component along the (x,y) axes of the *i*-th element.

The vertices in the array consist of four types, labeled as T-I to T-IV, in order of increasing energy (Fig. 2b). T-I and T-II obey the 2in/2out ice rules with a net vertex charge Q = 0, while T-III and T-IV do not, with Q = ± 2 and Q = ± 4, respectively, and are energetically unfavorable. The ground state of this system consists of a tiling of the lattice with T-I vertices, and due to the frustrated nature of such a system, has been difficult to obtain in experiment but of great interest for studying frustration physics in real space. A variety of methods have been used to achieve a pseudo-ground state – most commonly AC demagnetization protocols in which the sample is spun in a decaying AC magnetic field, or by utilizing the thermal ordering that occurs during growth.[23,28,30,34,37]

To reveal the switching process in the square ice lattice, the sample was subjected to a magnetic field along its [11] symmetry axis,[38] and the static state after field application was imaged. The [11] axis was chosen due to the nature of the symmetry, allowing for the formation of low energy vertices. Interactions and correlations between neighboring elements were studied by observing the in-plane component of the fringing fields of the individual magnetic islands as well as the local magnetization direction from Lorentz images (Fig. 3). The ability to image the field, unique from other common magnetic imaging techniques such as magnetic force microscopy or various X-ray techniques, allows for new insights in the physical processes governing the frustration to be obtained.

The reversal initiates with the switching of a single element, creating a pair of T-III vertices of opposite charge. As these charges propagate through the lattice, they leave behind a trail of switched elements between the respective charges. The string acts as a channel of magnetic flux between the charges (seen in Fig. 3b). Further increases lead to the nucleation of more defects. As the defects propagate, they tend to leave behind a trail of T-I vertices (Fig. 3c). The highest degree of ordering (as well as T-I vertex populations) happens near zero net magnetization. Furthermore, we quantify the correlations between neighboring elements, similar to that done in previous works,[23,35] with correlations of +1 for ground state interactions and -1 otherwise. We note three different types of correlations (D, T, L), defined in Fig. 4a. As the net magnetization decreases, these correlations increase significantly, suggesting a high degree of ground state ordering. However, longer range correlations are approximately 0 for low net magnetizations, an indication of the frustrated nature of the lattice.

The role of charge ordering becomes apparent when looking at the behavior of monopole defects as the net magnetization decreases. Below M/M$_s$ = 0.6, T-III populations saturate and are well below what would be expected in a non-interacting case. This can be explained through charge interactions. When defects of opposite charge occupy adjacent lattice sites, there is an attractive interaction. More evidence for charge interactions can be seen by tracking the charges between each field step. We find that neighboring charges of the opposite sign annihilate more than 70% of the time that this transition is topologically allowable. These observations, as well as the lack of T-IV vertices at any point of the reversal cycle, hint at the strong role that magnetic charge ordering plays in enforcing the ice rules in these frustrated systems.

5. Demagnetization cycling

Due to the high degree of ordering that was observed during a single reversal cycle along the [11] axis, we use simulations to study how a decaying AC field applied along the [11] axis further orders the frustrated system. In these simulations, following each semi-cycle of the hysteresis loop the maximal field is decreased. The demagnetization using a [11] cycling is distinctly different that of the more common AC demagnetization protocol. Whereas the standard AC demagnetization protocol alternatingly switches the [10] and

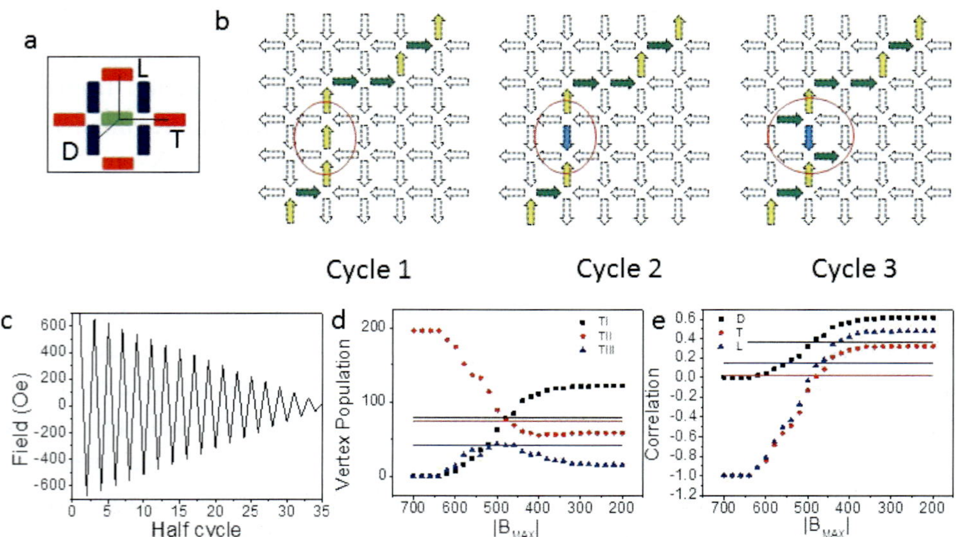

Fig. 4. (a) Definitions for diagonal (D), transverse (T), and Longitudinal (L) correlations in the square spin-ice lattice. (b) Schematic of the multiple step switching in a [11] demagnetization. A chain is broken by switching an unfrozen T-II vertex. This creates two new defects which may propagate, creating new T-I vertices. (c) The demagnetization field cycle protocol. Each half cycle the field is decreased by 2 mT until it falls below the field required to switch elements. (d) Vertex population vs. the amplitude of the field for each cycle. (e) Correlations vs. Field. It is apparent that the end of the cycling results in a high degree of ground state ordering, but complete ordering is blocked due to frustration effects. Lines in (d) and (c) correspond to vertex populations and correlations, respectively, at maximal ordering during a single experimental cycle as reference.

[01] axis for different angles during the sample rotation,[37] the [1̄1] cycling simultaneously switches both axis, exploiting the asymmetric coupling of the vertices to achieve local ordering. During successive cycles, low energy T-I vertices become frozen for subsequent field steps. Meanwhile, T-II vertex links between monopoles can split, separating the string into two sets of defect pairs, with distinct flux channels, that then may propagate through the lattice to form more T-I vertices (Fig. 4b).

Following the cycling, the majority of T-III vertices are the results of trapped defects, in which their motion is only possible through the destruction of a low energy T-I vertex or the creation of high energy T-IV vertices. This results in the low T-II and T-III populations, while achieving large populations of T-I vertices as well as large ground state correlations. For the case we simulated in this work, we used parameters identical to the [1̄1] reversal – a 30 mT switching field with 6 mT variation with the same misalignment from the [1̄1] axis. The field cycles such that the maximal amplitude decreases by 2 mT each half-cycle (Fig. 4c). Initial iterations show no change in correlations of vertex populations from the single cycle reversal, as the applied field is too large, leaving all T-I vertices unfrozen. However, for intermediate regions, T-I vertices become frozen, and do not change for subsequent cycling. This results in much larger final T-I populations and correlations at the end of the ordering process as compared to the simple reversal (Figs. 4d, e). This opens up a new method for achieving a quasi-ground state in which the behavior of emergent monopoles, and their relation to frustration physics, can be studied on the macroscale.

6. Outlook

Despite no observations of true Dirac monopoles in nature to this date, spin ices have opened up new avenues in which the properties of fractionalized magnet charges that behave as free monopoles may be studied. Furthermore, the flux channel analogs corresponding to the Dirac string are observable as they affect the phase of the electron due to the AB effect. Here, we have used magnetic imaging techniques based on the AB effect to link monopole-like defects in artificial spin-ices to field ordering. The defects locally enforce the ice rules during ordering processes, and these processes could give insights into the ordering that occurs in a variety of frustrated systems. Furthermore, exploitation of symmetries has been shown to suggest new methods of achieving highly ordered states, as was demonstrated here by cycling a field along the [1̄1] direction. As new advanced microscopy techniques are further developed, such as atomic resolution differential phase contrast imaging,[39] it may even be possible to study the atomic spin ices and the role that magnetic charges play in their ordering processes in real space.

Acknowledgments

This work was supported by U.S. Department of Energy, Office of Basic Energy Science, Material Sciences and Engineering Division, under Contract No. DE-AC02-98CH10886. The authors acknowledge V. Volkov for assistance in phase imaging.

References

1. A. Tonomura, *Electron Holography*, 2nd Ed. (Springer-Verlag, Berlin, Heidelberg, 1999), pp. 55-59.
2. C. N. Yang and R. L. Mills, Conservation of isotopic spin and isotopic gauge invariance, *Phys. Rev.* **96**(1), 191-195, (1954).
3. A. Tonomura, Applications of electron holography, *Rev. Mod. Phys.* **59**(3), 639-669, (1987).
4. P. J. Grundy and R. S. Tebble, Lorentz electron microscopy, *Adv. Phys.* **17**(66), 153-242, (1968).
5. V. V. Volkov and Y. Zhu, Phase imaging and nanoscale currents in phase objects imaged with fast electrons, *Phys. Rev. Lett.* **91**(4), 043904, (2003).
6. Y. Aharonov and D. Bohm, Significance of electromagnetic potentials in the quantum theory, *Phys. Rev.* **115**(3), 485-491, (1959).
7. W. Ehrenberg and R. E. Siday, The refractive index in electron optics and the principles of dynamics, *Proc. Phys. Soc. London, Sec. B* **62**(1), 8-21, (1949).
8. C. G. Kuper, Electromagnetic potentials in quantum mechanics: a proposed test of the Aharonov-Bohm effect, *Phys. Lett.* **79A**, 413-416, (1980).
9. P. Bocchieri, A. Loinger, and G. Siracusa, Nonexistence of the Aharonov-Bohm effect 2. Discussion of the experiments, *Il Nuovo Cimento A* **51**, 1-16, (1979).
10. P. Bocchieri and A. Loinger, Nonexistence of the Aharonov-Bohm effect, *Il Nuovo Cimento A* **47**, 475-482, (1978).
11. S. M. Roy, Condition for nonexistence of Ahronov-Bohm effect, *Phys. Rev. Lett.* **44**(3), 111-114, (1980).
12. A. Tonomura, N. Okasabe, T. Matsude, T. Kawasaki, J. Endo, S. Yano, and H. Yamada, Evidence for the Aharonov-Bohm effect with magnetic field completely shielded from electron wave, *Phys. Rev. Lett.* **56**(8), 792-795, (1986).
13. P. A. M. Dirac, Quantized singularities in the electromagnetic field, *Proc. Roy. Soc. London, Ser. A* **133**, 60-72, (1931).
14. P. A. M. Dirac, The theory of magnetic poles, *Phys. Rev.* **74**(7), 817-830, (1948).
15. K. A. Milton, Theoretical and experimental status of magnetic monopoles, *Rep. Prog. Phys.* **69**(6), 1637-1711, (2006).
16. M. Bertani, G. Giacomelli, M. R. Mondardini, B. Pal, L. Patrizii, F. Predieri, P. Serra-Lugaresi, G. Sini, M. Spurio, V. Togo, and S. Zucchelli, Search for magnetic monopoles at the Tevatron collider, *Europhys, Lett.* **12**(7), 613-616, (1990).
17. L. D. C. Jaubert and P. C. W. Holdsworth, Signature of magnetic monopole and Dirac string dynamics in spin ice, *Nature Phys.* **5**, 258-261, (2009).
18. T. Fennel, P. P. Deen, A. R. Wildes, K. Schmalzl, D. Prabhakaran, A. T. Boothroyd, R. J. Aldus, D. F. McMorrow, S. T. Bramwell, Magnetic Coulomb phase in the spin ice $Ho_2Ti_2O_7$, *Science* **326**(5951), 415-417, (2009).
19. C. Castelnovo, R. Moessner, and S. L. Sondhi, Magnetic monopoles in spin ice, *Nature* **451**, 42-45, (2008).

20. S. T. Bramwell, S. R. Giblin, S. Calder, R. Aldus, D. Prabhakaran, and T. Fennell, Measurement of the charge and current of magnetic monopoles in spin ice, *Nature* **461**, 956-959, (2009).

21. D. J. P. Morris, D. A. Tennant, S. A. Grigera, B. Klemke, C. Castelnovo, R. Moessner, C. Czternasty, M. Meissner, K. C. Rule, J. U. Hoffmann, K. Kiefer, S. Gerischer, D. Slobinsky, and R. S. Perry, Dirac strings and magnetic monopoles in the spin ice $Dy_2Ti_2O_7$, *Science* **326**(5951), 411-414, (2009).

22. C. Castelnovo, R. Moessner, and S. L. Sondhi, Spin ice, fractionalization, and topological order, *Annual Review of Condensed Matter Physics* **3**, 35-55, (2012).

23. R. F. Wang, C. Nisoli, R. S. Freitas, J. Li, W. McConville, B. J. Cooley, M. S. Lund, N. Samarth, C. Leighton, V. H. Crespi, and P. Schiffer, Artificial 'spin ice' in a geometrically frustrated lattice of nanoscale ferromagnetic islands, *Nature* **439**, 303-306, (2006).

24. Y. Qi, T. Brintlinger, and J. Cumings, Direct observation of the ice rule in an artificial kagome spin ice, *Phys. Rev. B* **77**(9), 094418, (2008).

25. Z. Budrikis, J. P. Morgan, J. Akerman, A. Stein, P. Politi, S. Langridge, C H. Marrows, and R. L. Stamps, Disorder strength and field-driven ground state domain formation in artificial spin ice: experiment, simulation, and theory, *Phys. Rev. Lett.* **109**(3), 037203, (2012).

26. Z. Budrikis, P. Politi, and R. L. Stamps, Diversity enabling equilibration: disorder and the ground state in artificial spin ice, *Phys. Rev. Lett.* **107**(21), 217204, (2011).

27. S. A. Daunheimer, O. Petrova, O. Tchernyshyov, and J. Cummings, Reducing disorder in artificial kagome ice, *Phys. Rev. Lett.* **107**(16), 167201, (2011).

28. J. P. Morgan, J. Akerman, A. Stein, C. Phatak, R. M. L. Evans, S. Langridge, and C. H. Morrows, Real and effective thermal equilibrium in artificial square spin ices, *Phys. Rev. B* **87**(2), 024405, (2013).

29. A. Schumann, B. Sothmann, P. Szary, and H. Zabel, Charge ordering of magnetic monopoles in triangular spin ice patterns, *Appl. Phys. Lett.* **97**(2), 022509, (2010).

30. J. P. Morgan, a. Stein, S. Langridge, and C. H. Marrows, Thermal ground-state ordering and elementary excitations in artificial magnetic square ice, *Nature Physics* **7**, 75-79, (2011).

31. U. B. Arnalds, A. Farhan, R. V. Chopdekar, V. Kapaklis, a. Balan, E. T. Papaioannou, M. Ahlberg, F. Nolting, L. J. Heytdermann, and B. Hjörvarsson, Thermalized ground state of artificial kagome spin ice building blocks, *Appl. Phys. Lett.* **101**(11), 112404, (2012).

32. S. Ladak, D. E. Read, G. K. Perkins,, L. F. Cohen, and W. R. Branford, Direct observation of magnetic monopole defects in an artificial spin-ice system, *Nature Physics* **6**, 359-363, (2010).

33. E. Mengotti, L. J. Heyderman, A. F, Rodoriguez, F. Nolting, R. V. Hügli, and H. B. Braun, Real-space observation of emergent magnetic monopoles and associated Dirac strings in artificial kagome spin ice, *Nature Physics* **7**, 68-74, (2011).

34. C. Patak, A. K. Petford-Long, O. Heinonon, M. Tanase, and M. De Graef, Nanoscale structure of the magnetic induction at monopole defects in artificial spin-ice lattices, *Phys. Rev. B* **83**(17), 174431, (2011).

35. S. D. Pollard, V. Volkov, and Y. Zhu, Propagation of magnetic charge monopoles and Dirac flux strings in an artificial spin-ice lattice, *Phys. Rev. B* **85**(18), 180402(R), (2012).

36. L. A. S. Mól, W. A. Moura-Melo, and A. R. Pereira, Conditions for free magnetic monopoles in nanoscale square arrays of dipolar spin ice, *Phys. Rev. B* **82**(5), 054434, (2010).

37. Z. Budrikis, P. Politi, and R. L. Stamps, Vertex dynamics in finite two-dimensional square spin ices, *Phys. Rev. Lett.* **105**(1), 017201, (2010).

38. V. V. Volkov, D. C. Crew, Y. Zhu, and L. H. Lewis, Magnetic field calibration of a transmission electron microscope using a permanent magnet material, *Rev. Sci. Instrum.* **73**(6), 2298-2304, (2002).

39. N. Shibata, S. D. Findlay, Y. Kohno, H. Sawada, Y. Kondo, and Y. Ikuhara, Differential phase-contrast microscopy at atomic resolution, *Nature Physics* **8**, 611-615, (2012).

Do Dispersionless Forces Exist?

Herman Batelaan and Scot McGregor

*Department of Physics and Astronomy, University of Nebraska—Lincoln,
208 Jorgensen Hall, Lincoln, Nebraska 68588-0299, USA,
e-mail: hbatelaan2@unl.edu*

A defining property of the Aharonov-Bohm effect is its dispersionless nature. This means that the response of a matter wave to external potentials of the type used in the A-B effect is frequency or, equivalently, velocity independent. In the classical limit the dispersionless nature is often equated with the absence of forces. But how is the classical limit defined in the context of the A-B effect? This is the question addressed in this paper and it is argued that the A-B physical system provides an interesting testing ground for the classical-quantum boundary.

1. Introduction

My first interaction with Dr. Akira Tonomura was just a few years ago through a simple email when I (H.B.) asked him if he wanted to collaborate on writing a paper for Physics Today on the Aharonov-Bohm effect. Dr. Tonomura had done one of the most beautiful experiments on this effect and my group had added to this a complementary measurement demonstrating the absence of force. It appeared to be a nice idea to put these two together and at the same time indicate some of the open physics questions in this research area. Dr. Tonomura's response was immediate and very positive. I was delighted with that response coming from such an established scientist, and I considered it an honor to collaborate. The collaboration was fruitful.[1] Not much later Dr. Tonomura asked me to come to a meeting at Okinawa he had organized and I had the great fortune to meet him in person. Dr. Tonomura was very encouraging on the scientific front. He also, with great enthusiasm, led the participants of the meeting to experience a bit of the Japanese culture. Even without meeting him, the elegance of his work had already set standards that I considered to be an example to be followed. After meeting him this now extends to the way he interacts with people. The work described below reconsiders a small part of the issues that we mentioned in our Physics Today paper, is in part motivated by some of the short discussions that I had with him, and would have been presented at the canceled Tokyo meeting, which was planned to be held in April 2011.

The Aharonov-Bohm effect is well known because it is thought to establish that the vector potential can cause measurable effects even when the fields (and thus the forces) are zero. It thus elevates the relevance of the vector potential from being a helpful mathematical construct to that of having direct physical reality associated with it. To highlight this it is interesting to combine two experimental results. The first is the demonstration of the Aharonov-Bohm effect. Tonomura's experiment[2] is not the first to do this, but certainly one of the most elegant ones. The second is the demonstration that forces are absent.[3] Both of these experiments together formed the center part of the Physics Today paper mentioned above.

An opposing view on the Aharonov-Bohm effect was provided in the previous decade. A force was proposed to explain the Aharonov-Bohm effect.[4] The x-component of the Lorentz force on the solenoid with cross-sectional area A and magnetic field B_0 is given by the expression

$$F_{sol}^x = \frac{B_0 A v_0}{4\pi c} \frac{4 x_e y_e}{\left(x_e^2 + y_e^2\right)^2},$$ (1)

where v_0 is the electron velocity along the x-direction and x_e and y_e are the xy-coordinates of the charge relative to the solenoid's z-axis. The supposed back-action force of the solenoid on the electron provided by Newton's third law can be integrated to yield a relative displacement between electrons passing on opposite sides of the solenoid of $\Delta x = eB_0 A / mv_0$. In a semi-classical approximation $\Delta\varphi = k\Delta x$. This phase turns out to be equal to the well-known Aharonov-Bohm phase shift $\Delta\varphi = eB_0 A/\hbar$. It should be emphasized that the fact that such a force can be formulated at all, is very surprising in view of the generally accepted interpretation of the effect. The proposed force was predicted to give rise to a time delay for electrons passing by a solenoid. This time delay was shown experimentally not to occur in the second experiment mentioned above and thus it may appear that this discussion is over. It is the purpose of this paper to revisit that apparent conclusion.

2. Statement of the problem

To start the discussion it may be useful to delineate between the classical, semi-classical and quantum-mechanical parts of the predictions. In the classical description it is noted that the force (1) has components along the direction of motion and thus may cause a time delay as compared to the free electron's motion. The delay can be estimated by making the impulse approximation. This means that we assume that the change in velocity is small compared to the electron's initial velocity v_0 and compute the displacement Δx. The semi-classical part consists of guessing what the associated phase shift is. A reasonable guess would be the use the phase factor e^{ikx} associated with a plane wave and assume that this factor changes by $e^{ik\Delta x}$. The quantum mechanical part is most readily attained by using the path integral approach and the phase shift accumulated over the electron's path as it passes by a solenoid is calculated as

$$\Delta\varphi_{AB} = \frac{e}{\hbar}\int_C \vec{A}\cdot d\vec{l} = \frac{e}{\hbar}\int \vec{B}\cdot d\vec{S}. \tag{2}$$

At this point it may appear convenient to simply rely on the fact that quantum mechanics is a superior theory, encompasses classical mechanics, and ignore the classical and semi-classical arguments. Such a convenient argument would neither do justice to the correspondence principle nor to the main reason why the A-B effect is famous as pointed out above. The question remains how to deal with classical forces in a quantum mechanical context.

2.1. Classical-Quantum deflection in a magnetic field. To answer this question it is perhaps useful to consider the simple deflection of an electron passing through a homogeneous magnetic field. Classical mechanics provides an answer that agrees with observation. Consider an electron entering a region with a homogeneous magnetic field (Fig. 1). The electron's velocity v is at right angle with the magnetic field. The classical deflection angle θ is given by $\theta = \Delta v/v = qvB\Delta t/mv = qBL/mv$.

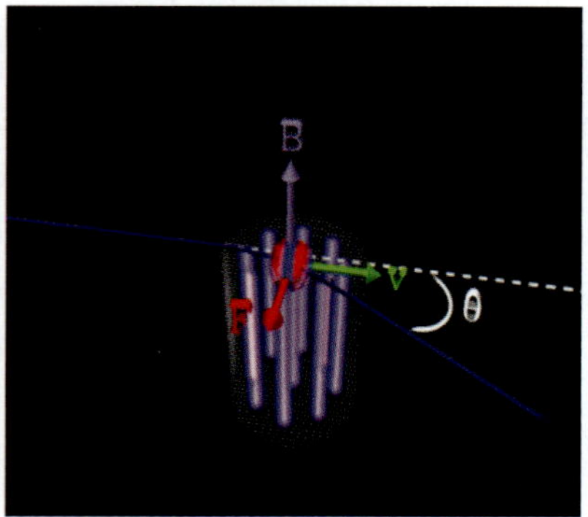

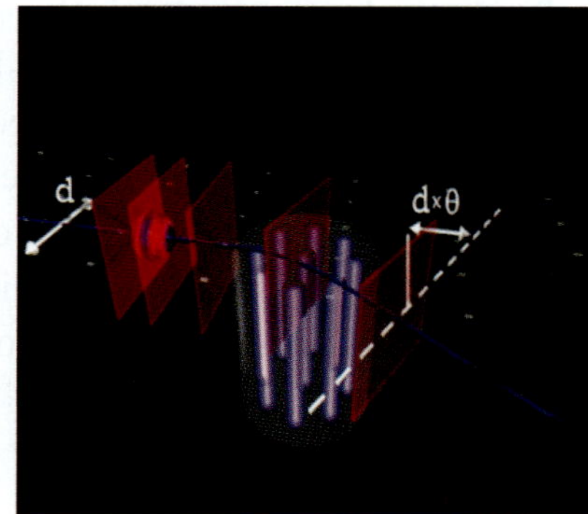

Fig. 1. Left: Electrons deflect by an angle θ after travelling through a region of space with a homogeneous magnetic field B and experiencing a Lorentz force F. Right: An electron wave accumulates a spatially dependent phase shift after travelling through a region of space with a spatially dependent vector potential. This deflects the electron wave by an angle θ.

Associated with the electron is a quantum mechanical wave. For a plane wave in free space the wave planes are at right angles to the direction of motion of the electron. If the planes of the wave tilt then the electron is deflected. Consider planes of constant phase of the wave while propagating through the homogeneous magnetic field.

The phase difference φ accumulated over the width d of a section of a plane wave, determines the tilt of the wavefront. The Lagrangian is given by $L = 1/2\,m_e\dot{x}^2 + q\vec{A}\cdot\dot{\vec{x}}$. The phase shift is $\varphi = 1/\hbar \int L\,dt = 1/\hbar \int qA\,dy \approx qAL/\hbar$ for a vector potential that corresponds to a homogeneous magnetic field $A_y = \int B_z\,dx \approx B_z\Delta x = B_z d$ in the z-direction. A wavefront section with a width d tilts by an angle $\theta = \left(\lambda_{dB}\varphi/(2\pi)\right)/d$, where λ_{dB} is the electron's de Broglie wavelength. This can be rewritten as $\theta = \Delta L/d = \left(\lambda_{dB}\varphi/(2\pi)\right)/d = \left(\lambda_{dB}\left(qB_z dL/\hbar\right)/(2\pi)\right)/d = qB_z L/mv$ and it is clear that the quantum and classical deflection are identical. In other words, the quantum-classical correspondence demands the presence of the phase shift.

The phase shift can be generalized for an arbitrary path to $\varphi = 1/\hbar \int q\vec{A}\cdot d\vec{l}$. For a closed path this is the A-B phase shift $\varphi = 1/\hbar \int_C q\vec{A}\cdot d\vec{l}$. Thus we can say that the deflection of a charged particle in a magnetic field is caused by the A-B phase shift. This should not be confused with the A-B effect which occurs when paths are considered through regions of space where the magnetic field is zero as would be the case when the solenoid in Fig. 2 would be extended in length to infinity. Returning to our main question we can conclude this paragraph by pointing out that the identical classical and quantum prediction may be pleasing but should not be mistaken for the classical quantum correspondence.

Fig. 2. An example of a current carrying solenoid with magnetic field lines (blue) and equi-(vector) potential lines (green) (see also cover article of Ref. 1.

2.2. Classical-Quantum correspondence

The correspondence principle demands that there is some limiting procedure by which we can recover from the quantum mechanical description the classical description. It is traditional to associate large quantum numbers or physically large systems with such a limit.[5] The textbook observation that Gaussian wave packets for particles of macroscopic mass (associated with large systems) have immeasurable small position and velocity spread is correct but does not represent an appropriate classical limit, after all, a wave packet for a large mass particle could still interfere with itself (in an interferometer type arrangement) and exhibit quantum mechanical behavior. Thus we would not expect a large mass to present a truly appropriate classical limit. Instead the capability to interfere must be removed. But what is the detailed description of this coherence removal? We could add an overall random phase factor to the wave packet, or we could instead add a random phase factor to each frequency component. Both modifications approach the classical limit in that the particle loses the capability to interfere with itself, but more than one choice is possible. One could attempt to describe the detailed underlying interaction with the environment. For example, large molecules lose their capability to interfere with themselves by interaction with a thermal background providing evidence for decoherence theory as a means to connect the quantum and classical world. This has been demonstrated in a beautiful controlled coherence experiment by the Arndt group in Vienna.[6] In this experiment it is thought that thermal excitations of internal molecular quantum states and thermal emission make the arms of the interferometer (in principle) distinguishable and thus taking a partial trace over the environment removes coherence. In the present context of the discussion of what types of forces exist, it is the external quantum states that are relevant. It the next section we do not describe decoherence theory for an A-B system, but do attempt to define the problem mathematically at a basic level.

3. Complete coherence and incoherence

Suppose we would like to experimentally test that quantum mechanics correctly describes a free particle. A short pulse could be made and its propagation studied. It is sufficient to investigate the propagation of two frequency components. Consider two plane waves of equal amplitude propagating along the positive x-axis with velocities $v+\Delta v$ and $v-\Delta v$. The wavefunction can be written as the sum of the two frequency components,

$$\psi(x,t) = \psi_{E_1}(x,t) + \psi_{E_2}(x,t) = \frac{1}{2}\sqrt{2}\,e^{i\left(\frac{2\pi m(v-\Delta v)}{h}x - \frac{\pi m(v-\Delta v)^2}{h}t\right)} + \frac{1}{2}\sqrt{2}\,e^{i\left(\frac{2\pi m(v+\Delta v)}{h}x - \frac{\pi m(v+\Delta v)^2}{h}t\right)} \qquad (3)$$

This wavefunction can be rewritten as the product of the frequency carrier and the envelope,

$$\psi(x,t) = \frac{1}{2}\sqrt{2}\,e^{i\left(\frac{2\pi m(v)}{h}x - \frac{\pi m(v)^2}{h}t\right)}\left(e^{i\left(\frac{2\pi m\Delta v}{h}(x-vt)+O((\Delta v)^2\right)} + e^{i\left(\frac{2\pi m\Delta v}{h}(x-vt)+O((\Delta v)^2\right)}\right), \qquad (4)$$

where the former has a phase velocity of v/2 and the latter travels at the group velocity of v, but only when the two components are coherently added. The probability distribution $\psi^*(x,t)\psi(x,t)$ follows the group velocity according to $x=vt$, in "correspondence" with the classical prediction. Such an argument can be generalized to a wave packet. If an interaction causes a phase shift that affects each frequency (or equivalently velocity) component in the same way; $\psi(x,t) = \psi_{E_1}(x,t)e^{i\varphi} + \psi_{E_2}(x,t)e^{i\varphi}$, then (3) and (4) are only modified by an overall phase factor that does not change the probability distribution. Thus dispersionless interactions do not cause a deviation from the classical path; hence we can state that a dispersionless interaction is associated with the absence of force.

If we assume that an underlying physical decoherence process removes all coherence then we can construct a density matrix and add $\psi_{E_1}(x,t)$ and $\psi_{E_2}(x,t)$ completely incoherently in an attempt to take a classical limit:

$$\rho = \left|\psi_{E_1}(x,t)\right\rangle\left\langle\psi_{E_1}(x,t)\right| + \left|\psi_{E_2}(x,t)\right\rangle\left\langle\psi_{E_2}(x,t)\right|. \tag{5}$$

Rewriting the density matrix as a product of the carrier wave and its envelope is now not possible. Instead we can calculate how the expectation value of the position propagates in time. The result is ill-defined because the expectation value for plane waves is ill-defined. This very basic simple step failed, and serves to illustrate that taking classical limits may be hard with and even without forces. Perhaps, we should not care about the correspondence principle and only demand that our best theory matches our experimental outcomes, and not that it should first match a presumably worse theory. So, let's next attempt to circumvent the classical limit and simply calculate the velocity dependence of the phase shift in the absence of force and when the force given by (1) is present.

4. Velocity dependent Phase shift with and without Forces

Using the path integral formulation of quantum mechanics, the phase shift is given by

$$\varphi = 1/\hbar \int L dt = 1/\hbar \int (1/2\, m_e \dot{x}^2 + q\vec{A} \cdot \dot{x}) dt, \tag{6}$$

where the integral is to be taken along a classical path that starts at (x_A, t_A) and ends at (x_B, t_B). For a particle that travels along a classical path that is free from any force, this expression can be simplified to

$$\varphi_{free} = 1/\hbar \int (1/2\, m_e \dot{x} + q\vec{A}) \cdot d\vec{x} = \frac{\pi(x_B - x_A)}{\lambda_{dB}} + 1/\hbar \int q\vec{A} \cdot d\vec{x}, \tag{7}$$

where the first term is similar to what is expected from the Huygens' principle for matter waves[7] except for a missing factor of two. It is straightforward to show that the factor of two can be recovered by considering only phase differences between paths that start and stop at the same time. The second term yields a phase that is velocity independent, and is thus dispersionless as expected.

For a particle that travels along a classical path that experiences a force given by (1), this expression has to be explicitly calculated. For our present discussion it will suffice to make a very crude approximation. Noting that the force is anti-symmetric under parity in x, we consider a simple piecewise constant force (Fig. 3) that modifies the velocity to $v - \Delta v$ when $x < 0$, and $v + \Delta v$ when $x > 0$.

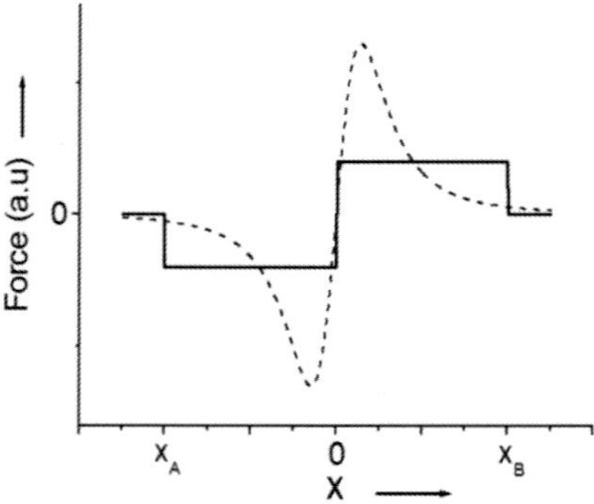

Fig. 3. The x-dependence of the force given by equation (1) is given (dashed line). A crude estimate (solid line) is used to estimate the phase dependence on velocity.

Further, consider a particle that starts at location $x_A = -L/2$ and ends at $x_B = L/2$. In this case the phase shift can be approximated by

$$\varphi_{force} = \frac{\pi(0-x_A)m(v-\Delta v)}{h} + \frac{\pi(x_B-0)m(v+\Delta v)}{h} + 1/\hbar \int q\vec{A}\cdot d\vec{x}$$

$$= \frac{\pi(x_B-x_A)}{\lambda_{dB}} + 1/\hbar \int q\vec{A}\cdot d\vec{x},$$

(8)

which is identical to the phase shift for the free particle. The reason that the result with force is not the same as the semi-classical phase $e^{ik\Delta x}$ is caused by the demand that the path has the same start and stop time as in free particle case. A key feature of the force is that it is linear in the velocity itself, which results in a phase that is velocity independent. *In other words it appears possible to construct forces that are dispersionless.*

5. Approximately dispersionless

If the magnitude of the force is large in the sense that the change in velocity Δv is not small compared to the initial velocity v, then the demand that the start and end time should not change leads to dispersion. In specific, for the conditions $(v-\Delta v_1)t_1 + (v+\Delta v_2)t_2 = x_B - x_A$ and $t_1 + t_2 = t_B - t_A$, the decrease in velocity for the region $x < 0$ does not equal the increase in velocity in the region $x > 0$. The result is that the cancelation of the Δv terms in (8) is removed, which results in a velocity dependent phase shift in (8). This leads to a time delay and it may be possible to falsify such a prediction experimentally with a refined version of the experiment reported in Ref. 3.

6. Summary and Conclusion

To identify if there is a force, one can measure a time delay of a pulse or a deflection of a beam of particles. This experimental definition appears to be very clear. But can we conclude that if there is no deflection or delay that no forces acted? This is not obvious. Nevertheless, that is the operational definition for the claim that the Aharonov-Bohm effect occurs in the absence of any force. A counter argument based on the non-zero force (1) is hard to rule out. In our first attempt to do so (section 3) by demanding that the correspondence principle should hold, we find that it is hard to find an appropriate classical limit. In our second attempt to rule out this force (section 4) it turns out to be dispersionless. However, dispersionless interaction is considered to be a defining property of the A-B effect. This leads to the question raised in the title: "Do dispersionless forces exist?" A potential way to resolve this issue presents itself when we realize that the force is only dispersionless for small changes in velocity. A re-analysis of the experimental data for small delays may rule out the approximately dispersionless forces. A complicating factor for large changes in velocity is the issue to what extent decoherence and "the classical limit" can be avoided for such conditions.

Acknowledgment

This material is based upon work supported by the National Science Foundation under Grant No. 2505210148001.

References

1. H. Batelaan and A. Tonomura, The Aharonov–Bohm effects: Variations on a subtle theme, *Phys. Today* **62**(9), 38-43, (September 2009).
2. A. Tonomura, N. Osakabe, T. Matsuda, T. Kawasaki, J. Endo, S. Yano, and H. Yamada, Evidence for Aharonov-Bohm effect with magnetic field completely shielded from electron wave, *Phys. Rev. Lett.* **56**(8), 792-795, (1986).
3. A. Caprez, B. Barwick, and H. Batelaan, Macroscopic test of the Aharonov-Bohm effect, *Phys. Rev. Lett.* **99**(21), 210401, (2007).
4. T. H. Boyer, Semiclassical explanation of the Matteucci-Pozzi and Aharonov-Bohm phase shifts, *Foundations of Physics* **32**(1), 41-49, (2002).
5. J. R. Nielsen (Ed.), *Niels Bohr, Collected Works, Volume 3, The Correspondence Principle (1918–1923)*, (North-Holland, Amsterdam, 1976).
6. L. Hackermüller, K. Hornberger, B. Brezger, A. Zeilinger, and M. Arndt, Decoherence of matter waves by thermal emission of radiation, *Nature* **427**, 711-714, (2004).
7. R. P. Feynman, Space-time approach to non-relativistic quantum mechanics, *Rev. Mod. Phys.* **20**(2), 367-387, (1948).

Aharonov-Bohm Effect and Geometric Phases — Exact and Approximate Topology

Kazuo Fujikawa

RIKEN Nishina Center
Wako, Saitama 352-0198, Japan
E-mail: k-fujikawa@riken.jp

By analyzing an exactly solvable model in the second quantized formulation which allows a unified treatment of adiabatic and non-adiabatic geometric phases, it is shown that the topology of the adiabatic Berry's phase, which is characterized by the singularity associated with possible level crossing, is trivial in a precise sense. This topology of the geometric phase is quite different from the topology of the Aharonov-Bohm effect, where the topology is specified by the external local gauge field and it is exact for the slow as well as for the fast motion of the electron.

1. Aharonov-Bohm effect and geometric phases

Akira Tonomura made important contributions to the studies of the Aharonov-Bohm effect and the double-slit experiment which is closely related to the analysis of geometric phases. Both of these effects are related to phases and interference in quantum mechanics. The phase in quantum mechanics is also closely related to the notion of topology in mathematics.

The topology of the Aharonov-Bohm effect is provided by the external boundary condition for the gauge field,[1] and the Aharonov-Bohm effect is best described by the path integral representation

$$\langle \vec{x}_f, T | \vec{x}_i, 0 \rangle = \int D\vec{x} \exp\left\{ \frac{i}{\hbar} \int_0^T \left[\frac{m\dot{\vec{x}}^2}{2} - e\vec{A}(\vec{x}) \frac{d\vec{x}}{dt} \right] dt \right\} \tag{1}$$

for the propagation of an electron.

On the other hand, the geometric phase, or Berry's phase,[2-5] for the electron placed in a rotating magnetic field $\vec{B}(t)$, which is solved exactly as shown below, is given by

$$\psi_{\pm}(T) = w_{\pm}(T)\exp\left[-\frac{i}{\hbar}\int_0^T dt w_{\pm}^{\dagger}(t)(\hat{H} - i\hbar\partial_t)w_{\pm}(t)\right]$$

$$= w_{\pm}(T)\exp\left[-\frac{i}{\hbar}\int_0^T dt w_{\pm}^{\dagger}(t)\hat{H}w_{\pm}(t)\right]$$

$$\times\exp\left[-\frac{i}{\hbar}\int_0^T \vec{A}_{\pm}(\vec{B})\frac{d\vec{B}}{dt}dt\right], \tag{2}$$

where

$$\vec{A}_{\pm}(\vec{B}) \equiv w_{\pm}^{\dagger}(t)\left(-i\hbar\frac{\partial}{\partial\vec{B}}\right)w_{\pm}(t) \tag{3}$$

gives an analogue of the gauge potential (or connection).

These two expressions are very similar, but the important difference is that the electron moves outside the magnetic field in the case of the Aharonov-Bohm effect while the electron moves inside the magnetic field in the case of geometric phases. This difference suggests that the topology of these two phases, though similar, is fundamentally different. In fact, the topology of the Aharonov-Bohm effect is precise for the non-adiabatic as well as adiabatic motion of the electron. On the other hand, it is shown that the topology of Berry's phase is valid only in the ideal adiabatic limit and it is lost once one moves away from ideal adiabaticity. We explain this difference in the following since it is not widely recognized.

2. Second quantization and hidden gauge symmetry

To analyze the topology of the geometric phase, one needs a formulation which treats the adiabatic[2-5] and non-adiabatic phases[6-8] in a unified manner.[9,10] We start with the action

$$S = \int dt d^3x\left[\hat{\psi}^{\dagger}(t,\vec{x})\left(i\hbar\frac{\partial}{\partial t} - \hat{H}(t)\right)\hat{\psi}(t,\vec{x})\right] \tag{4}$$

For a time-dependent Hamiltonian $\hat{H}(t)$. We then expand

$$\hat{\psi}(t,x) = \sum_n \hat{c}_n(t)v_n(t,\vec{x}) \tag{5}$$

with $\int d^3 x v_n^*(t,\vec{x}) v_m(t,\vec{x}) = \delta_{n,m}$. For the fermion, we impose anti-commutator $\{\hat{c}_l(t), \hat{c}_m^\dagger\} = \delta_{l,m}$. The Fock states are defined by $|l\rangle = \hat{c}_l^\dagger(0)|0\rangle$.

By inserting the expansion into the action S, we have

$$
S = \int dt \left\{ \sum_n \hat{c}_n^\dagger(t) i\hbar \partial_t \hat{c}_n(t) \right.
$$

$$
\left. - \sum_{n,m} \int d^3 x \left[\hat{v}_n^*(t,\vec{x}) \hat{H}(t) \hat{v}_m(t,\vec{x}) - v_n^*(t,x) i\hbar \partial_t v_m(t,\vec{x}) \right] \hat{c}_n^\dagger(t) \hat{c}_m(t) \right\} , \tag{6}
$$

and the appearance of "geometric phase" in the last term is automatic.

The solution of the conventional Schrödinger equation with the initial condition $\psi(0,\vec{x}) = v_n(0,\vec{x})$ is given by $\psi_n(t,\vec{x}) = \langle 0|\hat{\psi}(t,\vec{x})\hat{c}_n^\dagger(0)|0\rangle$. The second quantized formulation contains the following *gauge* (or redundant) freedom[11]

$$
\hat{c}_n(t) \to e^{-i\alpha_n(t)} \hat{c}_n(t), \, v_n(t) \to e^{i\alpha_n(t)} v_n(t) \tag{7}
$$

which keeps $\hat{\psi}(t,\vec{x})$ in (5) invariant, with the phase freedom $\{\alpha_n(t)\}$ being arbitrary functions of time. Under this *hidden gauge transformation*, the Schrödinger amplitude is transformed as

$$
\psi_n(t,\vec{x}) = \langle 0|\hat{\psi}(t,\vec{x})\hat{c}_n^\dagger(0)|0\rangle \to e^{i\alpha_n(0)} \psi_n(t) , \tag{8}
$$

namely, the ray representation of the state vector is induced. One may ask what is the physical implication of this hidden gauge symmetry? The answer is "it controls all the geometric phases, either adiabatic or non-adiabatic."[12] In the analysis of geometric phases, it is crucial to note that the combination $\psi_n^*(0,\vec{x})\psi_n(t,\vec{x})$ is manifestly gauge invariant.

3. Exactly solvable example and geometric phase

We consider the motion of a spin inside the rotating magnetic field

$$
\vec{B}(t) = B\left(\sin\theta \cos\varphi(t), \sin\theta \sin\varphi(t), \cos\theta \right) \tag{9}
$$

and $\varphi(t) = \omega_0 t$ with a constant ω_0 . The action is written as

$$
S = \int dt \left[\hat{\psi}^\dagger(t) \left(i\hbar \frac{\partial}{\partial t} + \vec{B}\cdot\vec{\sigma}/2 \right) \hat{\psi}(t) \right], \tag{10}
$$

with $\vec{\sigma}$ Pauli matrices. The field operator is expanded as

$$\hat{\psi}(t,\vec{x}) = \sum_{l=\pm} \hat{c}_l(t) w_l(t)$$

with the anti-commutation relation, $\left\{\hat{c}_l(t),\hat{c}_m^\dagger\right\} = \delta_{lm}$.

The effective Hamiltonian for the above spin system is exactly diagonalized if one defines

$$w_+(t) = \begin{pmatrix} e^{-i\varphi(t)}\cos\frac{\vartheta}{2} \\ \sin\frac{\vartheta}{2} \end{pmatrix}, w_-(t) = \begin{pmatrix} e^{-i\varphi(t)}\sin\frac{\vartheta}{2} \\ -\cos\frac{\vartheta}{2} \end{pmatrix} \tag{11}$$

with $\vartheta = \theta - \theta_0$, and the constant parameter θ_0 is defined by

$$\tan\theta_0 = \frac{\hbar\omega\sin\theta}{B + \hbar\omega\cos\theta}. \tag{12}$$

The effective Hamiltonian is then written as $\hat{H}_{eff}(t) \equiv \sum_l E_l \hat{c}_l^\dagger(0)\hat{c}_l(0)$ with time-independent effective energy eigenvalues

$$E_\pm = w_\pm^\dagger(t')(\hat{H} - i\hbar\partial_{t'})w_\pm(t')$$

$$= \mp\frac{1}{2}B\cos\theta_0 - \frac{1}{2}\hbar\omega_0\left[1 \pm \cos(\theta - \theta_0)\right]. \tag{13}$$

The exact solution of the Schrodinger equation is given by[12]

$$\psi_\pm(t) = \left\langle 0 \middle| \hat{\psi}(t)\hat{c}_\pm^\dagger(0) \middle| 0 \right\rangle$$

$$= w_\pm(t)\exp\left[-\frac{i}{\hbar}\int_0^t dt' w_\pm^\dagger(t')(H - i\hbar\partial_{t'})w_\pm(t')\right]$$

$$= w_\pm(t)\exp\left[-\frac{i}{\hbar}E_\pm t\right], \tag{14}$$

where the exponent has been calculated in Eq. (13).

The basis vectors satisfy $w_\pm(T) = w_\pm(0)$ with the period $T = 2\pi/\omega_0$. The solution is thus cyclic, namely periodic up to a phase freedom, and, as an exact solution, it is also applicable to the non-adiabatic case.

3.1. Adiabatic limit

The adiabatic limit is defined by $|\hbar\omega_0 / B| \ll 1$ for which the parameter $\theta_0 \to 0$ in Eq. (12), and the exact Schrödinger amplitude (14) approaches

$$\psi_\pm(t) \Rightarrow w_\pm(t)\exp\left[\frac{i}{2}\left[\omega_0\left(1\pm\cos\theta\right)t\right]\right]\exp\left[-\frac{i}{\hbar}\left[\mp\frac{1}{2}B\right]t\right] \tag{15}$$

where the first phase factor is called *geometric phase* and the second phase factor is called *dynamical phase*.

The conventional geometric phase or "Berry's phase"

$$\exp\left[i\pi\left(1\pm\cos\theta\right)\right] \tag{16}$$

is recovered after one cycle $t = T = 2\pi / \omega_0$ of the motion. The Berry's phase is known to have a *topological meaning* as the phase generated by a magnetic monopole located at the origin of the parameter space $\mathbf{B}$.[5]

Note that the dynamical phase in (15) vanishes at $\mathbf{B} = 0$, namely the *level crossing* appears in the conventional adiabatic approximation.

We note that, in the generic case (14) with the period $T = 2\pi / \omega_0$, one can in principle measure $\psi_+^\dagger(0)\psi_+(T)$ by looking at the interereference[12]

$$|\psi_+(T)+\psi_+(0)|^2 = 2|\psi_+(0)|^2 + 2\,\mathrm{Re}\,\psi_+^\dagger(0)\psi_+(T)$$

$$= 2+2\cos\left[\left(\mu B\cos\theta_0\right)T - \frac{1}{2}\Omega_+\right], \tag{17}$$

where the geometric phase

$$\Omega_+ = 2\pi\left[1-\cos\left(\theta-\theta_0\right)\right] \tag{18}$$

stands for the solid angle drawn by $w_+^\dagger(t)\vec{\sigma}w_+(t)$.

3.2. Non-adiabatic limit

The non-adiabatic limit is defined by $|\hbar\omega_0 / B| \gg 1$, and thus $\theta_0 \to \theta$ in Eq. (12) so that the geometric phase vanishes in the exact amplitude (14),

$$\exp\left[-\frac{i}{\hbar}\left[-\frac{1}{2}\hbar\omega_0\left(1\pm\cos\left(\theta-\theta_0\right)\right)\frac{2\pi}{\omega_0}\right]\right] = 1. \tag{19}$$

The formal gauge connection in (3) also vanishes. Namely, the *adiabatic Berry's phase is smoothly connected to the trivial phase inside the exact solution* and thus the topology of Berry's phase is trivial in a precise sense. In our unified formulation of adiabatic and non-adiabatic phases, we can analyze a transitional region from the adiabatic limit to the non-adiabatic region in a reliable way, which was not possible in the past formulation.

4. Conclusion

The present second quantized approach allows a unified treatment of all the geometric phases, either adiabatic or non-adiabatic, and thus one can analyze the transitional region from adiabatic to non-adiabatic phases in a reliable way. One then recognizes that the topology of the adiabatic Berry's phase is actually trivial, in contrast to the topology of the Aharonov-Bohm effect, which is exact for the fast as well as slow motion of the electron.

The analyses of other aspects of geometric phases from a point of view of second quantization are found in the review[13] with further references.

References

1. Y. Aharonov and D. Bohm, Significance of electromagnetic potentials in the quantum theory, *Phys. Rev.* **115**(3), 485-491, (1959).
2. H. C. Longuet-Higgins, Intersection of potential energy surfaces in polyatomic molecules, *Proc. Roy. Soc. (London) A* **344**, 147-156, (1975).
3. B. Simon, Holonomy, the quantum adiabatic theorem, and Berry's phase, *Phys. Rev. Lett.* **51**(24), 2167-2170, (1983).
4. M. V. Berry, Quantal phase factors accompanying adiabatic changes, *Proc. Roy. Soc. (London) A* **392**, 45-57, (1984).
5. M. V. Berry, Quantum phase corrections from adiabatic iteration, *Proc. Roy. Soc. (London) A* **414**, 31-36, (1987).
6. Y. Aharonov and J. Anandan, Phase change during a cyclic quantum evolution, *Phys. Rev. Lett.* **58**(16), 1593-1596, (1987).
7. J. Samuel and R. Bhandari, General setting for Berry's phase, *Phys. Rev. Lett.* **60**(23), 2339-2342, (1988).
8. J. Anandan, The geometric phase, *Nature* **360**, 307-313, (1992), and references therein.
9. K. Fujikawa, Topological properties of Berry's phase, *Mod. Phys. Lett. A* **20**, 335-344, (2005).
10. S. Deguchi and K. Fujikawa, Second-quantized formulation of geometric phases, *Phys. Rev. A* **72**(1), 012111, (2005).
11. K. Fujikawa, Geometric phases and hidden local gauge symmetry, *Phys. Rev. D* **72**(2), 025009, (2005).
12. K. Fujikawa, Geometric phases for mixed state and decoherence, *Ann. of Phys.* **322**, 1500-1517, (2007).
13. K. Fujikawa, Geometric phases and hidden gauge symmetry, Bulletin of Asia-Pacific Center for Theoretical Physics (APCTP), 23-24, (2009). 29. arXiv:0910.0396[quant-ph].

A Brief Overview and Topological Aspects of Gaseous Bose-Einstein Condensates

Masahito Ueda

Department of Physics, The University of Tokyo,
7-3-1 Hongo, Bunkyo-ku, Tokyo 113-0033, Japan
ueda@phys.s.u-tokyo.ac.jp

This is a reproduction of a seminar I presented at Hitach Central Research Laboratory for Dr. Tonomura late 2011 on a brief overview of Bose-Einstein condensates of dilute atomic gases with a particular emphasis on symmetry breaking and topological excitations.

1. Introduction

It is still hard for me to face too-early death of Dr. Akira Tonomura at the age of 70. My first personal encounter with Dr. Tonomura was at a conference "Frontiers in Quantum Physics" in Malaysia in 1997, where he was among the main speakers together with Klaus von Klitzing and Gerard 't Hooft. As usual, he made a very lucid account of how James Clerk Maxwell identified his vector potential with Faraday's electro-tonic state, and then showed his spectacular movie on the vortex-antivortex pair annihilation. It was at an early stage of my career and I told him some idea about his experiments. I only vaguely remember what I actually talked about. What I can vividly recall is his keen interest in what I was talking about. In retrospect, this is very typical of Dr. Tonomura who always encouraged young researchers by showing his genuine interest in what they study. Since then, Dr. Tonomura had been the behind-the-scenes supporter of me. What he kept telling me was, "You may be involved in many responsibilities, but you should stay focused on your research."

My last meeting with Dr. Tonomura came in late November 2011 when he was struggling to rehabilitate from surgery of pancreas cancer. He asked me to give a talk on superfluidity, vortices, and symmetry breaking in gaseous Bose-Einstein condensates by saying "We want to do interesting experiments with our electron microscope." It was not easy for me to prepare my seminar to accomodate such a request. After some thoughts, I decided to construct my seminar so that the bare essentials of gasous Bose-Einstein condensates would be presented in a concise yet stimilating manner.

The seminar was very private and I was not able to go through all slides because Dr. Tonomura's condition was obviously not good. But it seemed that the seminar gave him some vigor, as I later learned from his secretary, Ms. Matsuyama, which was reassuring to me. I will here try to reproduce the main contents of that seminar which, hopefully, give readers a brief overview about the fundamentals of gaseous Bose-Einstein condensation.[1]

2. A New Path to Ultralow Temperatures

The creation of a gaseous Bose-Einstein condensate is based on a series of ground-breaking techniques to achieve ultralow temperatures such as laser cooling and evaporative cooling.[2] Laser cooling exploits the quantized nature of an atom and a laser light. If the laser frequency is tuned slightly below a resonant frequency of the atom, every time the atom absorbs and emits a photon, there is an energy deficit because the frequency of the emitted photon is on average equal to the resonant frequency. This energy deficit is compensated for by the kinetic-energy loss of the atom caused through recoil. Although the recoil momentum per one photon absorption-emission cycle is minute because photons are massless, an enormous spontaneous emission rate at frequencies above the near-infrared regime makes the deceleration of the atom about one hundred thousand times as large as the gravitational acceleration g. (Note that the acceleration of a rocket is only a few times g.) Thus, very fast room-temperature atoms can be stopped over a tiny cross-sectional volume of three pairs of mutually orthogonal laser beams. Laser cooling thus enables us to control translational degrees of freedom of atoms by cooling them below 1mK. Then, the thermal de Broglie length of an atom becomes of the order of 0.1 μm and the Young-type interference experiment becomes possible[3] without resorting to a conventional collimation method.[4]

To reach quantum degeneracy, evaporative cooling is often invoked. Here, laser-cooled atoms are confined in a trapping potential, and by lowering the height of the potential, the most energetic atoms are selectively ejected or "evaporated" from the edges of the potential and the remaining atoms thermalize at progressively lower temperatures through elastic collisions. Evaporation is extremely efficient in cooling the system as we know from our daily experience. When a hot water cools down in a cup, only a negligible fraction of water molecules will evaporate. To reach quantum degeneracy, however, roughly 99% of the atoms must be forcibly evaporated to reach the critical phase-space density with the temperature of the remaining atoms going down to 1μK or even lower.

The atomic cloud is usually probed by the time-of-flight (TOF) method. When the trapping potential is turned off, atoms will expand according to their initial velocities. The faster the initial velocities are, the farther atoms can travel until the density distribution of atoms is photographed with a CCD camera. Thus, by measuring the density distribution of atoms, one can find the velocity distribution of the trapped atomic cloud. When a Bose-Einstein condensate emerges, the TOF image shows a bimodal distribution in which the thermal component displays a broadly expanded isotropic distribution due to the equipartition law in statistical mechanics, while the condensate component peaks at the center because it can expand only slowly due to the interparticle interaction and zero-point kinetic pressure. Such a dramatic change in the TOF image from a single isotropic density profile above the Bose-Einstein transition temperature to the bimodal structure below it is a consequence of the highly compressible nature of dilute atomic systems. This makes a sharp contrast with the liquid helium which is incompressible and shows no obvious change in the density distribution at the transition temperature. Such a distinction explains why it took almost a quarter of a century for the anomalous behavior of liquid helium-4 to be identified as due to Bose-Einstein condensation and why dilute-gas Bose-Einstein condensation commanded instant recognition.

3. Some Fundamental Properties of Bose-Einstein Condensates

Bose-Einstein condensation is a macroscopic manifestation of a single-particle state and as such it exhibits a number of macroscopic quantum phenomena. A remarkable consequence of the macroscopic occupation of a single-particle state is that a single-shot measurement is sufficient to provide the probability distribution of a given few-body observable. For example, TOF images taken at varying times show how the initial condensate evolves in time. If the condensate is initially confined in an elongated potential, the condensate expands faster in the tightly confined direction than in other directions. From single-shot images, one can thus find how the density distribution evolves in time due to Heisenberg's uncertainty relation and interparticle interactions.[5] One can also follow how symmetry breaking proceeds in a Bose-Einstein condensate that is subject to external perturbations such as rotation[6] or sudden changes in system's parameters (quench experiments).[7]

Another remarkable experiment is the observation of an interference between two independently prepared condensates.[8] Because there is no initial phase relationship between the two condensates, the observed interference pattern should not be interpreted as a Young-type interference but as a macroscopic analogue of the Hanbury Brown-Twiss (HBT) experiment of neutral atoms.[9] In fact, the interference pattern exhibits shot-to-shot fluctuations in the peak positions[10] and disappears upon ensemble averaging. Here, the interference pattern manifests itself in individual images because a large number of particles share the same single-particle state in each sample. Note that a Young-type interference pattern does not average out upon ensenmbe averaging because the peak positions are determined by the geometrical path difference and not by the phase difference between the two condensates.

An infallible hallmark of macrosopic occupation of a single-particle state is the nonclassical response of a condensate to an external rotation. The single-valuedness of the wave function dictates that the angular momentum of the system be quantized in units of $\hbar$. If N particles share the same single-particle state, the angular momentum should be quantized in units of $N\hbar$ as observed experimentally,[11] and the system develops a quantized vortex if the frequency of an external rotation exceeds a critical frequency.[6] What I am personally very interested in is the fact that the superfluid system is at rest below the critical frequency (Hess-Fairbank effect). The crucial question is with respect to which frame of reference the condensate is at rest? It is certainly not the laboratory frame because one can use the Hess-Fairbank effect as the principle for a gyroscope to sense the rotation of the Earth. Perhaps, one might use the condensate at rest to define the local inertial frame over a small length scale of μm.

Finally, the defining character of a Bose-Einstein condensate is the off-diagonal long-range order,[12,13] which is said to exist if the largest eigenvalue of the reduced single-particle density matrix is of the order of the total number of particles. In this case, the correlation function $\langle \hat{\Psi}^\dagger(\mathbf{r}')\hat{\Psi}(\mathbf{r})\rangle$ remains nonvanishing in the limit of $|\mathbf{r}-\mathbf{r}'| \to \infty$.[12] That is, if one takes out a particle at $\mathbf{r}$ and returns it at a distant place $\mathbf{r}'$, the probability amplitude of the state to remain unaltered is finite. This implies an ability of the system to travel a long distance without changing its state and explains why Bose-Einstein condensation and superfluidity occur simultaneously in many systems, despite the fact that they are neither necessary nor sufficient to each other.

4. Symmetry Breaking

Gaseous Bose-Einstein condensates offer a cornucopia of symmetry breaking. The primary reason is that the system involves several well-separated energy scales that extend over five orders of magnitude. Apart from the Zeeman energy which depends on an external magnetic field, the largest energy scale is the spin-independent Hartree interaction ($\sim \mu$K), which together with a trapping potential determines the density distribution of particles. Three orders of magnitude smaller than this is the spin-dependent or spin-exchange interaction ($\sim$nK) which determines the magnetism or symmetry of the spinor order parameter. By far the smallest is the magnetic dipole-dipole interaction (~ 0.1nK). Albeit its minuteness, it plays a dominant role in creating spin textures which describe spatiotemporal variations of the spin state.

A crucial question is why these minute nano-Kelvin phenomena can be observed at much higher temperatures, say 0.1μK. The answer is Bose enhancement: once the system undergoes Bose-Einstein condensation, all condensed particles behave in exactly the same manner and the resultant collective behavior is robust against thermal fluctuations. For example, the energy scale of the magnetic dipole-dipole interaction is of the order of 0.1nK and 10nK for ^{87}Rb and ^{52}Cr, respectively. However, once one hundred thousand atoms undergo Bose-Einstein condensation, the collective energy amounts to 1μK for ^{87}Rb and 100μK for ^{52}Cr, both of which are well above the usual temperature of the system which is 0.1μK or lower. According to the tim-energy uncertainty principle, we may expect that the magnetic dipole-dipole interaction cannot be ignored when we consider the dynamics of the system longer than 100 ms for ^{87}Rb and 1 ms for ^{52}Cr.

An obvious but very important feature of gaseous condensates is their extreme diluteness with the atomic density being five orders of magnitude more dilute than that of the air. This makes the collision time very long, of the order of milliseconds and therefore kinetics becomes important. In particular, if the system is driven out of equilibrium or if some parameter of the system is suddenly changed as in quench experiments, the ensuing nonequilibrium relaxation and thermalization processes can be observed in real time. In particular, the dynamics of phase transitions and defect necleation have been successfully observed in real time. Examples include the phase transition of Bose-Einstein condensation,[14] vortex nucleation,[6] and spontaneous formation of scalar vortices[15] and spin vortices.[7]

The full symmetry group of the system at high temperature is $G = U(1) \times SO(3) \times \mathbb{R}^3$, each of which describes the gauge symmetry, spin-rotation symmetry, and translation symmetry in real space. As the temperature of the system is lowered, the system undergoes a symmetry breaking transition into a phase with lower symmetry. The isotropy group H describes the remaining symmetry of the broken symmetry phase. For example, the ferromagnetic phase breaks the spin-rotation symmetry but it is invariant under an arbitrary rotation about the magnetization axis, and hence $H = SO(2)$. The isotropy groups of other phases involve discrete subgroups of $SO(3)$ such as the $\mathbb{Z}_2$ symmetry for the polar phase[16] and the tetrahedral symmetry for the cyclic phase.[17,18] The order parameter manifold is thus given by the coset $M = G/H$. Possible topological excitations can be found by examining the homotopy group for each order parameter manifold.

5. Topological Excitations

There are several distinctive features of gaseous Bose-Einstein condensates that are instrumental in studying topological excitations. In alkali atoms, the magnetic moment originates primarily from electronic rather than nuclear spin. This allows us to make local control of spin textures. In contrast, in superfluid helium-3, local manipulation of spin textures is difficult because the magnetic moment of the nuclear spin is too small.

Order parameters of spinor condensates feature discrete symmetries except for the ferromagnetic phase. As mentioned above, the polar phase of a spin-1 condensate such as ^{23}Na has a two-fold symmetry and the cyclic phase of a spin-2 condensate such as ^{87}Rb has a tetrahedral symmetry. The spin-3 condensate such as ^{52}Cr has even richer symmetries.[19] These discrete symmetries combined with the continuous U(1) gauge symmetry lead to fractional vortices[16–18] and non-Abelian vortices.[20] The spin-1 polar phase hosts a one-half vortex because of the combined spin-gauge Z_2 symmetry. It also accomodates rather exotic topological objects such as knots[21] in three dimensions because the underlying order parameter manifold involves the Hopf map $S^3 \rightarrow S^2$.

On the other hand, ferromagnetic condensates possess continuous spin-gauge symmetry that couples supercurrent with spin textures. A general form of the ferromagnetic order parameter involves the gauge angle and the spin rotation angle about its direction as a linear combination. Because a local change of the gauge angle induces a supercurrent, the spin-gauge symmetry implies that if you rotate spin locally, the system tries to undo it by flowing supercurrent, and vice versa. The spin-gauge symmetry can be used to convert synthetic gauge fields, Berry phase, etc. to superflow. A related implication of the spin-gauge symmetry is that the spin texture acts as an effective gauge field and contributes to the Berry phase. Consequently, the circulation alone is not quantized, but the sum of the circulation and the contribution from the Berry phase is quantized.[22] In contrast, superfluid helium-3 has the orbital-gauge symmetry.

The ground state of a spin-2 Bose-Einstein condensate of ^{87}Rb is predicted to be near the boundary between the antiferromagnetic and cyclic phases. Interestingly, vortices created in either case can be non-Abelian.[20] When the system accomodates non-Abelian vortices, the collision dynamics of vortices becomes highly nontrivial. In general, there are three different types of collision dynamics between vortices: reconnection, passing through, and formation of a rung vortex. In superfluid helium-4, the collision dynamics is usually dominated by reconnection because it is energetically most favorable. However, for non-Abalian vortices, neither reconnection nor passing through is topologically allowed, and therefore the formation of a rung vortex is topologically imposed. A rung vortex is formed every time two non-Abelian vortices collide. We thus expect that the system will develop a large scale of vortex network. This makes a sharp contrast with the case of Abelian vortices which become smaller every time they collide via reconnection and cascade processes.

This distinction should have an impact on the statistical properties of quantum turbulence.

6. Conclusions

The preparation of the seminar for Dr. Tonomura was an invaluable occasion for me to reexamine the bare essentials of dilute-gas Bose-Einstein condensation. It seems that they are all about low density, electronic spin, and precision control. The extreme diluteness of the particle density implies long lifetime and high compressibility. The ground states of alkali atoms are metals, yet Bose-Einstein condensation has been created in a metastable state because the three-body recombination, which is the first step to solidification, is strongly suppressed due to the low atomic density. The high compressibility of the gaseous phase has been instrumental for detecting the onset of Bose-Einstein condensation and also for manipulation of the system via laser light. The electronic spin implies high succeptibility to an external magnetic field. The ability to control spin textures locally is indispensable to create topological objects. The laser control to prepare, manipulate, and probe ultracold atomic systems has given us numerous new possibilities, and enables us to directly see macroscopic quantum phenomena which stimulates our imagination and inspiration. It then seems apparent why last fall Dr. Tonomura wanted to know what's going on in ultracold atomic gases: to push forward his lifelong aspiration of directly seeing what everyone once envisaged invisible to the extreme through the newly constructed electron microscope.

Acknowledgements

This work was supported by Grants-in-Aid for Scientific Research (Kakenhi Grants No. 22340114 and No. 22103005), the Global Center of Excellence Program "The Physical Science Frontier," and the Photon Frontier Network Program of the Ministry of Education, Culture, Sports, Science, and Technology of Japan.

References

1. M. Ueda, *Fundamentals and New Frontiers of Bose-Einstein Condensation*. (World Scientific, Singapore, 2010).
2. W. Ketterle, D. S. Durfee, and D. M. Stamper-Kurn, *Making, probing and understanding Bose-Einstein condensates*, In eds. M. Inguscio, S. Stringari, and C. E. Wieman, *Bose-Einstein condensation in atomic gases*, pp. 67–176. Proceedings of the International School of Physics "Enrico Fermi," Course CXL. IOS Press, Amsterdam, (1999).
3. F. Shimizu, K. Shimizu, and H. Takuma, Double-slit interference with ultracold metastable neon atoms, *Phys. Rev. A* **46**(1), R17–R20, (1992).

4. O. Carnal and J. Mlynek, Young's double-slit experiment with atoms: A simple atom interferometer, *Phys. Rev. Lett.* **66**(21), 2689–2692, (1991).

5. M.-O. Mewes, M. R. Andrews, N. J. van Druten, D. M. Kurn, D. S. Durfee, and W. Ketterle, Bose-Einstein condensation in a tightly confining dc magnetic trap, *Phys. Rev. Lett.* **77**(3), 416–419, (1996).

6. K. W. Madison, F. Chevy, V. Bretin, and J. Dalibard, Stationary states of a rotating Bose-Einstein condensate: Routes to vortex nucleation, *Phys. Rev. Lett.* **86**(20), 4443–4446, (2001).

7. L. E. Sadler, J. M. Higbie, S. R. Leslie, M. Vengalattore, and D. M. Stamper-Kurn, Spontaneous symmetry breaking in a quenched ferromagnetic spinor Bose-Einstein condensate, *Nature.* **443**, 312–315, (2006).

8. M. R. Andrews, C. G. Townsend, H.-J. Miesner, D. S. Durfee, D. M. Kurn, and W. Ketterle, Observation of interference between two Bose condensates, *Science.* **275**(5300), 637–641, (1997).

9. M. Yasuda and F. Shimizu, Observation of two-atom correlation of an ultracold neon atomic beam, *Phys. Rev. Lett.* **77**(15), 3090–3093, (1996).

10. Z. Hadzibabic, S. Stock, B. Battelier, V. Bretin, and J. Dalibard, Interference of an array of independent Bose-Einstein condensates, *Phys. Rev. Lett.* **93**(18), 180403, (2004).

11. F. W. Chevy, K. W. Madison, and J. Dalibard, Measurement of the angular momentum of a rotating Bose-Einstein condensate, *Phys. Rev. Lett.* **85**(11), 2223–2227, (2000).

12. O. Penrose and L. Onsager, Bose-Einstein condensation and liquid helium, *Phys. Rev.* **104**(3), 576–584, (1956).

13. C. N. Yang, Concept of off-diagonal long-range order and the quantum phases of liquid He and of superconductors, *Rev. Mod. Phys.* **34**(4), 694–704, (1962).

14. H.-J. Miesner, D. M. Stamper-Kurn, M. R. Andrews, D. S. Durfee, S. Inouye, and W. Ketterle, Bosonic stimulation in the formation of a Bose-Einstein condensate, *Science.* **279**(5353), 1005–1007, (1998).

15. C. N. Weiler, T. W. Neely, D. R. Scherer, A. S. Bradley, M. Davis, and B. P. Anderson, Spontaneous vortices in the formation of Bose-Einstein condensates, *Nature.* **445**, 948–951, (2008).

16. F. Zhou, Spin correlation and discrete symmetry in spinor Bose-Einstein condensates, *Phys. Rev. Lett.* **87**(8), 080401, (2001).

17. H. Mäkelä, Y. Zhang, and K.-A. Suominen, Topological defects in spinor condensates, *J. Phys. A: Math. Gen.* **36**(32), 8555, (2003).

18. G. W. Semenoff and F. Zhou, Discrete symmetries and 1/3-quantum vortices in condensates of F=2 cold atoms, *Phys. Rev. Lett.* **98**(10), 100401, (2007).

19. Y. Kawaguchi and M. Ueda, Symmetry classification of spinor Bose-Einstein condensates, *Phys. Rev. A.* **84**(5), 053616, (2011).

20. M. Kobayashi, Y. Kawaguchi, M. Nitta, and M. Ueda, Collision dynamics and rung formation of non-Abelian vortices, *Phys. Rev. Lett.* **103**(11), 115301, (2009).

21. Y. Kawaguchi, M. Nitta, and M. Ueda, Knots in a spinor Bose-Einstein condensate, *Phys. Rev. Lett.* **100**(18), 180403, (2008).

22. M. Ueda, Bose gases with nonzero spin, *Annual Review of Condensed Matter Physics.* **3**, 263–283, (2012).

Mapping Electric Fields with Inelastic Electrons in a Transmission Electron Microscope

Christian Colliex

Laboratoire de Physique des Solides (UMR CNRS 8504), Bldg 510,
Université Paris Sud 11, 91405 Orsay, France
E-mail: colliex@lps.u-psud.fr

A tribute to the work of Akira Tonomura

This text is dedicated to the memory of Akira Tonomura. Akira and myself came across in the early seventies when, during his period with Prof. G. Möllenstedt in Tübingen, he organized a visit to my place in Orsay. We discovered that we were both very motivated by the building of cold field emission electron guns of intrinsic high brightness. He succeeded beautifully at Hitachi. This proved to be the key for having access to the highly coherent beam necessary for realizing great achievements in phase contrast electron microscopy. It resulted in particular into the superb experimental confirmation of the Aharonov-Bohm effect and into the direct visualization of magnetic vortices in superconductors. On my own side, I later developed the tools and applications of electron energy loss spectroscopy under the impact of a very tiny electron probe. Over the years, we have built a strong and permanent friendship and have remained enthusiastic to follow at regular intervals, the successes brought respectively by the use of the elastic and inelastic scattering processes in electron microscopy.

1. Introduction

The electrons in the electron microscope have demonstrated for a long time, their efficiency in analyzing and mapping with very high spatial resolution, the position and the nature of atoms, ions and electron clouds in thin objects as a consequence of their strong scattering probabilities. The structural information is mostly conveyed in high resolution imaging and diffraction relying, in both cases, on elastic scattering, phase contrast or momentum transfer. The analytical information on its side, is brought by spectroscopies such as electron energy-loss spectroscopy (EELS) or photon emission (energy dispersive X-ray spectroscopy (EDX) for X rays or cathodoluminescence (CL) in the near-visible domain), resulting from inelastic scattering and transfer of energy between the incident electron and those in the target.

When considering electric $(\vec{E})$ and magnetic fields $(\vec{B})$, as a consequence of the Lorentz forces, there is no exchange of energy for the magnetic case (the resulting force is perpendicular to the velocity $\vec{v}$ of the incoming electron) while the component of the electric field parallel to the velocity of the incoming electron generates a transfer of energy of the type $(\vec{v}.\vec{E})$. Consequently magnetic fields can best be visualized using modes (phase contrast, electron holography) resulting from elastic scattering. Among his most successful achievements, Akira Tonomura has spectacularly visualized the magnetic flux lines of vortices in superconductors. Using the defocus contrast, also named Lorentz microscopy mode, he has imaged the lattice distribution of the vortices, their dynamics, and their interactions with pinning defect centers, first in type II superconductors,[1] then in high Tc superconductors with new vortex configurations.[2] See Ref. 3 for an up-to-date review of the contribution of Akira Tonomura to the field of vortex physics using a transmission electron microscope (TEM) with a bright and coherent illumination system.

Let us turn now to the possibility of visualizing photons or, equivalently, their associated electromagnetic (EM) fields in a TEM. It is well known that, based on elementary arguments of energy and momentum conservation laws, electrons and photons cannot directly couple in free space. The key ingredient for making it possible is to confine the electric field, thus providing as a consequence of the Heisenberg's principle, the momentum which is necessary for the inelastic scattering to occur. Such confined fields can be created in the vicinity of nano-objects, as the evanescent fields generated by surface plasmons. These plasmons are oscillations of surface charge densities at the boundaries between metallic and dielectric media, which induce electric fields decaying exponentially with the distance to the interface. In the following, we will focus on localized surface plasmons (LSP) of interest for structures of size small with respect to the wavelength of the associated EM radiation, in contrast with surface plasmons propagating along unbounded interfaces (also known as surface plasmon polaritons: SPP). Figure 1 illustrates the physical nature of these two modes (oscillating charges at frequency ω and associated evanescent electric fields $\vec{E}$) and how they can be coupled to external photons (impossible without an extra momentum, see orange dotted arrows) or to impinging electrons without any problem.

Since the earliest experiments measuring energy losses in the TEM environment, the existence and the physical nature of bulk, surface and interface plasmons have been abundantly investigated. Over the past few years, however, the field of surface plasmons, in particular, has been extraordinarily revived, or "resurrected" as suggested by the title of a recent editorial in the Nature Photonics November 2012 issue (over 6,000 papers have been published on plasmons in the year 2011!!). This is largely due to the huge increase of research stimulated by the possible tools for guiding, enhancing and

                    C. Colliex

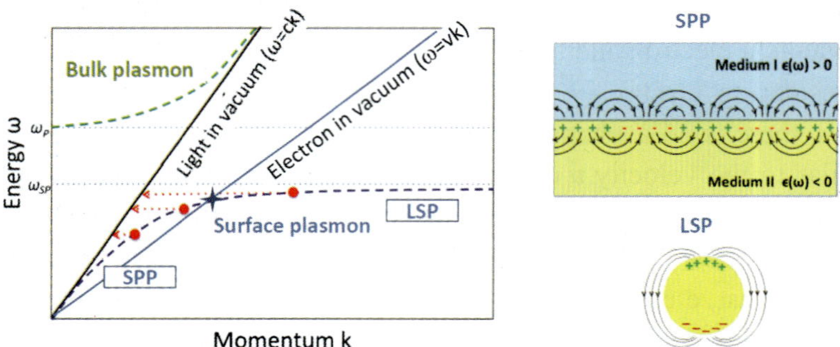

Fig. 1 Left: dispersion curves (frequency-momentum $\varphi(k)$) for plasmon modes in and at the surface of a material of dielectric constant $\varepsilon\ (\omega)$, together with the corresponding curves for electron and light in vacuum. We have neglected the case of light in a dielectric and the associated Cerenkov effect. Right: scheme of electric charge density distributions and associated electric fields for an unbounded interface (SPP) and a small object (LSP).

controlling light at the much sub-wavelength scale which could be offered by a deeper knowledge of the interaction processes between charges and light at surfaces, generally designed as plasmonics. Of direct interest for the present text, the use of electron probes, such as involved in S(canning)TEMs with sub-nm spatial resolution, has really opened new possibilities to look closer at these subtle interactions, for which the study of the triangular nanoplatelets of silver has constituted a clear demonstration of its potential impact.[4] First, we will describe recent instrumental and methodological developments which have been responsible for the explosion of nano-plasmonics studies with electron beams observed over the past few years. Then, practical cases ranging from test situations to a practical example of meta-atom and meta-material will be briefly considered. Finally, new challenges for the future will be identified.

2. The multi-signal strategy in a STEM to track the optical response of individual metallic nanoparticles

In a STEM, a very narrow electron beam of high primary energy electrons (typically from 60 to 200 keV), is travelling at a well-controlled position with respect to the investigated target. With a field emission source, currents of the order of 10 to 100 pA can now be focused in a sub-nm probe, which can be even as small as 1 Å when the microscope is equipped with an aberration corrector (Cs corrector). One can consider that on average one electron per ns impinges onto the target. Figure 2 is a simplified representation of the interaction between such an incident electron travelling in the vacuum at a distance of a few nm outside the external surface of an isolated metallic particle of diameter in the 10 to 100 nm range. This swift point charge can be regarded as a source of white light,[5] which induces the oscillation of charges in the object, i.e. the creation of a LSP, with a resonance frequency depending on its nature (through its dielectric constant ε), its shape and its size. This surface plasmon generates an evanescent induced field, which the electron travelling at velocity $\vec{v}$ feels, thus giving rise to an energy loss measured by the electron energy loss spectrometer (EELS). It is important to point out significant orders of magnitude: the interaction duration between the electron and the target is typically $2b/v \approx 10^{-16}$ s with b being the distance between the electron and the target, while the plasmon decay time is on the 10^{-15} s scale. When compared with the rate of arrival of electrons (10^9/s), it is obvious that the incident electrons interact individually with the target and that it is the same electron which generates and detects the surface plasmon mode. The induced EM field can also be directly detected by an optically coupled spectrometer and photon detector, in which case the signal is named cathodoluminescence (CL).

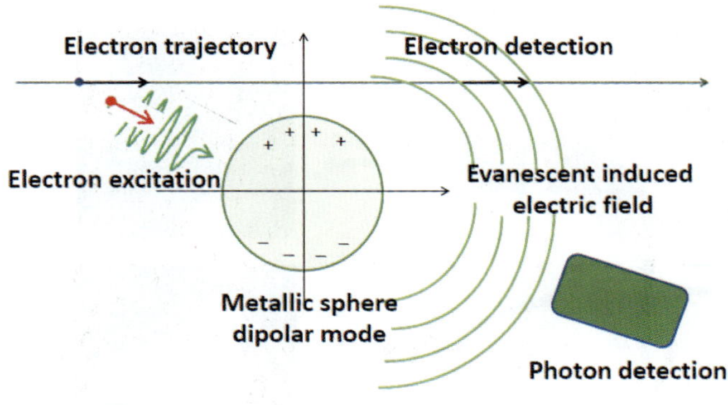

Fig. 2. Schematics of the generation and detection channels for localized EM fields induced by an impinging swift electron on a metallic nanoparticle.

An alternative way to stimulate the creation of LSPs on a metallic nanoparticle is to shine it directly with a photon beam. The induced evanescent field similar to that shown in Fig. 2 can also be detected by a swift electron travelling aloof the particle or directly with the photon spectrometer-detector device (in this case, this is a photoluminescent signal and there is no spatial resolution at the nm level). As the decay time of the plasmon is much shorter than the time elapsed between the arrival of two electrons, it is necessary to correlate in time both photon and electron impacts. This direct visualization of near-EM fields (called PINEM for photon induced near-field electron microscopy) has been beautifully demonstrated by the group of Zewail,[6,7] who have used ultra-fast electron microscopy techniques in a pump-probe approach. However, beyond the demonstration of a strong imaging process between inelastic electrons and photons produced by LSPs, this approach does not provide rich information on the specimen. This is why the remaining part of this paper will be devoted to the discussion of multi-signal STEM approach.

Practically, in a STEM microscope (see Fig. 3), the incident electron beam is scanned over the specimen. Several detectors are positioned so that their signals can be acquired simultaneously, pixel after pixel, and provide complementary views: (i) the light emitted in the near-visible spectral range is collected by an elliptic mirror and transferred to the photon analyzer and detector device in order to provide a CL spectrum; (ii) the high angle annular dark field (HAADF) detector collects all electrons scattered at large angles, so that a topographic view, in practice a mass-thickness one, can be displayed; (iii) the transmitted inelastic electrons are analyzed by a magnetic spectrometer in order to provide an EELS spectrum. For each probe position (or image pixel), one has thus access

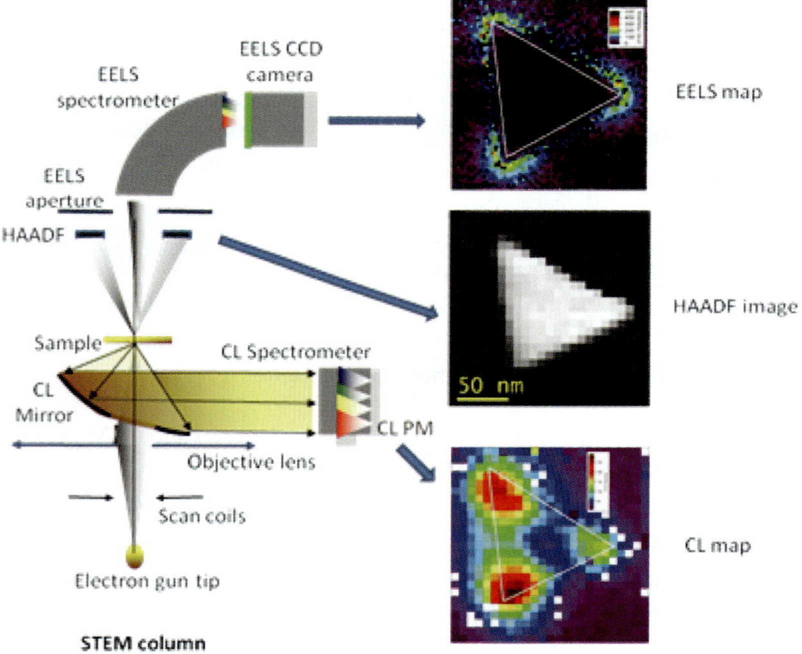

Fig. 3. Schematic diagram of a STEM microscope running in the multi-signal mapping mode. For each position of the incident probe on the specimen (a gold nanoprism deposited on a thin amorphous carbon layer), one simultaneously records three signals (HAADF for topographical and structural mapping; EELS spectrum in the visible domain, from 2 to 4 eV; optical emission –CL– spectrum between 300 and 700 nm in wavelength). The two maps (EELS and CL) on the right display the distribution in intensity of the major peaks identified in the spectra (diagram courtesy of M. Kociak and L. Zagonel).

to a topographic signal and to two spectra. Those are then processed to extract the main parameters (energy or frequency, line width, intensity) of major peaks and lines and displayed in maps of the different excitations.[8,9] Comparison of EELS and CL maps provides a good view of the complete optical properties at the nanoscale.[10]

The spatial resolution, mainly set by the diameter of the primary beam, is typically in the sub-nm region when dealing with a low energy-loss range (between 1 and 4 eV). The energy resolution in the EELS spectrum-images, on the other hand, is mostly governed by the spectral width of the primary electron beam (of the order of 100 meV in most recent experiments realized with a monochromatized electron source and 200 meV when using deconvolution techniques and cold field-emission electron sources). For CL maps, sub-10 meV spectral resolution together with a few nanometer spatial resolution has been demonstrated, when mapping the emission spectrum of semiconducting quantum disks.[11]

It must be added for completeness that an alternative way for providing plasmon maps down to below 1 eV is to record them in the TEM energy-filtering mode with the use of a monochromator. In this case, full images are realized at once with electrons contained in an energy width, typically 0.2 eV wide defined by a slit (energy-filtering TEM images). After having recorded a series of such images while ramping the energy loss with 0.2 eV increments, maps displaying different spatial distributions of intensities as a function of energy loss can be displayed. The spectral range which can thus be investigated now extends down to about 0.2 eV, i.e., clearly in the IR domain.[12]

3. A rapid journey through the blossoming world of nanoplasmonics explored with electron beams

As a matter of fact, the present explosion of electron beam studies of plasmonic modes and of their induced EM fields around metallic nanoparticles results from a timely combination of diversified factors. First, the global potentialities they open for manipulating light at very local scales with respect to its wavelength offer rich perspectives in broad domains of applications such as energy, chemistry and therapy. Then, the instrumental developments in energy-loss analysis and filtering described above coincide with parallel progress in adequate specimen preparation, numerical modeling and innovative theory.

Specimen preparation: Most nanostructures investigated up to now for plasmonics applications have been realized in Ag, Au or a combination of both. Although numerous syntheses or fabrication approaches have been used for growing individual or assemblies of these metallic nanoparticles, they can be classified into two families: the bottom-up type, which relies on colloidal chemistry and the top-down type, which uses the most modern tools of nanolithography. The major parameters to be controlled are the shapes (spheres, rods, triangular platelets, cubes, stars, icosahedra) and dimensions (lengths, diameters, thicknesses, aspect ratios between two dimensions), and the substrate nature. Generally, chemical processes provide random collections of individual particles which can be diluted in solutions and spread over a substrate (amorphous carbon, mica, graphitic flakes, SiO_2 thin foil, and so on), and in some cases when the concentration is higher, self-assembled distributions can be obtained. On the contrary, electron and/or ion-beam lithography techniques generate well-controlled distributions of individual or interacting particles.

Numerical simulations: The theory underlying the numerical tools required for simulating the EM response of a metallic nanoparticle to the impact of a primary swift electron basically relies on Maxwell equations. The material itself is characterized by its dielectric constant $\varepsilon\,(\omega,\,\vec{q})$ depending on the energy loss $\hbar\omega$ and transferred wave vector $\vec{q}$, which reduces to $\varepsilon(\omega)$ in the local approximation valid for small angle inelastic scattering conditions of EELS experiments. The calculation of the energy loss implies the integration of $(\vec{v}.\vec{E})$ along the trajectory of the primary electron where $\vec{E}$ is the induced electric field. This classical theory has been reviewed by Garcia de Abajo[5] for many cases (planar interfaces, spheres, cylinders) where the solutions can be found analytically. For more complex shapes, numerical methods have been implemented, the two mostly used being the boundary element method (BEM) and the discrete dipole approximation (DDA). They differ by the fact that BEM looks for the solutions at discrete points on the surface of the object, while DDA introduces a discretization in the volume as well as on the surface. In both cases, these calculations provide access to the electrostatic potential on the boundaries, and from there to the distribution of surface charge densities and of the generated electric field. When both BEM and DDA numerical simulations of the EM response have been performed on the same nanostructure, they do not exhibit significant differences.

New theoretical developments: It has been demonstrated[13] that EELS and CL can probe a quantity named the electromagnetic-local density of states (EM-LDOS): $\rho_{ph}(\vec{r},\,\omega)$ which can be calculated as $\approx \sum_n |E^z_n(\vec{r},\,\omega)|^2.\delta(\omega_n-\omega)$. Here $E^z_n(\vec{r},\,\omega)$ is the component projected along the electron axis z of the eigen-field generated by the plasmonic eigen-mode peaking at the eigen-energy ω_n. The EM-LDOS, filtered at a given energy, is thus a map of the variation of the related plasmon, and has been used in the past to interpret scanning near-field optical microscopy (SNOM) experiments.[14] It nicely justifies EELS and CL as alternative candidates to near field optical methods to investigate optical phenomena, yet with improved spatial resolution. As pointed out in Ref. 15, neither EELS nor SNOM directly measure the eigencharges but Fig. 4 illustrates the comparison between the simulated SNOM and EELS map for a given frequency, and how they result

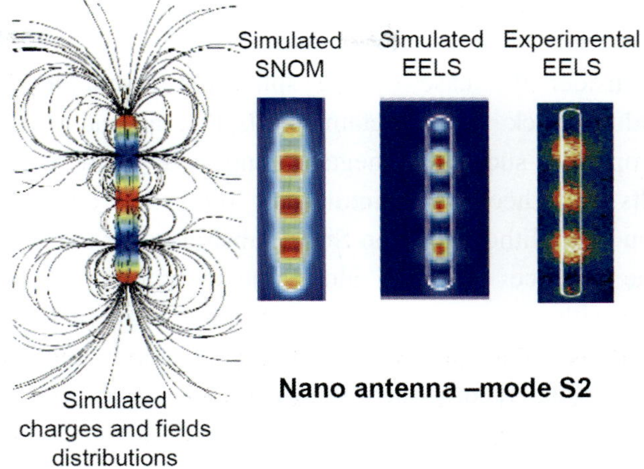

Fig. 4. Non retarded BEM simulation of the charge and electric field distributions for a silver nanorod (aspect ratio 15 between length 300 nm and diameter 20 nm), together with simulated EELS and SNOM maps, as compared with an experimental EELS map. The S2 mode (i.e., the second order symmetric one) corresponds to a symmetric charge distribution with maxima at the center and at the extremities. Let us recall that both SNOM and EELS vary as the square modulus of E_z and therefore exhibit maxima where the magnitude of E_z is maximum. The z direction is perpendicular to the figure: for a nanorod, one assumes a revolution symmetry for charges and fields around the main axis of the particle (figure courtesy of G. Boudarham and M. Kociak; experimental EELS is from Ref. 12).

from a given distribution of eigencharges on the surface and eigenfields outside the nanoparticle.

This evolution in our understanding of the observed EELS plasmon maps in the electron microscope, i.e., that they can be linked to a generic optical quantity, has induced a noticeable change of standpoint: we are really measuring the properties of EM fields and not directly those of the nanoparticles.

From model objects to metamaterials: The new concepts and tools introduced above have been tested over the past five years by different teams around the world on a range of specimens of different shapes and sizes: spheres, elongated spheres, dumbbells, rods, antennas, triangles, decahedra, cubes, stars, and also complementary holes and slits. It is not in our scope here to go into detail through this rapidly emerging field of literature. Let us mention only two clear results at that point: (i) EELS can reveal all modes, should they be bright or dark under more traditional optical illumination[16]; (ii) plasmon coupling between two nanoparticles in close interaction (split nanowires or aligned dimers, triangular platelets in bow-tie configuration) has been shown to induce a mode splitting into a bonding and an anti-bonding one with separation in energies as small as a few hundreds of meV.[17]

Let us now consider the case of the split ring resonator (SRR) which is a representative building brick for a metamaterial, i.e., a material exhibiting unusual electromagnetic properties such as a negative index of refraction. Figure 5 gathers experimental results and theoretical simulations published in two studies of SRRs prepared by electron beam lithography on Si_3N_4 substrates, for the first plasmon mode at lowest energy. The absence of signal along the symmetry axis corresponds to an antisymmetric mode with opposite charges on the tips of the SRR (see Fig. 5(a)). The resulting oscillating current along the entire ring from one tip to another is responsible for the occurrence of the strong B_z component in its center as confirmed in Fig. 5(b).

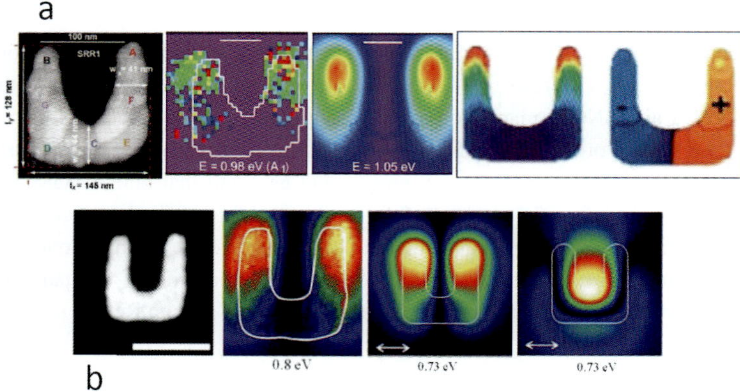

Fig. 5. Map of the lower energy plasmon mode named A1 in isolated SRRs: (a) ADF image, EELS map of the A1 mode at 0.98 eV, BEM simulation of the EELS map recorded in the spectrum-image STEM mode, calculation of the electric charge density (amplitude and phase) confirming that the charges at the two extremities are of different sign[18]; (b) ADF image, EELS map recorded in a monochromated STEM-EELS mode at an energy of 0.8 eV, calculated field distribution of E_z and B_z in a plane located 20 nm above the nanostructure, when optically excited by a plane wave, with polarization shown by the white arrows.[19]

4. New challenges for future studies

These few examples have clearly demonstrated that EELS studies constitute quite efficient probes of the optical response of nanostructured materials with unique spatial resolution. They will surely generate further studies on a broader range of geometries involving nm-sized features to be built by the newly available fabrication tools.

Up to now, our identification of the observed plasmon modes has mostly been based on a classical electromagnetic description. When considering nanostructures with at least one dimension of size smaller than the skin depth (typically 20 nm for silver in the visible range), one can consider[20] the mapped LSPs as stationary waves of oscillating electron densities, respectively along 1D (the longitudinal axis of the nanorod shown in Fig. 4) or 2D (in plane for the triangular platelets shown in Fig. 3). They correspond to the response of the whole nanostructure to the external perturbation brought by the incoming electron at a given position. Varying a global parameter modifies the local response, such as the energy and the intensity of the plasmon peaks, which means that the "coherence length" of these LSPs is larger than their sizes.

In contrast to this global response situation, there exist cases where the detailed geometry can break it down. One of these has been identified in gold nanostars with individual tips protruding out of a common core.[21] A core mode of weak intensity can be detected, however overwhelmed by plasmon resonance peaks of different energies and intensities located at the tips. This behavior has been attributed to a loss of spatial coherence between the different tips linked by a core larger than the optical skin depth, thus impeding cross-talk between the surface electron oscillations of neighboring tips. For sure, these possibilities of local modification and control of plasmon resonance will generate a strong interest in the nano-optics community for manipulating and localizing light on shortest sizes, at "hot spots" for instance.

Several TEM studies aim at further exploring the concept of "coherence length" for plasmons, and surface plasmons in particular. Röder and Lichte (Ref. 22 and references therein) have described a scheme for measuring the extension of coherence in an inelastic process, such as a plasmon, by using an electron holographic method. In these experiments, the contrast of the interference fringes corresponding to variable shear distances defined by the biprism voltage is measured in energy filtered images corresponding to the inelastic events under investigation. The first experiments that probe the modulation of the coherence of the incident electron beam induced by surface plasmon scattering and measured at a given distance outside an Al bar would suggest that it could be associated to the density-density correlation function of the excited mode.[22] They also clearly plead in favor of further investigations along this direction.

Let us conclude with a few words on "quantum plasmonics". As highlighted by Z. Jacob in a recent review,[23] a new frontier needs be fully explored when critical dimensions of the nanostructure become smaller and smaller, when the quantum nature of light and matter induces non-classical behavior. As a first striking result, Scholl *et al.*[24] have reported the results of an EELS investigation of the surface plasmon energy and lifetime in a set of silver nanoparticles free of stabilizing ligands, with diameters ranging typically from 2 to 20 nm. The theoretical model that they propose introduces a discretization of the individual electron levels when their number decreases, as responsible for the observed blue shift and increased width of the measured surface plasmon line. Future research effort is obviously required in cases involving coupling of metallic particles between themselves, or with quantum emitters and/or molecular sensors, i.e., when the smallest dimensions are essential to create quantum effects such as quantum tunneling.

This list of new challenges opens up extensive studies of the interactions between electrons and photons, mediated by matter at the nm scale, and is far from being exhaustive. There is no doubt that in the near future, many more results will appear, such as those incorporating time-resolved measurements.

Acknowledgements

The present work is indebted to many collaborators and sponsors over years. I wish to point out the key role of M. Kociak for promoting these new and successful fields of nanoplasmonics and nanophotonics in our research group at Orsay. Fruitful discussions with O. Stéphan, L. Zagonel, L. Tizei, J. Garcia de Abajo are also acknowledged. The contribution of the Ph.D. students (J. Nelayah, D. Taverna, S. Mazzucco, G. Boudarham, A. Losquin, Z. Mahfoud) has been essential. The support of the European Union projects NMP 4-2006-SPANS and I3 2006 ESTEEM 026019 has been very helpful.

References

1. K. Harada, T. Matsuda, J. Bonevich, M. Igarashi, S. Kondo, G. Pozzi, U. Kawabe, and A. Tonomura, Real-time observation of vortex lattices in a superconductor by electron microscopy, *Nature* **360**, 51-53, (1992).
2. A. Tonomura, H. Kasai, O. Kamimura, T. Matsuda, K. Harada, T. Yoshida, T. Akashi, J. Shimoyama, K. Kishio, T. Hanaguri, K. Kitazawa, T. Matsui, S. Tajima, N. Koshizuka, P. L. Gammel, D. Bishop, M. Sasase, and S. Okayasu, Observation of structures of chain vortices inside anisotropic high-Tc
2. superconductors, *Phys. Rev. Lett.* **88**(23), 237001, (2002).
3. K. Harada, N. Osakabe, and Y. A. Ono, Electron microscopy study on magnetic flux lines in superconductors: Memorial to Akira Tonomura, *IEEE Trans. Appl. Superconductivity* **23**(1), 8000507, (2012).
4. J. Nelayah, M. Kociak, O. Stéphan, F. J. García de Abajo, M. Tence, L. Henrard, D. Taverna, I. Patoriza-Santos, L. M. Liz-Marzán, and C. Colliex, Mapping surface plasmons on a single metallic nanoparticle, *Nature Phys.* **3**, 348-353, (2007).
5. F. J. Garcia de Abajo, Optical excitations in electron microscopy, *Rev. Mod. Phys.* **82**(1), 209-275, (2010).
6. S. T. Park, M. Lin, and A. H. Zewail, Photon-induced near-field electron microscopy (PINEM): theoretical and experimental, *New J. Phys.* **12**, 123028, (2010).
7. A. Yurtsever and A. H. Zewail, Direct visualization of near-fields in nanoplasmonics and nanophotonics, *Nano Lett.* **12**(6), 3334-3338, (2012).
8. C. Colliex, N. Brun, A. Gloter, D. Imhoff, M. Kociak, K. March, C. Mory, O. Stéphan, M. Tencé, and M. Walls, Multi-dimensional and multi-signal approaches in scanning transmission electron microscopes, *Phil. Trans. R. Soc. A* **367**, 3845-3858, (2009).
9. C. Colliex, From electron energy-loss spectroscopy to multi-dimensional and multi-signal electron microscopy, *J. Electron Microsc. (Tokyo)* **60**, Suppl. 1: S161-171, (2011).

10. For a recent review, see M. Kociak and F. J. Garcia de Abajo, Nanoscale mapping of plasmons, photos and excitons, in MRS Bulletin, Vol. **37**, Special issue on "Spectroscopy imaging in electron microscopy, eds. S. J. Pennycook and C. Colliex, pp. 39-46, January 2012.

11. L. F. Zagonel, S. Mazzucco, M. Tencé, K. March, R. Bernard, B. Laslier, G. Jacopin, M. Tchernycheva, L. Rigutti, F. H. Julien, R. Songmuang, and M. Kociak, Nanometer scale spectral-imaging of quantum emitters in nanowires and its correlation to their atomically resolved structure, *Nano Lett.* **11**(2), 568-573, (2011).

12. D. Rossouw, M. Couillard, J. Vickery, E. Kumacheva, and G. A. Botton, Multipolar plasmonic resonances in silver nanowire antennas imaged with a subnanometer electron probe, *Nano Lett.* **11**(4), 1499-1504, (2011); D. Rossouw and G. A. Botton, Plasmonic response of bent silver nanowires for nanophotonic subwavelength waveguiding, *Phys. Rev. Lett.* **110**(6), 066801, (2013).

13. F. J. Garcia de Abajo and M. Kociak, Probing the photonic local density of states with electron energy-loss spectroscopy, *Phys. Rev. Lett.* **100**(10), 106804, (2008).

14. C. Girard, Near fields in nanostructures, *Rep. Prog. Phys.* **68**(8), 1833, (2005).

15. G. Boudarham and M. Kociak, Modal decompositions of the local electromagnetic density of states and spatially resolved electron energy loss probability in terms of geometric modes, *Phys. Rev. B* **85**(24), 245447, (2012).

16. M.-W. Chu, V. Myroshnychenko, C. H. Chen, J.-P. Deng, C.-Y. Mou, and F. J. García de Abajo, Probing bright and dark surface-plasmon modes in individual and coupled noble metal nanoparticles using an electron beam, *Nano Lett.* **9**(1), 399-404, (2009).

17. I. Alber, W. Single, S. Müller, R. Neumann, O. Picht, M. Rauber, P. A. van Aken, and M. E. Toimil-Molares, Visualization of multipolar longitudinal and transversal surface plasmon modes in nanowire dimers, *ACS Nano* **5**(12), 9845-9853, (2011).

18. G. Boudarham, N. Feth, V. Myroshnychenko, S. Linden, J. Garcia de Abajo, M. Wegener, and M. Kociak, Spectral imaging of individual split-ring resonators, *Phys. Rev. Lett.* **105**(25), 255501, (2010).

19. F. von Cube, S. Irsen, J. Niegemann, C. Matyssek, W. Hergert, K. Bush, and S. Linden, Spatio-spectral characterization of photonic meta-atoms with electron energy-loss spectroscopy, *Optical Materials Express* **1**(5), 1009-1018, (2011).

20. J. Nelayah, M. Kociak, O. Stéphan, N. Geuquet, L. Henrard, F. J. García de Abajo, I. Pastoriza-Santos, L. M. Liz-Marzán, and C. Colliex, Two-dimensional quasistatic stationary short range surface plasmons in flat nanoprisms, *Nano Lett.* **10**(3), 902-907, (2010).

21. S. Mazzucco, O. Stéphan, C. Colliex, I. Pastoriza-Santos, L. M. Liz-Marzán, J. García de Abajo, and M. Kociak, Spatially resolved measurements of plasmonic eigenstates in complex-shaped, asymmetric nanoparticles: gold nanostars, *Eur. Phys. J. Appl. Phys.* **54**(3), 33512, (2011).

22. F. Röder and H. Lichte, Inelastic electron holography–first results with surface plasmons, *Eur. Phys. J. Appl. Phys.* **54**(3) 33504, (2011).

23. Z. Jacob, Quantum plasmonics, *MRS Bulletin* **37**(08), 761-767, (August 2012).

24. J. Scholl, A. L. Koh, and J. A. Dionne, Quantum plasmon resonances of individual metallic nano-particles, *Nature* **483**(7390), 421-427, (2012) and associated comment by F. J. Garcia de Abajo, Microscopy: Plasmons go quantum, *Nature* **483**(7390), 417-418, (2012).

"The Picture is My Life"

Shuji Hasegawa

Department of Physics, The University of Tokyo,
7-3-1 Hongo, Bunkyo-ku, Tokyo 113-0033, Japan
E-mail: shuji@phys.s.u-tokyo.ac.jp

This article is dedicated to the memory of the late Dr. Akira Tonomura, with whom I worked at Hitachi Advanced Research Laboratory (HARL) from 1985 to 1990. I report my research experience at HARL and at the University of Tokyo after 1990 together with intriguing experimental photographs from my favorite collection taken by myself and my colleagues. These photographs include electron holography micrographs, scanning tunneling micrographs, electron diffraction patterns, photoemission band dispersion, and images of four-tip scanning tunneling microscopes. I conclude with a discussion on my 'beyond seeing' study.

1. Introduction

After my graduation from Graduate School of Science, The University of Tokyo with MS, I spent five years as a researcher in Tonomura group at Hitachi Advanced Research Laboratory (HARL). When I joined his group, they were at the final stage of verification experiments of the Aharonov-Bohm effect using a ring magnet covered with a toroidal superconductor. I participated in the sample preparation and optical reconstruction processes. In addition, I engaged in the development of a new computer-controlled optical reconstruction interferometer to enhance the sensitivity in phase measurement of electron waves, which was later used for detecting very weak magnetic fields, such as leakage fields from magnetic recording media and quantized magnetic fluxes penetrating superconductors. Research results from these topics constituted my Ph.D. thesis in 1991.

When we received the galley proof of our Phys. Rev. Lett. (PRL) paper[1] reporting the first observation of quantized magnetic flux lines penetrating a superconducting Pb film, we found out that the holographic micrograph of the flux was in a half-column size. The contents of the micrograph were simple so that I thought the half-column size was acceptable. However, Tonomura-san was not satisfied and wrote to the PRL editor asking for an enlargement to the double-column size with a comment "Since the picture is my life, please show the micrograph as large as possible in the printed paper." I was strongly impressed by his words at that time that really symbolized his attitude toward research (See Fig. 1 in Ref. 1 for the double-column figure).

After moving to The University of Tokyo as an assistant professor at Department of Physics, I had to give up the electron holography research because I could not purchase an expensive electron microscope. Instead, I started research on surface physics using simple electron diffraction and later extended to the study of electronic transport on crystal surfaces. Just after publishing my first PRL paper on the surface transport[2], someone remarked that it was the least expensive PRL paper but interesting one. Later I purchased a scanning tunneling microscope (STM) and a photoemission spectroscopy instrument that enabled me to explore the surface physics from different aspects. After establishing my own research group, I started to develop a world-first four-tip STM.[3] Tonomura-san's passion for developing new equipment has been carried on to my research.

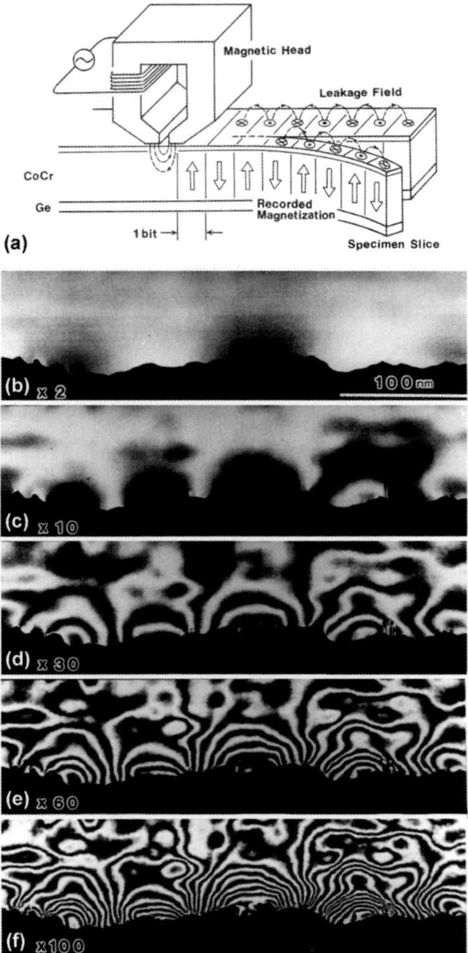

Fig. 1. Digital magnetic recording on a CoCr film in the perpendicular magnetization mode, observed by electron holography with phase amplification (Ref. 4).

2. Electron holography results

Figure 1 shows reconstructed holographic interference micrographs[4] with its sample structure, showing leakage fields from a CoCr thin film for a digital magnetic recording medium. The bit width was ca. 100 nm and the magnetizations were alternately up or down along the perpendicular direction. The phase sensitivity is enhanced with the new interferometer I developed: single interference fringes in units of h/e ($= 4 \times 10^{-15}$ Wb) are as follows: 1/2 in Fig. 1b, 1/10 in Fig. 1c, 1/30 in Fig. 1d, 1/60 in Fig. 1e, and 1/100 in Fig. 1f. Here h is Planck's constant and e is the elementary charge. The measurement resolution of the phase of electron wave is as small as $2\pi/100$, which is roughly an order of magnitude improvement compared with the previous measurements.

Figure 2 shows holographic interference micrographs of quantized magnetic fluxes penetrating superconducting Pb films.[5] A single interference fringe corresponds to a single flux quantum $h/2e$. The magnetic flux structures vary depending on the film thickness d and the applied magnetic field H. For thin films, individual fluxes are made of single quanta, while for thicker films, every flux is made of multi-quanta. While the quantized magnetic fluxes at the surface of superconductors are described only by schematic illustrations in many of the text books of solid-state physics, they should be replaced by the present pictures.

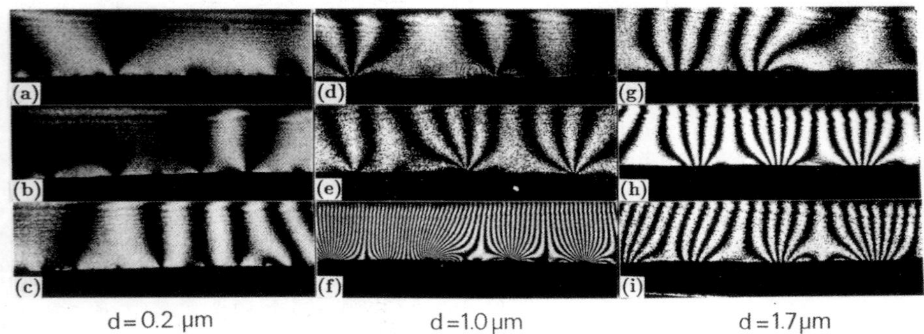

Fig. 2. Quantized magnetic fluxes penetrating superconducting Pb thin films with different thickness *d*, observed by electron holography under an applied transverse magnetic field *H*. The field is increased from the top panel to the bottom panel at each thickness. The dark area in the lower part in each micrograph is the Pb film and the upper part is vacuum (Ref. 5).

3. Scanning tunneling microscopy experiment

Figure 3 shows an STM image of a Si crystal surface partially covered with a monolayer of Ag film.[6] The Ag-covered area is denoted by $\sqrt{3}\times\sqrt{3}$, while the uncovered area is indicated by 7×7. Fine and regular arrays of protrusions correspond to atomic lattices with different periodicities in two areas. There are meandering atomic steps running from left to right in the center (indicated by (A)) and straight domain boundaries (indicated by (B)). In addition to the regular lattice pattern, we find wavy patterns with about 3 nm wavelength near the peripheries of the $\sqrt{3}\times\sqrt{3}$ area, which are just like ripples on water surface. These wavy patterns are electron standing waves. Electrons in the monolayer-Ag film freely moves around along the surface and they are reflected by defects such as steps and domain boundaries. The incident wave and reflected wave interfere with each other to form the standing wave. The STM enables imaging not only geometric protrusions at an atomic scale, but also distributions of electron density. Although I had learned the wave nature of electron through electron holography study in the Tonomura group, I was really impressed by this STM picture of electron standing waves on the crystal surface. Electrons confined in narrow domains make concentric standing waves, revealing quantum-well states. This is nothing but a visualization of quantum mechanics.

4. Electron diffraction experiment

The wave nature of electron was historically verified as a form of electron diffraction from crystals by

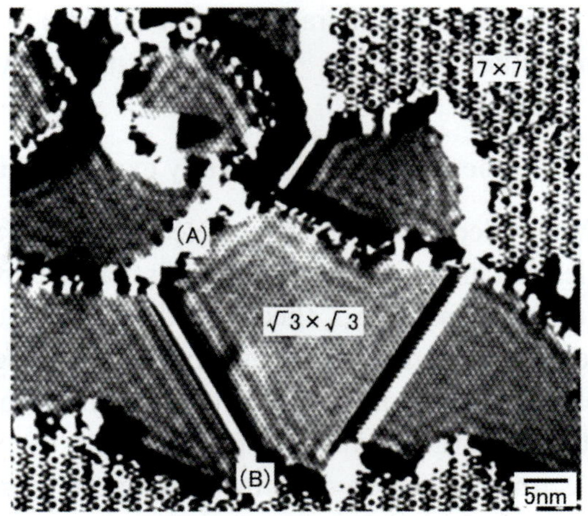

Fig. 3. STM image of electron standing waves in a monatomic-layer Ag on a Si crystal surface (Ref. 6).

Davisson and Germer, Thomson, and Kikuchi. Electron diffraction is now widely used to investigate atomic structures on crystal surfaces. Figure 4 shows reflection-high-energy electron diffraction (RHEED) patterns taken from a single sample, a monatomic-layer In adsorbed on a Si surface, at room temperature (RT) (left figure) and at 100 K (right figure).[7] Diffraction spots in the RT pattern indicate a regular atomic arrangement of a so-called 4×1 surface superstructure. This surface structure is now explained by massive arrays of In atomic chains aligned along a particular crystal orientation of the Si substrate, resulting in a quasi-one-dimensional metallic surface. By cooling the sample down to 100 K, several additional spots (indicated by black arrowheads) and streaks (indicated by arrow) appear, showing a surface superstructure of so-called 8×'2'. This is a temperature-induced surface phase transition we have found; the nature is believed to have a Peierls-

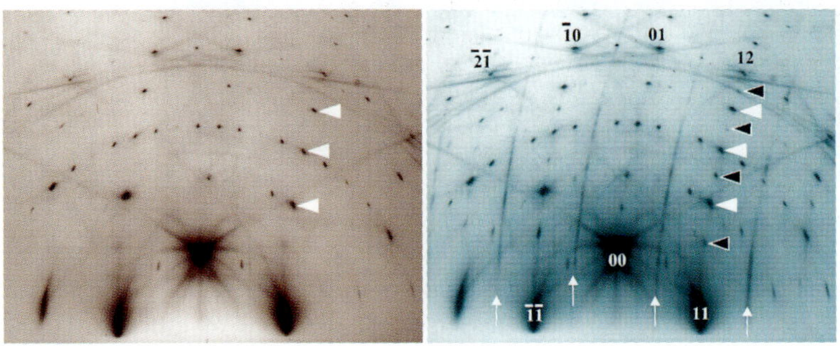

Fig. 4. Electron diffraction patterns taken from a monatomic-layer In adsorbed on a Si surface, at room temperature (left) and at 100 K (right), showing a phase transition (Ref. 7).

like meta-insulator transition accompanied by a charge-density wave (CDW) formation and lattice distorsion.[7] Its mechanism, however, is not clarified yet.

5. Photoemission spectroscopy experiment

While electron microscopy, electron holography, and STM enable real-space imaging, electron diffraction and photoemission spectroscopy enable reciprocal-space imaging.

From Einstein's theory of photoelectric effect, electrons in crystals and on the surfaces are emitted outside by absorbing the energy of irradiated photons. This phenomenon is now widely used as photoemission spectroscopy for analyzing electronic states in/on crystals. We can estimate the initial-state energy levels and momenta of electrons in/on crystals by measuring the kinetic energy and emission angle of the emitted photoelectrons. Then, we can deduce the energy-wavenumber relation (band dispersion), since the wavenumber k is defined as momentum p divided by $\hbar$.

Figures 5(a)-(c) show the band dispersion of Bi thin films of different thicknesses, measured by angle-resolved photoemission spectroscopy (ARPES).[8] The vertical axis expresses the binding energy E with respect to the Fermi energy (E_F). The horizontal axis expresses the component of the wavenumber parallel to the surface. We notice that several bands are dispersed, and that with the increase in film thickness, the number of bands increases and the energy separation between the bands becomes smaller. As schematically shown in Fig. 5(d), the electrons are confined in the direction perpendicular to the film surface, which results in formation of quantum-well states and discrete energy levels. The energy separation between the quantized energy levels becomes smaller as the width of quantum well (thickness of the film) becomes larger. On one hand, the electrons in the film can move around freely along the surface, resulting in dispersive bands in the

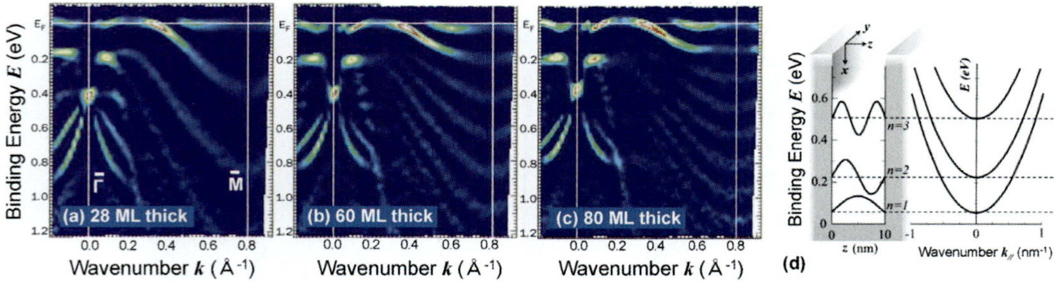

Fig. 5. Bismuth single-crystalline ultrathin films grown on Si, observed by angle-resolved photo-emission spectroscopy. (a)-(c) Band dispersion of the films of different thickness, 28 ML (monolayer), 60 ML, and 80 ML thick, respectively. (d) Schematics showing quantum-well states and sub-bands in a thin film (Ref. 8).

surface-parallel direction. The bands observed in Figs. 5(a)-(c) are sub-bands of the quantum-well states formed inside the film.

However, the bands crossing E_F do not change with the film thickness changes, indicating that these metallic states do not come from the inside of the film, but come from the surface of the film. This means that the Bi crystal has a metallic surface though the inside is a semi-metal. Since the metallic surface states are found to have a high electrical conductivity[9] and a spin-split characteristic,[10] existence of the surface state might lead to a possibility of spin-polarized current flowing at the non-magnetic surface.

6. Four-tip STM experiment

A single-tip STM enables atomic-scale imaging of structures and electronic density. A multi-tip STM enables us to measure electrical conductivity at a nanometer scale by utilizing these tips as electrodes. Figure 6 shows scanning electron micrographs of tips of the developed four-tip STM.[3,11] These tips are mechanically and electrically independent of each other and can be controlled by a single PC. Two of the four tips are used for source and drain electrodes and the other two are used for voltage measurement. The tip separation and arrangement can be changed arbitrarily under scanning electron microscope observation. With carbon nanotube tips, the tip separation can be reduced to as small as 20 nm[11] (See the right figure of Fig. 6). This instrument is now commercially available from Unisoku Co. and used world-wide to measure electronic transports on surfaces of nano-materials such as nanowires, graphene, and molecule assemblies.

7. Beyond seeing – electron/spin transports in relation to structures

Microscopy, diffraction, and spectroscopy analyses provide fundamental information on structures and electronic states so that it becomes possible to interpret various kinds of

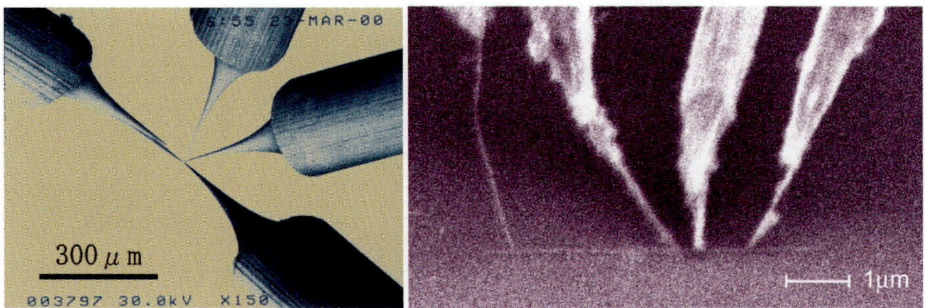

Fig. 6. Scanning electron microscope images of the four tips in the developed four-tip scanning tunneling microscope. Left figure shows W tips (Ref. 3) and right figure shows carbon nanotube tips (Ref. 14).

physical and chemical properties. My strategy in research is to combine electronic and spin transport studies with microscopy/diffraction/spectroscopy studies. Before my first paper on this subject[2] in 1992, no one considered electrical conduction at surfaces in relation to surface structures. Twenty years later, no one now doubts the importance of surface-state transport, i.e., electrical conduction only at the topmost atomic layers.

Figure 7 shows resistance of a Si crystal measured as a function of temperature.[12] The surface is covered by (sub)monolayer-indium, and different surface superstructures are created depending on the In coverage; $\sqrt{3} \times \sqrt{3}$ (1/3 ML), $\sqrt{31} \times \sqrt{31}$ (2/3 ML), 4×1 (1 ML), and $\sqrt{7} \times \sqrt{3}$ (2 ML). It should be noted that the resistance of a Si crystal having a macroscopic thickness (0.5 mm) changes many orders of magnitude just by adsorption of monoatomic-layer In on the surface. These findings can be obtained only through the surface-sensitive measurements using the developed microscopic four-point probes as shown in Fig. 6. The temperature dependence of resistance is also quite different depending on the surface structures. These properties can be explained from the surface-state bands revealed by photoemission spectroscopy. The $\sqrt{7} \times \sqrt{3}$ surface superstructure shows superconducting transition[13] at 2.8 K (See the right figure in Fig. 7). This is the first example of surface-state superconductivity.

8. Concluding remarks

Electron holography visualizes magnetic field in a form of distribution of magnetic lines of force. Scanning tunneling microscope visualizes local electronic density of states. Photoemission spectroscopy visualizes energy levels and Fermi surfaces. As indicated in

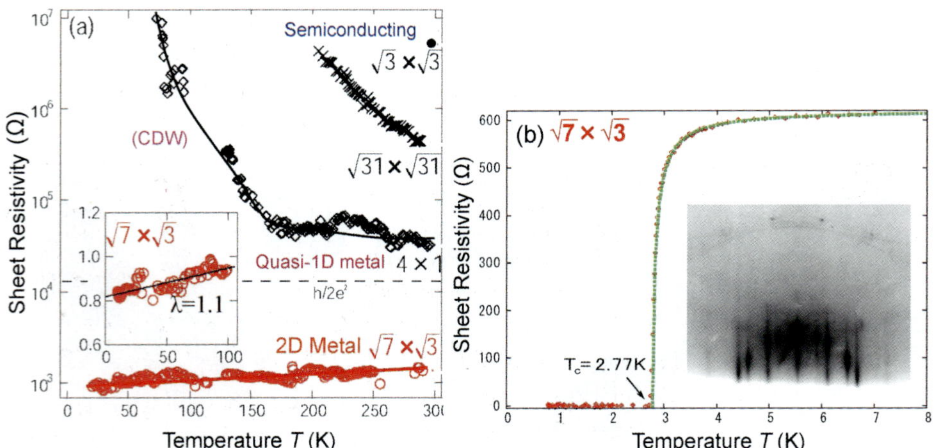

Fig. 7. Temperature-dependent electrical resistance of different surface superstructures formed with (sub)monolayer In adsorption on a Si crystal (Refs. 12, 13). The inset in (b) is a RHEED pattern showing Si(111)-$\sqrt{7} \times \sqrt{3}$ -In surface superstructure.

these examples, epoch-making experimental methods and instruments visualize what used to be invisibles just as if they were real objects. This is a real pleasure of science. Tonomura-san's words 'The picture is my life.' shows the heart of science. Four-tip STM, my life work, can be used for visualizing electronic transport in many ways. I will continue to struggle to obtain pictures that visualize hitherto invisible quantities.

I dedicate the pictures shown in this article to Tonomura-san in token of gratitude for his warm encouragement throughout my research life.

References

1. T. Matsuda, S. Hasegawa, M. Igarashi, T. Kobayashi, M. Naito, H. Kajiyama, J. Endo, N. Osakabe, A. Tonomura, and R. Aoki, Magnetic field observation of a single flux quantum by electron holographic interferometry, *Phys. Rev. Lett.* **62** (21), 2519-2522, (1989).
2. S. Hasegawa and S. Ino, Surface structures and conductance at epitaxial growths of Ag and Au on the Si(111) surface, *Phys. Rev. Lett.* **68** (8), 1192-1195, (1992).
3. I. Shiraki, F. Tanabe, R. Hobara, T. Nagao, and S. Hasegawa, Independently driven four-tip probes for conductivity measurements in ultrahigh vacuum, *Surf. Sci.* **493** (1-3), 633-643, (2001).
4. S. Hasegawa, T. Kawasaki, J. Endo, A. Tonomura, Y. Honda, M. Futamoto, K. Yoshida, F. Kugiya, and M. Koizumi, Sensitivity-enhanced electron holography and its application to magnetic recording investigation, *J. Appl. Phys.* **65** (5), 2000-2004, (1989).
5. S. Hasegawa, T. Matsuda, J. Endo, N. Osakabe, M. Igarashi, T. Kobayashi, M. Naito, A. Tonomura, and R. Aoki, Magnetic flux quanta in superconducting thin films observed by electron holography and digital phase analysis, *Phys. Rev. B* **43** (10), 7631-7650, (1991).
6. N. Sato, T. Nagao, S. Takeda, and S. Hasegawa, Electron standing waves on the Si(111)-$\sqrt{3}\times\sqrt{3}$-Ag surface, *Phys. Rev. B* **59** (3), 2035-2039, (1999).
7. H. W. Yeom, S. Takeda, E. Rotenberg, I. Matsuda, K. Horikoshi, J. Schaefer, C. M. Lee, S. D. Kevan, T. Ohta, T. Nagao, and S. Hasegawa, Instability and charge density wave of metallic quantum chains on a silicon surface, *Phys. Rev. Lett.* **82** (24), 4898-4901, (1999).
8. T. Hirahara, T. Nagao, I. Matsuda, G. Bihlmayer, E. V. Chulkov, Yu. M. Koroteev, and S. Hasegawa, Quantum-well states in ultrathin Bi films: angle-resolved photoemission spectroscopy and first-principles calculations study, *Phys. Rev. B* **75** (3), 035422, (2007).
9. T. Hirahara, I. Matsuda, S. Yamazaki, N. Miyata, T. Nagao, and S. Hasegawa, Large surface-state conductivity in ultrathin Bi films, *Appl. Phys. Lett.* **91** (20), 202106, (2007).
10. T. Hirahara, K. Miyamoto, I. Matsuda, T. Kadono, A. Kimura, T. Nagao, G. Bihlmayer, E. V. Chulkov, S. Qiao, K. Shimada, H. Namatame, M. Taniguchi, and S. Hasegawa, Direct observation of spin splitting in Bismuth surface states, *Phys. Rev. B* **76** (15),153305, (2007).
11. S. Yoshimoto, Y. Murata, K. Kubo, K. Tomita, K. Motoyoshi, T. Kimura, H. Okino, R. Hobara, I. Matsuda, S. Honda, M. Katayama, and S. Hasegawa, Four-point probe resistance measurements using PtIr-coated carbon nanotube tips, *Nano Letters* **7** (4), 956-959, (2007).
12. S. Yamazaki, Y. Hosomura, I. Matsuda, R. Hobara, T. Eguchi, Y. Hasegawa, and S. Hasegawa, Metallic transport in a monatomic layer of In on a Silicon surface, *Phys. Rev. Lett.* **106** (11), 116802, (2011).
13. M. Yamada, T. Hirahara, S. Hasegawa, H. Mizuno, Y. Miyatake, and T. Nagamura, Surface electrical conductivity measurement system with micro-four-point probes at sub-Kelvin temperature under high magnetic field in ultrahigh vacuum, *e-J. Surf. Sci. Nanotech.* **10** (1), 400-405, (2012).

Direct Observation of Electronically Phase-Separated Charge Density Waves in Lu₂Ir₃Si₅ by Transmission Electron Microscopy[*]

Cheng-Hsuan Chen

*Center for Condensed Matter Sciences and Department of Physics, National Taiwan University,
Taipei 10617, Taiwan
E-mail: chchen35@ntu.edu.tw*

Transmission electron microscopy has been shown to be a powerful experimental tool for studies of charge-density-wave phase transitions. The capabilities of real-space imaging can provide details of the phase transitions not obtainable from diffraction experiments. Charge density waves have been found predominantly in low dimensional materials where strong nesting effect occurs due to favorable topological features of the Fermi surface. In this article, I'll describe transmission electron microscopy studies of an unusual charge density wave phase transition in a three-dimensional material Lu₂Ir₃Si₅, which exhibits unprecedented electronic phase separations in the charge-density-wave state at low temperatures.

1. Introduction

Charge density waves (CDWs), an electronically driven instability, are normally found in low-dimensional metallic systems due to favorable nesting features of the Fermi surface that greatly soften the electronic state and result in a periodic electron density fluctuation in real space. The gain in electronic energy by the formation of electron density fluctuation is, however, always offset by the presence of crystalline elastic energy due to accompanied lattice distortions. In a typical CDW phase transition, the system first goes to an incommensurate state followed by an incommensurate to commensurate (lock-in) transition as the temperature is lowered from the normal state above the CDW transition.[1-4] The CDW phase transition, signified by the presence of anomalies in various transport measurements, is further characterized by the appearance of superlattice reflection spots observable in electron and X-ray diffraction experiments. Transmission electron microscopy (TEM) has been shown to be a powerful experimental tool for

---

[*] This work was done in collaboration primarily with Dr. M. H. Lee, Center for Condensed Matter Sciences and Department of Physics, National Taiwan University.

studies of charge density waves phase transitions.[3-5] Electron diffraction capabilities in TEM allow single-crystalline diffraction patterns to be obtained from an area smaller than a single grain in a polycrystalline sample, whereas the Xx-ray diffraction experiments on CDW materials often require single crystals of much larger sizes. Furthermore, the capabilities of real-space imaging in TEM can provide details of the phase transitions not obtainable from diffraction experiments. In real space, an incommensurate CDW phase is often characterized by the presence of discommensurations (DC) or domain walls and it is well known that nucleation, growth, and annihilation of DCs play the most critical role for the incommensurate-commensurate transition.[1-4] When the temperature is lowered toward the commensurate phase, the density of DC continues to decline through motions and annihilations of DCs leading to a gradual decrease of incommensurability for the CDW superlattice diffraction spots. It should be noted that CDW is an electronically homogeneous phase transition that the entire sample would undergo the transition as it reaches the critical temperature.

Rare-earth transition-metal ternary silicides, such as the types of $R_5T_4Si_{10}$ and $R_2T_3Si_5$, despite their seemingly three-dimensional crystallographic structures, have been shown to exhibit CDW phase transitions with remarkable anomalies observable in the thermal and electrical transport measurements.[6-10] In this article, I'll describe TEM studies of the CDW phase transition in $Lu_2Ir_3Si_5$ by electron diffraction and dark-field imaging using the CDW superlattice diffraction spots. Most interestingly, the CDW state at low temperatures is found to be inhomogeneous and electronically phase-separated into a coexistence of CDW domains and low-temperature normal phase domains. Upon change of temperatures, unlike other typical incommensurate CDW systems in which commensurability varies continuously with temperatures, we find that commensurability remains unchanged in the present case and the predominant change takes place in the redistribution of the area ratio of the two coexisted phases. The electronic phase separation in the CDW state of $Lu_2Ir_3Si_5$ is unprecedented in CDW systems.

2. Experimental details

The polycrystalline sample $Lu_2Ir_3Si_5$ was prepared by arc-melting high-purity elements under argon atmosphere. The sample has been re-melted several times under the identical preparation conditions to improve the homogeneity and purity of phase. The resulting ingot was then sealed in a quartz ampoule with about 160 Torr of argon and annealed at 1250°C for one day followed by fourteen days at 1050°C. Transport measurements of this sample have been reported elsewhere.[11] Samples for the TEM studies were prepared by mechanical polishing followed by ion milling at liquid nitrogen temperature. Our electron diffraction and TEM dark-field imaging experiments were carried out in a JEOL 2000FX transmission electron microscope (operating at 200 kV) equipped with a low temperature sample holder capable of reaching 20 K and a 14 bit CCD imaging detector.

3. Results and discussions

Figure 1 compares [101] zone-axis electron diffraction patterns taken at room temperature and 95 K below the CDW phase transition at 140 K. The room temperature diffraction pattern shown in Fig. 1(a) is consistent with the known orthorhombic lattice structure with space group Ibam. Presence of CDW superlattice spots can be easily seen in Figs. 1(b) and 1(c) obtained at 95 K, and these superlattice spots are characterized by modulation wave vectors $\bar{q}_1 = \delta_1(\bar{1}21)$ and $\bar{q}_2 = \delta_2(12\bar{1})$ with δ_1 and δ_2, which, in the range of 0.23–0.25, are usually not the same. We note that Figs. 1(b) and 1(c) were obtained from different areas of the same grain showing incommensurate and commensurate CDW modulations in Fig. 1(b) and Fig. 1(c), respectively.

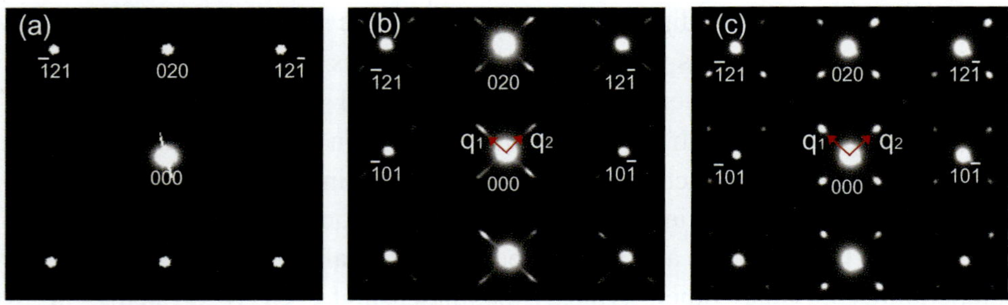

Fig. 1. (a) shows the [101] zone-axis electron diffraction pattern obtained at the room temperature. (b) and (c) are similar patterns obtained at 95 K which show the presence of CDW superlattice reflections and the systematic (10-1) reflections which are absent at room temperature. CDW superlattice modulation in (b) is incommensurate whereas (c) obtained from a different area is commensurate. We marked the modulation wave vectors $\bar{q}_1 = \delta_1(\bar{1}21)$ and $\bar{q}_2 = \delta_2(12\bar{1})$ with δ_1 and δ_2 varying ~0.23–0.25 from area to area.

The commensurability of CDW superlattice spots is found to vary from area to area with most areas exhibiting incommensurate modulations. This is unusual compared with other known CDW systems in which global homogeneity of modulations are commonly observed. It is also noted that the incommensurate superlattice reflections are somewhat diffusive and elongated along the $(\bar{1}21)$ or $(12\bar{1})$ planar directions indicating a rather short coherence length ~10 nm. The width of superlattice reflections in the orthogonal direction, however, is much narrower. Commensurate CDW superlattice spots, on the other hand, are much sharper in both directions. Electron diffraction patterns obtained from other zones axes such as [11-1] and [210] also yield the same CDW modulation wave vectors. The appearance of CDW superlattice spots becomes discernible at ~140 K on cooling and, upon warming from low temperatures, they disappear ~180 K with a 40 K thermal hysteresis, consistent with the recent transport measurement.[11,12] In addition to the CDW superlattice spots, forbidden Bragg reflections (*h* 0 *l*), which violate the extinction condition of *h, l* = 2n, such as (-1 0 1), (-3 0 3) and so forth, can also be seen in Figs. 1(b) and 1(c) indicating a structural phase transition associated with the CDW transition. This is consistent with previous X-ray diffraction results obtained at the liquid nitrogen temperature showing a cell doubling along the c-axis.[12] In our experiment, the structural phase transition temperature was found to be ~ 15 K higher than the CDW superlattice formation for both cooling and warming. The higher structural transition temperature is also consistent with a recent thermal conductivity measurement which found that the anomaly from lattice contribution was ~15 K higher than its electronic counterpart.[11]

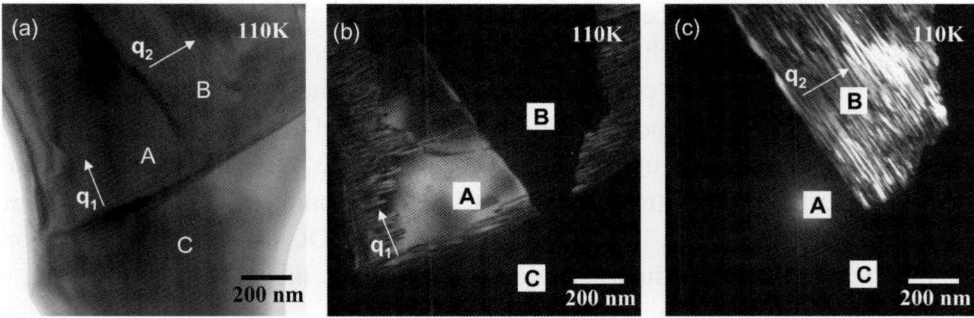

Fig. 2. (a) Bright-field image at 110 K of an area showing three domains marked as A, B, and C. (b) and (c) are dark-field images obtained from $\bar{q}_1$ and $\bar{q}_2$ superlattice spots, respectively. It is clear that modulation wave vectors $\bar{q}_1$ and $\bar{q}_2$ indicated in Fig. 1 actually come from different domains. It is noted that $\bar{q}_1$ is nearly commensurate and $\bar{q}_2$ is incommensurate.

TEM dark-field imaging using the CDW superlattice spots further reveals that the $\bar{q}_1 = \delta_1(\bar{1}21)$ and $\bar{q}_2 = \delta_2(12\bar{1})$ modulations as shown in Figs. 1(b) and 1(c) actually come from separate domains (but still in the same grain) marked as A and B, respectively, as shown in Fig. 2. Note the big difference in DC density in these two domains, with domain A (thinner area) nearly commensurate with few residual DCs and domain B (thicker region) obviously incommensurate with very high density of DCs. Chemical microanalysis with electron nano-probe revealed no inhomogeneity between these two regions. The variations of local strains may have played an important role for the disparity of DC density in regions of different thicknesses. Crystallographically, $(\bar{1}21)$ and $(12\bar{1})$ reciprocal lattice vectors are equivalent and related by simple symmetry operations. This implies that the CDW phase transition is only characterized by one modulation wave vector, not two. The formation of crystallographically equivalent CDW domain structure is a natural consequence for a phase transition adopting a lower crystallographic symmetry at low temperatures. We note that the $\vec{q}_1$ and $\vec{q}_2$ domain structure shown in Fig. 2 can change during different thermal cycles in which areas originally characterized with $\vec{q}_1$ modulation can change into $\vec{q}_2$ modulation and vice versa. This indicates a weak pinning effect in $Lu_2Ir_3Si_5$. With the absence of band structure calculations in this class of materials, it is not totally clear why the modulation wave vector adopts an uncommon $(\bar{1}21)$ reciprocal lattice direction in the present case. Nevertheless, some clues can be gathered from the crystallographic structure of $Lu_2Ir_3Si_5$. It is noted that $2b^* \sim c^*$ and this makes the b and c components of the modulation wave vector nearly equal, which probably enhances the nesting geometry at the Fermi surface.

In the following, we'll discuss the changes of the CDW superlattice spots and its domain structure as a function of temperature. Figure 3(a) shows an intensity profile of an incommensurate CDW superlattice reflection as a function of temperature. It is clear that the position and the width of the superlattice peak remain nearly unchanged in this temperature range, in striking contrast with typical incommensurate CDW systems which normally exhibit gradual sharpening and movement toward the commensurate position of the superlattice peak as the temperature is lowered.[13,14] This very unusual observation sets the CDW in $Lu_2Ir_3Si_5$ apart from the rest of known incommensurate CDW systems in which, when the temperature is lowered, the density of DC decreases through motions and annihilations of DCs leading to a gradual decrease in incommensurability and peak width of the superlattice diffraction spots.[3-5]

The mystery of this anomalous CDW behavior in $Lu_2Ir_3Si_5$ is unveiled when we examine the CDW domain structure as a function of temperature in real space by the satellite dark-field imaging. Close examination of satellite dark-field images shown in Fig. 2 reveals that an electronic phase separation actually takes place during the CDW phase transition. In addition to the CDW domains A and B, domain C is present without CDW satellite modulations. Domain C represents the normal low-temperature phase which is different from the high temperature phase with the structural distortion mentioned earlier. In other words, the entire sample is phase separated into domains of CDW and regions without CDW.

It should be noted that these phase coexistence appears within a single grain of the sample. The spatial ratio of the CDW phase to the low-temperature normal phase is about 1:3 at 95 K. There exists a distinct, rather straight boundary between these two phases. When temperature is varied, no changes of the density or detailed configuration of DCs within the CDW domains are observed. This is evidenced in Figs. 3(b) and 3(c) which display two dark-field images taken at 110 K and 146 K, respectively. Several distinct features of DCs are marked from 1 to 5 in both figures for comparison. It is clear that no discernible changes or movements of DCs have taken place. The only change observed is

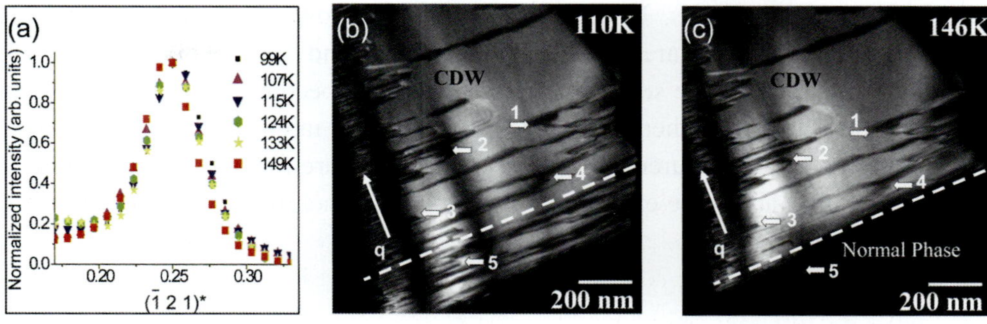

Fig. 3. (a) Intensity profile of CDW superlattice reflections showing that the position and width remain largely unchanged as a function of temperature. (b) and (c) show the CDW domains recorded at 110 K and 146 K, respectively. With increasing temperatures, the domain boundary (dashed line) separating the CDW phase (area with brighter contrast) and the low temperature normal phase (area with dark contrast) moves in a direction along the CDW modulation wave vector such that the area of non-CDW phase grows at the expense of the CDW domain. It is noted that detailed configurations of discommensurations in the CDW domain, marked 1 through 5 for reference, remain unchanged with changing temperatures.

the motion of the boundary separating the two phases leading to an expansion of CDW domain with decreasing temperature and vice versa. The motion of the boundary, which is thermally reversible, is indicated by the dashed line in Fig. 3(b) when temperature is changed. The distribution of these two coexisted phases and their spatial ratio are therefore temperature dependent. We have also cooled the sample down to 20 K and have found that the CDW phase has now become a predominant phase covering ~ 90 % of the area. It should be reminded that the initial CDW transition occurs at a much higher transition temperature ~140 K. The transition appears to be a very sluggish one.

We emphasize that the unprecedented phase-separated CDW state in Lu₂Ir₃Si₅ described above is an intrinsic electronic inhomogeneity, similar to the electronic phase separation in manganites[15,16] in which coexistence of charge-ordered insulating domains and ferromagnetic metallic domains is a common phenomenon. The electronic phase separation in manganites, which arises from the intricate interplay of charge, spin, orbital, and lattice degrees of freedom, plays the most crucial role for the colossal magneto-resistance effect through the percolative transport in the phase-separated state.[16] However, the spatial ratio of these two phases in manganites below the transition temperature is largely determined by the hole-carrier concentration and rather insensitive to the temperature changes, unlike the case for Lu₂Ir₃Si₅. It is not clear why the delicate balance and competition between the CDW phase and the low temperature normal phase are manifested so remarkably in Lu₂Ir₃Si₅. Throughout our experiments, we find that incommensurability tends to vary among different grains in the same sample and the pattern of phase co-existence in the same grain can also vary among different thermal cycles. It is then plausible that minute variations of local strains play a critical role for the mixture of these two phases. Under this scenario, the low temperature normal phase would be construed to be areas under higher strain and the development of CDW modulations naturally relieve some strain and eventually become the predominant phase at lower temperatures. It is then plausible that the system might be sensitive to external pressure. However, no measurements under applied pressure for the present system have been reported so far and these experiments could be very helpful to understand the CDW phase transition in $Lu_2Ir_3Si_5$.

4. Conclusions

The CDW state in $Lu_2Ir_3Si_5$ was found to be electronically phase-separated into regions exhibiting CDW modulations and regions without it. Changing temperature only causes the motion of the distinct boundary separating these two phases and therefore changes the ratio of their spatial extent with the CDW phase prevailing at lower temperatures. The incommensurability and the width of the CDW superlattice reflections are therefore largely unchanged as the temperature varies. Detailed configurations of DC are found to be intact and do not change with temperature, unlike typical CDW systems in which motions, nucleations, and annihilations of DCs play the most critical role for the CDW phase transitions. This article was based largely on our recently published work.[17]

When Dr. Akira Tonomura came to visit me at Bell Labs at Murray Hill, NJ sometime around 1982, I showed him my newly discovered CDW domain walls (discommensurations) in $2H\text{-}TaSe_2$, a compound in the family of transition metal dichalcogenides exhibiting series of intricate and novel CDW phase transitions. Akira, being an ardent explorer of advanced techniques of electron microscopy and its novel applications to frontier physical problems, immediately expressed strong interest in my new results and appreciated the profound significance of these new findings to condensed matter physics made by transmission electron microscopy. Akira would certainly be very pleased to know that, after 30 years, I'm still using electron microscopy to explore new frontiers in CDW and making new advances in the field.

References

1. W. L. McMillan, Landau theory of charge-density waves in transition-metal dichalcogenides, *Phys. Rev. B* **12**(4), 1187-1196, (1975).
2. W. L. McMillan, Theory of discommensurations and the commensurate-incommensurate charge-density-wave phase transition, *Phys. Rev. B* **14**(4), 1496-1502, (1976).
3. C. H. Chen, J. M. Gibson, and R. M. Fleming, Microstructure in the incommensurate and the commensurate charge-density-wave states of $2H\text{-}TaSe_2$: A direct observation by electron microscopy, *Phys. Rev. B* **26**(1), 184-205, (1982).

4. K. K. Fung, S. McKernan, J. W. Steeds, and J. A. Wilson, Broken hexagonal symmetry in the locked-in state of 2Ha-TaSe$_2$ and the discommensurate microstructure of its incommensurate CDW states, *J. Phys. C: Solid State Phys.* **14**(35), 5417, (1981).

5. C. M. Tseng, C. H. Chen, and H. D. Yang, Direct observation of charge-density waves in Ho$_5$Ir$_4$Si$_{10}$, *Phys. Rev. B* **77**(15), 155131, (2008).

6. Y. K. Kuo, Y. Y. Chen, L. M. Wang, and H. D. Yang, Multiple and weak-coupling charge-density-wave transitions inY$_5$Ir$_4$Si$_{10}$, *Phys. Rev. B* **69**(23), 235114, (2004).

7. Y. K. Kuo, F. H. Hsu, H. H. Li, H. L. Huang, C. W. Huang, C. S. Lue, and H. D. Yang, Ionic size and atomic disorder effects on the charge-density-wave transitions in R_5Ir$_4$Si$_{10}$ (R=Dy-Lu), *Phys. Rev. B* **67**(19), 195101, (2003).

8. R. Tediosi, F. Carbone, A. B. Kuzmenko, J. Teyssier, D. van der Marel, and J. A. Mydosh, Evidence for strongly coupled charge-density-wave ordering in three-dimensional R_5Ir$_4$Si$_{10}$ compounds from optical measurements. *Phys. Rev. B* **80**(3), 035107, (2009).

9. S. van Smaalen, The Peierls transition in low-dimensional electronic crystals, *Acta Cryst. A* **61**, 51-61, (2005).

10. K. Tsutsumi, S. Takayanagi, K. Maezawa, and H. Kitazawa, Low-temperature specific heat study of antiferromagnetic transition in ternary rare-earth metal silicide R$_5$Ir$_4$Si$_{10}$ (R=Tb, Dy, Ho, Er), *J. Alloys Compd.* **453**(1-2), 55-57, (2008).

11. Y. K. Kuo, K. M. Sivakumar, T. H. Su, and C. S. Lue, Phase transitions in Lu$_2$Ir$_3$Si$_5$: An experimental investigation by transport measurements, *Phys. Rev. B* **74**(4), 045115, (2006).

12. Y. Singh, D. Pal, S. Ramakrishnan, A. M. Awasthi, and S. K. Malik, Phase transitions in Lu$_2$Ir$_3$Si$_5$, *Phys. Rev. B* **71**(4), 045109, (2005).

13. S. van Smaalen, M. Shaz, L. Palatinus, P. Daniels, F. Galli, G. J. Nieuwenhuys, and J. A. Mydosh, Multiple charge-density waves in R_5Ir$_4$Si$_{10}$ (R=Ho, Er, Tm, and Lu), *Phys. Rev. B* **69**(1), 014103, (2004).

14. F. Galli, R. Feyerherm, R. W. A. Hendrikx, E. Dudzik, G. J. Nieuwenhuys, S. Ramakrishnan, S. D. Brown, S. van Smaalen, and J. A. Mydosh, Coexistence of charge density wave and antiferromagnetism in Er$_5$Ir$_4$Si$_{10}$, *J. Phys.: Condens. Matter* **14**(20), 5067, (2002).

15. S. Mori, C. H. Chen, and S-W. Cheong, Pairing of charge-ordered stripes in (La,Ca)MnO$_3$, *Nature* **392**, 473-476, (2 April 1998).

16. M. Uehara, S. Mori, C. H. Chen, and S.-W. Cheong, Percolative phase separation underlies colossal magnetoresistance in mixed-valent manganites, *Nature* **399**, 560-563, (10 June 1999).

17. M. H. Lee, C. H. Chen, M. W. Chu, C. S. Lue, and Y. K. Kuo, Electronically phase-separated charge-density waves in Lu$_2$Ir$_3$Si$_5$, *Phys. Rev. B* **83**(15), 155121, (2011).

Basic Discoveries in Electromagnetic Field Visualization

Daisuke Shindo

*Institute of Multidisciplinary Research for Advanced Materials, Tohoku University,
Katahira 2-1-1, Sendai 980-8577, Japan
RIKEN Center for Emergent Matter Science, Wako, Saitama 351-0198, Japan
shindo@tagen.tohoku.ac.jp*

Basic discoveries in the electromagnetic field visualization are presented, mentioning the late Dr. A. Tonomura's significant achievements in this field. First, the discovery of the electron biprism interferences by G. Möllenstedt and his colleagues was noted. Having studied Möllenstedt's interference experiments, A. Tonomura and his colleagues have extended the electron holography system to clearly prove the physical reality of vector potentials, the so-called Aharonov-Bohm effect. They also succeeded in observing the dynamic motions of magnetic flux quanta (fluxons) in a superconducting Nb film. In a joint research with A. Tonomura, we succeeded in visualizing a fluxon pinned by an insulating particle in a high-Tc Y-Ba-Cu-O superconductor by combining electron holography and scanning ion microscopy. As the study of a scalar potential, the visualization of the orbits of electron-induced secondary electrons around positively charged biological specimens was noted. Finally, although the electromagnetic field analysis using electron holography on the basis of Maxwell's equations seems to be promising, it is pointed out that there have been some controversies on the interpretation and treatment of electromagnetic field.

1. Introduction

The author attended the "Tonomura FIRST International Symposium on Electron Microscopy and Gauge Fields" held in Tokyo, May 9-10, 2012, and delivered a talk entitled "Basic Discoveries with the Electron Microscope." In the talk, basic discoveries related to field visualization were discussed while referring to the late Dr. Tonomura's significant achievements in this field. In this article, on the basis of this talk, the basic discoveries in the electromagnetic field visualization are presented along with the author's joint research work with A. Tonomura. Although there have been many achievements in electromagnetic field visualization, only some basic discoveries especially relating to the work of A. Tonomura have been discussed in this article. Other achievements may be found in the review articles on the analysis of electric[1] and magnetic[2] fields provided as references.

2. Importance of field and its visualization

First, it is important to note the field concept and its visualization. In this context, an excerpt from the book[3] written by Einstein and Infeld would be pertinent.

> A new concept appears in physics, the most important invention since Newton's time : the field. It needed great scientific imagination to realize that it is not the charges nor particles but the field in the space between the charges and particles which is essential for the description of physical phenomena. The field concept proves most successful and....
>
> The theory of relativity arises from the field problems....

In general, it is known that the physical phenomena basically result from the four fundamental forces in nature, i.e., gravitational, strong, weak and electromagnetic forces. Among these, we are most familiar with electromagnetic forces resulting from the electromagnetic field. Thus, its visualization using electron holography is of significant importance in understanding and clarifying the mechanisms of various physical phenomena.

3. Discoveries of electron biprism interference fringes

For actualization of the electromagnetic field visualization, the discovery of the electron biprism interferences by G. Möllenstedt and his colleagues in Tübingen University[4] should be first noted. Figure 1 shows an illustration of the principles of electron interference, depicting the so-called hologram obtained by the Möllenstedt-type biprism consisting of a fine filament bridged in the center and two plate-shaped electrodes both being at ground potential. Under the kind encouragement of D. Gabor, the system was steadily improved upon. The presence of distinct interference fringes verifying de Broglie's relation $\lambda = h/(mv)$ was communicated to de Broglie

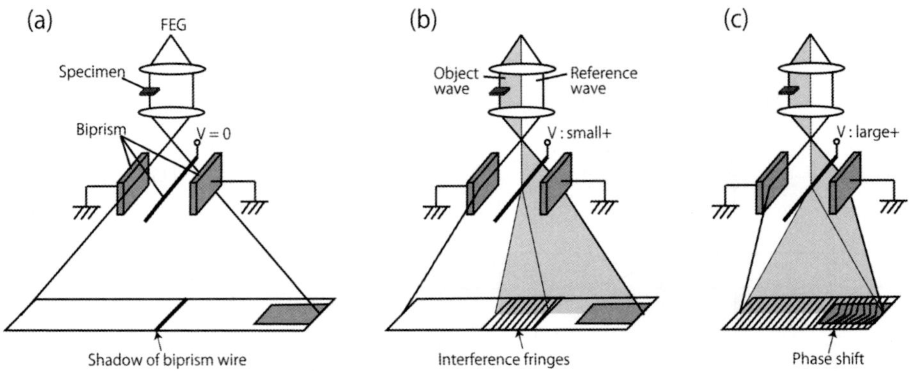

Fig. 1.   Principles of electron interference with Möllenstedt-type biprism. Voltage of a central filament is (a) zero, (b) small plus, and (c) large plus.

who was happy with the result. Having visited Tübingen University where he learned the electron interferometry,[5] A. Tonomura worked on the accuracy and resolution of electron holography and succeeded in observing magnetic fields in various materials.

4. Proof of physical reality of vector potential

One of the greatest achievements of A. Tonomura is the clear proof of the existence of vector potential through electron interference experiments. The characteristic feature of the vector potential is explained and compared with scalar (electric) potential in Fig. 2. If the specimen is homogeneous and has the same thickness, the phase shifts of the incident electron between positions C and D ($\Delta\phi(r)$) given by

$$\Delta\phi(r) = \sigma \int_0^t \varphi(r_C) - \varphi(r_D)dz$$

$$\sigma = \frac{2\pi}{\lambda V(1 + \sqrt{1 - \beta^2})}, (\beta = v/c) \tag{1}$$

are equal. Here, V is the accelerating voltage. On the other hand, if the specimen is a magnetic material with saturation magnetization $\boldsymbol{B}$, the phase shift obtained from the vector potential $\boldsymbol{A}$ is given as

$$\Delta\phi(r) = \phi(r_C) - \phi(r_D)$$
$$= \frac{e}{\hbar} \oint_{ABDC} \boldsymbol{A}(r)d\boldsymbol{s}$$
$$= \frac{e}{\hbar} \iint B_n dS = \frac{e}{\hbar} \Phi \tag{2}$$

Being different from the scalar potential, the phase shifts between positions C and D in this case are different for the same specimen thickness (Fig. 2(b)). Furthermore, even if the electron beams do not pass through the specimen or its magnetic field $\boldsymbol{B}$, as shown in Fig. 2(c), there is a difference in the phase shifts between positions C and D because $\boldsymbol{B}$ exists between the two beam paths A-C and B-D. In other words,

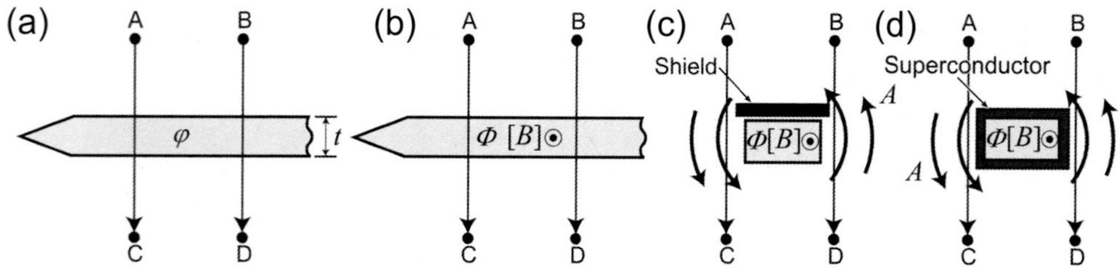

Fig. 2.   Illustration of phase shift due to (a) scalar and (b-d) vector potentials. See details in the text.

Fig. 3.   Photo of the late Dr. A. Tonomura (front left), J. J. Kim (back left), K. Matsuyama who was Dr. Tonomura's secretary (back right) and the author (front right) taken in Sendai on June 15, 2007.

this difference results from the existence of the vector potential $\boldsymbol{A}$ shown in Fig. 2(c). This is the so-called Aharonov-Bohm effect.[6] A. Tonomura and his colleagues have extended the electron holography system using a toroidal ferromagnet covered with a superconductor (see Fig. 2(d)) with which they were able to clearly prove the physical reality of vector potentials.[7] They also succeeded in observing the dynamic motions of magnetic flux quanta (fluxons) in a superconducting Nb film: the details are available in A. Tonomura's book.[8]

5. Joint research on magnetic field analysis

Our group has been conducting joint research with Tonomura's team since the summer in 2007, when J.J. Kim was awarded the doctorate degree under my supervision and moved to Tonomura's group as a post-doctoral researcher (see Fig. 3). In the beginning of the joint project, we visualized magnetic flux distribution around a magnetic recording head in a magnetic field (Fig. 4).[9]

In collaboration with A. Tonomura and his colleagues, we also succeeded in observing magnetic flux distribution around a high-Tc Y-Ba-Cu-O superconductor at low temperatures. Magnetic flux distribution under an external magnetic field whose direction is indicated in Fig. 5(a) shows the Meissner effect where the magnetic flux detours around the superconductor (as seen in the top edge region (T) and the insulating particle regions (I) in the side surfaces). On the other hand, when the external magnetic field becomes zero, the fluxon pinned by an insulating particle in a high-Tc Y-Ba-Cu-O superconductor can be visualized by combining electron holography and scanning ion microscopy (Fig. 5(b)).[10] It should be noted that the visualization of the pinning behavior of fluxons is important in improving the critical current density of the superconductors. (Details are given in Ref. 10).

176   *D. Shindo*

Although his health condition deteriorated at the beginning of this year (2012), A. Tonomura devoted himself to complete the manuscript of his work on the newly developed "split-illumination electron holography."[11] In the split illumination method, in which the additional biprisms are inserted into the condenser lens system (Fig. 6(a)), the reference wave is obtained from a region far from the specimen, being free from various stray fields around the specimen. As indicated in Fig. 6(b), the magnetic flux distribution around a submicron hole located at a distance of 9 μm from the specimen edge is clearly visualized, and thus this method is expected to widen the application of electron holography extensively. A. Tonomura was so delighted with this technology that he prepared the experimental data very carefully, revised the text so intensely, and finally completed the manuscript. In his e-mail on the completeness of the manuscript, he stated "I may not be able to do such a thing forever..." He passed away without seeing the paper published; the figures of the paper found place on the cover of Applied Physics Letters and the paper was included in "Editor's Picks" of the special edition (50th Anniversary) of Applied Physics Letters.

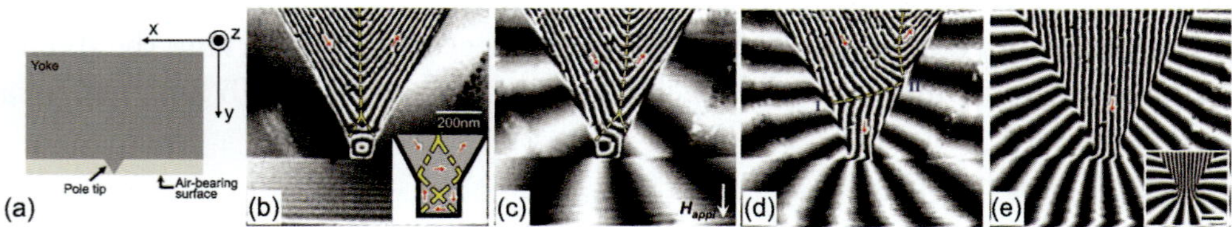

Fig. 4.   (a) Illustration of writer pole. (b)-(e) Reconstructed phase image exhibiting magnetization process of Ni-Fe pole tip. (b) Remanent state. Direction of external magnetic field is indicated in (c), and its magnitude is (c) 3.2 kA/m, (d) 5.6 kA/m, and (e) 12.0 kA/m. Dotted lines indicate domain walls. The inset shown in (e) is a simulation. Ref. 9.

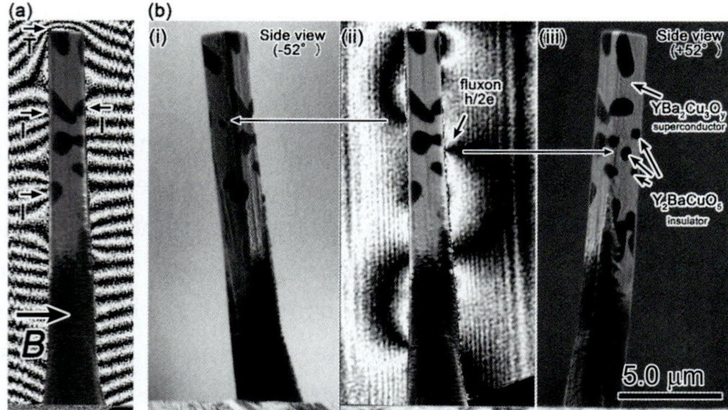

Fig. 5.   (a) Magnetic flux distribution around a high-Tc Y-Ba-Cu-O superconductor under external magnetic field 8.0 kA/m at 12 K. (b-ii) Magnetic flux distribution without external magnetic field at 13 K. (b-i) and (b-iii) are scanning ion microsope images showing left and right side views of the specimen. Ref. 10.

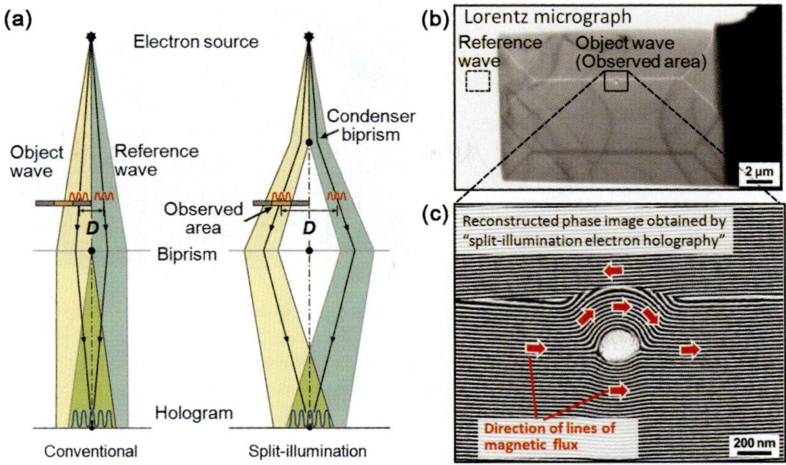

Fig. 6. (a) Schematic diagrams of electron-optical methods. Conventional (left) and split-illumination (right) electron holography. (b) Lorentz micrograph of electrical steel sheet. (c) Reconstructed phase image showing the magnetic flux distribution around the hole in a rectangular region in (b). Ref. 11.

6. Analysis of scalar potential for detecting electron motion

As studies of the scalar potential, trials of visualizing the motion of electrons through the electric field variations by electron holography are noted. In the experiment of field emission with a single $TaSi_2$ nanowire shown in Figs. 7(a) and (b), the electric potential fluctuation in the nanowire due to the ballistic emission was found (note the comparison of the observed electric field (c) and its simulation (d)).[12] However, the current of emitted electron current was about 1nA, which was too small for directly detecting the motion of the electrons. On the other hand, in the study of the charging effect in biological specimens such as a sciatic nerve tissue, we found a characteristic feature, i.e., there exist the electric potential fluctuations that correspond to the orbits of electron-induced secondary electrons around positively charged microfibrils being situated around the sciatic nerve tissue. The change in the size of the orbit was clarified by inserting the probe near the specimen and applying the voltage (see Fig. 8).[13] On the basis of Maxwell's equations, a computer simulation of the orbits of secondary electrons confirmed the results of the electron holography study when the specimen drift was small.[14] We also indicated that the location of electron orbits by electron holography is the so-called disturbance-free observation.[13] The author presented the holography data of the electron orbital locations to several distinguished researchers. Among them, A. Tomonura showed the most noteworthy reaction. When I displayed the reconstructed amplitude image showing the orbit of the secondary electrons around microfibrils in the screen, he came to the screen and

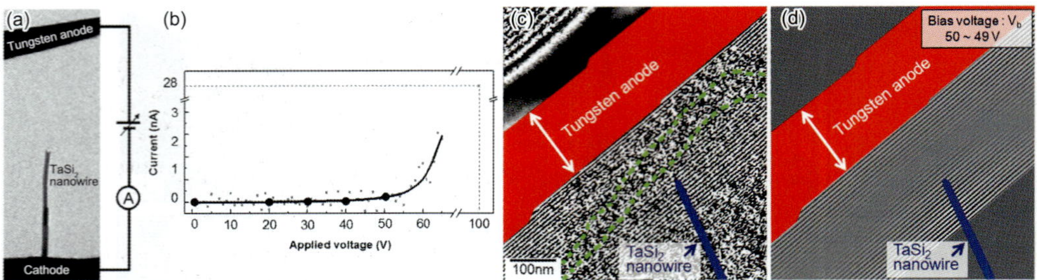

Fig. 7.   (a) Transmission electron microscope image showing geometrical configuration of nanowire and W anode. (b) Electric current observed as a function of applied voltage. (c) Reconstructed phase image showing an irregular contrast region with broken lines. (d) Simulated phase image obtained by taking into account the change of electric potential in the nanowire during field-emission process. Ref. 12.

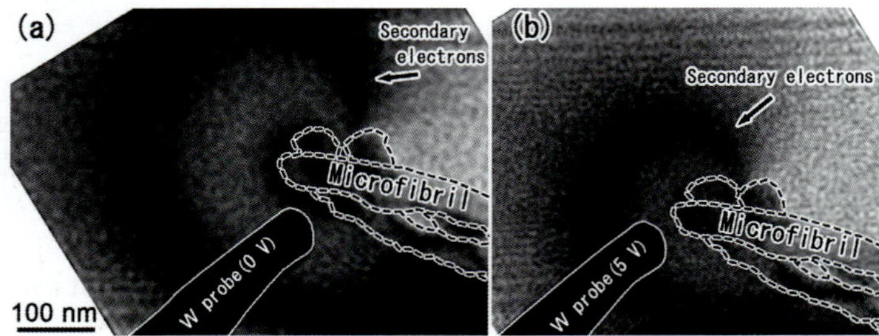

Fig. 8.   Reconstructed amplitude images showing the orbit of secondary electrons around a positively charged microfibril. Voltages of inserted W probe are (a) zero and (b) 5 V. Ref. 13.

observed the details, saying, "Location of the orbit of electrons will turn out to have a significant importance...."

7. Concluding remarks

Thus far, hardware and software for electron holography have been well established, and thus, both electric and magnetic fields can be analyzed quantitatively on the basis of Maxwell's equations. Therefore, the wide application of electron holography seems to be imminent. However, it should be noted here that there have been some controversies on the interpretation and treatment of the electromagnetic field. For example, two important remarks can be cited. The first remark is that of A. Einstein when referring to experiences with the theory of gravitation:[15]

> One may, for example, call on Maxwell's equations of empty space by way of comparison. These are formulations which coincide with the experiences of infinitely weak electro-magnetic fields. This empirical origin already determines their linear form; it has, however, already been emphasized above that the true laws cannot be linear. Such linear laws fulfill the super-position-principle for their solutions, but contain no assertions concerning the interaction of elemental bodies.

The second remark made by P.A.M. Dirac on "Quantum Electrodynamics"[16] is

> Hence most physicists are very satisfied with the situation. They say: "Quantum electrodynamics is a good theory, and we do not have to worry about it any more." I must say that I am very dissatisfied with the situation, because this so-called "good theory" does involve neglecting infinities which appear in its equations, neglecting them in an arbitrary way.

Finally, it should be noted that by taking into account these controversies carefully, we have to consider the field further on the basis of data obtained from electron holography.

References

1. D. Shindo and Y. Murakami, Electron holography study of electric field variations, *J. Electron Microscopy.* **60(Supplement 1)**, S225–S237, (2011).
2. D. Shindo and Y. Murakami, Electron holography of magnetic materials, *J. Phys. D: Appl. Phys.* **41**, 183002(1)–183002(21), (2008).
3. A. Einstein and L. Infeld, *The Evolution of Physics*, p. 244. Cambridge University Press, Cambridge, 2nd edition, (1978).
4. G. Möllenstedt, *THE HISTORY OF THE ELECTRON BIPRISM*, In *Introduction to Electron Holography*, pp. 1–15. Kluwer Academic, New York, (1999).
5. A. Tonomura, Obituary: Prof Dr Gottfried MÖLLENSTEDT, *J. Electron Microscopy.* **47**(5), 363–364, (1998).
6. Y. Aharonov and D. Bohm, Significance of electromagnetic potentials in the quantum theory, *Phys. Rev.* **115**(3), 485–491, (1959).
7. A. Tonomura, N. Osakabe, T. Matsuda, T. Kawasaki, J. Endo, S. Yano, and H. Yamada, Evidence for Aharanov-Bohm effect with magnetic field completely shielded from electron wave, *Phys. Rev. Lett.* **56**(8), 792–795, (1986).
8. A. Tonomura, *Electron Holography*. (Springer-Verlag, Berlin, 1999), 2nd edition.
9. J. J. Kim, K. Hirata, Y. Ishida, D. Shindo, M. Takahashi, and A. Tonomura, Magnetic domain observation in writer pole tip for perpendicular recording head by electron holography, *Appl. Phys. Lett.* **92**(16), 162501(1)–162501(3), (2008).
10. Z. Akase, H. Kasai, S. Mamishin, D. Shindo, M. Morita, and A. Tonomura, Imaging of magnetic flux distribution in vicinity of insulating particles in high-Tc superconductor by electron holography, *J. Appl. Phys.* **111**(3), 033912(1)–033912(5), (2012).
11. T. Tanigaki, Y. Inada, S. Aizawa, T. Suzuki, H. S. Park, T. Matsuda, A. Taniyama, D. Shindo, and A. Tonomura, Split-illumination electron holography, *Appl. Phys. Lett.* **101**(4), 043101(1)–043101(4), (2012).
12. J. J. Kim, D. Shindo, Y. Murakami, W. Xia, L. J. Chou, and Y. L. Chueh, Direct observation of field emission in a single $TaSi_2$ nanowire, *Nano Lett.* **7**(8), 2243–2247, (2007).
13. D. Shindo, J. J. Kim, K. H. Kim, W. Xia, N. Ohno, Y. Fujii, N. Terada, and S. Ohno, Determination of orbital location of electron-induced secondary electrons by electric field visualization, *J. Phys. Soc. Jpn.* **78**, 104802(1)–104802(8), (2009).
14. M. Inoue, S. Suzuki, Z. Akase, and D. Shindo, Computer simulation of electric field variations due to movements of electric charges, *J. Electron Microscopy.* **61**(4), 217–222, (2012).
15. A. Einstein, *Einstein's Autobiography*, In *Albert Einstein : Philosopher-Scientist*, p. 89. Cambridge University Press, London, 3rd edition, (1969).
16. P. A. M. Dirac, *Directions in Physics*, p. 36. A Wiley-Interscience Publication, USA, (1978).

Nanomagnetism Visualized by Electron Holography

Hyun Soon Park

RIKEN Center for Emergent Matter Science,
Hirosawa, Wako, Saitama 351-0198, Japan
E-mail: hspark@riken.jp

Electron holography, with its high phase and spatial resolutions, is a powerful tool enabling direct visualization of the phase shift of electron waves due to electromagnetic fields inside and/or outside emergent matters. Observing and characterizing the magnetization distributions on a nanometer scale are of vital importance for understanding nanomagnetism and its application to spintronics. For the studies of their magnetic properties and behavior, nanoscale imaging of magnetic field is indispensable. The applications discussed here, using electron holography and Lorentz microscopy, are selected to highlight the visualization of magnetization distribution that has not yet been investigated on the nanometer scale due to device complexity and/or the insufficient spatial resolution of the microscopes available. These studies demonstrate the promise of electron holography not only for real-space, quantitative magnetic-field imaging but also for detecting the ensemble of spins in emergent matter systems.

Commemoration of Dr. Akira Tonomura

I (Hyun Soon Park) would like to express my sincere condolence on the passing of Dr. Akira Tonomura, a pioneer in the fields of electron microscopy and fundamental quantum physics experiments, who entered into eternal rest on the dawn of May 2, 2012. He had been not only the director of Single Quantum Dynamics Research Group but also the team leader of Quantum Phenomena Observation Technology Team at the Advanced Science Institute of RIKEN since 2001. My first impression of him was his lecture at Tohoku University in 2004, still unforgettable, where he presented several beautiful experimental results observed by electron holography and introduced 1-MV holography electron microscope.[1-6] At that time (my doctoral course period), I had explored several magnetic materials *in situ* using electron holography and Lorentz microscopy under the guidance of Prof. Daisuke Shindo at Tohoku University. In his lecture, he had encouraged us, young students and researchers, to fall deeply into to the interesting fields of electron holography, quantum mechanics and magnetism. Six years later in March

2011, I was lucky to have an opportunity to work with him for the FIRST Tonomura Project,[7] after I spent five years from January 2006 to February 2011 as a postdoc at California Institute of Technology working on development of 4D electron microscopy and its applications to materials under the guidance of Prof. Ahmed H. Zewail, 1999 Nobel Laureate in Chemistry.

The aim of the FIRST Tonomura project is to develop a "holography electron microscope" capable of observing quantum phenomena in the microscopic world that he had continued exploring. A dream microscope turns out to be the last masterpiece of him, which is supposed to have the world's highest spatial and phase resolutions. Now, he left a powerful electron microscope for enthusiastic young researchers not only in Japan but also in the world. I, electron microscopist and materials scientist, will continuously recall his great enthusiasm, challenging and coherent thinking towards science that he showed us despite the course of medical treatment on pancreatic cancer. Described here are the results on the visualization of nanomagnetism which came to fruition through his encouragement, although those do not hold a candle to his beautiful experiments.

1. Introduction

Understanding magnetism on a nanometer scale, mainly associated with magnetic moments (or spin), plays an essential role in both fundamentals in multiferroics as quantum electromagnets and a wide range of applications to spin-electronic devices.[8-10] Visualization of magnetism, with appropriate spatial (or temporal) and phase resolutions, is of vital importance and critical to our understandings of the behavior of electron spin in emergent matters; that is, seeing and observing magnetism is understanding. Thus far, a number of microscopy techniques have been developed for observing the magnetization state of magnetic materials.[11-15]

Among the magnetic imaging techniques, electron holography using coherent electron waves provides unique capabilities for detecting and visualizing the phase shift of electron waves due to the electromagnetic fields. Electron holography has thus paved a new way for visualizing and measuring nanoscopic objects and electromagnetic fields that were previously inaccessible employing other techniques. The fields of science explored by holography are quantum mechanics, superconductivity, magnetism with many beautiful experiments of the late Dr. Akira Tonomura and his colleagues.[1-6] With advancement of nanoscience and nanotechnology, emergently needed are higher spatial and phase resolutions of electron holography since device structures suffer from more complicated fabrication scaled down to a nanometer (nm) scale and the spin orders in interface regions should be examined with high precision. Here, I highlight recent application results[16-18] of electron holography and Lorentz microscopy in fundamental materials and physical sciences as well as industrial applications. Using a holography electron microscope with relativistic electron wavelength of 2.0 pm at 300 kV, we have demonstrated that the magnetization distribution in real device structures and the topological spin textures, such as skyrmion with a lattice parameter of 18 nm, can be investigated on a nanometer scale.

2. Experimental procedure

A holography electron microscope (HF-3300 series, Hitachi, Ltd.) at an acceleration voltage of 300 kV was utilized for all experiments. This microscope has two positions in which the specimen can be placed, i.e., the field-free position and the objective lens position. Electron holograms (under no applied field) were observed in the field-free position where the residual magnetic field was approximately 0.01 mT. In the field-free position, we are able to apply a magnetic field up to 50 mT to the specimen in three dimensions using a direction-free magnetic field application system.[19] A liquid helium cooling holder was used to investigate the temperature dependence of the magnetic structure. It enabled the specimen temperature to be reduced to ~5 K. An objective lens current from 0 to 12 A was used to create a magnetic field perpendicular to the thin sample, by which magnetic phase transitions or magnetization process can be triggered.

Electron holography is an imaging technique that records the electron interference pattern of an object using electron biprisms and then reconstructs an image by optical or numerical methods. The holograms are formed when coherent electrons illuminating the thin specimen (object wave) were interfered with those passing through the vacuum (reference wave) using the biprisms as illustrated in Fig. 1. For recording the holograms, a slow-scan charge-coupled-device camera (UltraScan[®]4000, Gatan Inc.) was utilized. Using the double-biprism electron interferometry, we controlled the fringe spacing and the width of interference region independently.[20] The first biprism was installed in the image plane of the objective lens and the second biprism was set behind the first projector (magnifying) lens, inside the shadow area of the first biprism. To increase the phase resolution, we averaged phase images reconstructed from a few tens of holograms consecutively taken under the same conditions.

To visualize the electron phase shifts due to electromagnetic field, we show an example in Figures 1a-1c, *i.e.*, Lorentz image, electron hologram, and its reconstructed phase image of a sintered $Nd_2Fe_{14}B$ permanent magnet. In the Lorentz image in Fig. 1a, magnetic domain walls (DW) appear as white and black lines indicated by the arrowheads. In the hologram corresponding to the rectangular region in (a), the electron interference fringes appeared to curve from place to place due to strong magnetic field of this material. In the reconstructed phase image of the hologram, the lines of magnetic flux changed the direction at the domain walls inside the specimen and also the magnetic flux leakage was visualized outside the specimen: this is the case for the strong phase objects.

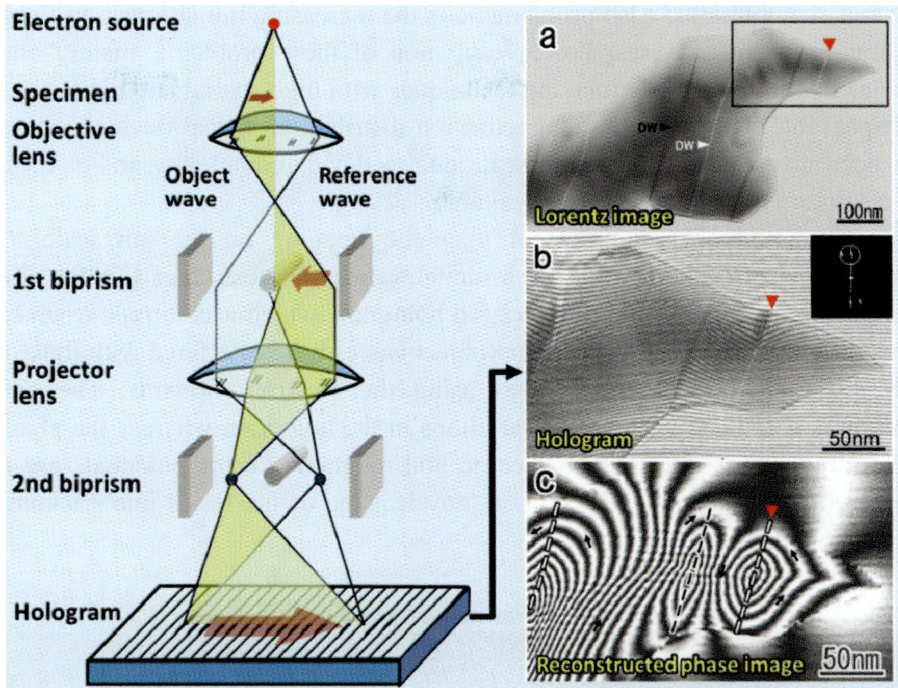

Fig. 1. Schematic illustration of double-biprism interferometry and the information obtained by electron holography. a) Lorentz image of a sintered $Nd_2Fe_{14}B$ permanent magnet. b) Electron hologram of the rectangular region in (a). Inset in (b) indicates a digital diffractogram obtained from the electron hologram through the Fourier transformation. c) Reconstructed phase image. The inverse Fourier transformation was carried out after selecting the scattering amplitude of the circled region in the digital diffractogram [inset in (b)]. Black arrows in (c) indicate the direction of lines of magnetic flux.

3. Applications of electron holography and Lorentz microscopy

3.1. *Nanoscale magnetic characterization of real device structures*

Ultrahigh density recording in hard disk drives (HDDs) is a production challenge for future advanced storage technology. With the discovery of giant magnetoresistance (GMR),[8] read heads and magnetic sensors for HDDs have been developed, not only in design but also in more complicated fabrication, which is scaled down to nanometers (nm). This has led to an increase in areal recording density by three orders of magnitude (from 0.1 to 100 Gbit/in^2) and the emergence of spin-electronic devices such as magnetic random access memory.[21] However, there are many factors that are still unclear and that have a negative effect on the MR effect and spin electronics.[22,23] They include interfacial and barrier spin scattering, magnetic flux leakage, changes in the output voltage, and smaller signal amplitudes. Also problematic is the increasing miniaturization from μm to nm in spin-electronic devices. For investigation of these problems, mainly associated with magnetization, magnetic imaging techniques with high spatial and phase resolutions are indispensable. However, the magnetization distribution in real device structures has not yet been investigated on the nm scale due to device complexity and/or insufficient spatial resolution of the microscopes available.

Here is one significant example[16] of magnetic imaging on the nm scale, *i.e.*, the application of electron holography to a tunneling magnetoresistance (TMR) spin valve head. Shown in the upper panel of Fig. 2 is a hologram, which was formed when coherent electrons illuminating the thin specimen (object wave) were interfered with those passing through the vacuum (reference wave) using the electron biprisms. The amplitude information is contained in intensity variations in the hologram whereas the phase shifts of the electron waves due to both electric and magnetic vector potentials are directly detected from the phase images. Without any leakage of the fields into vacuum, phase change $\Delta\varphi$ is expressed as follows:

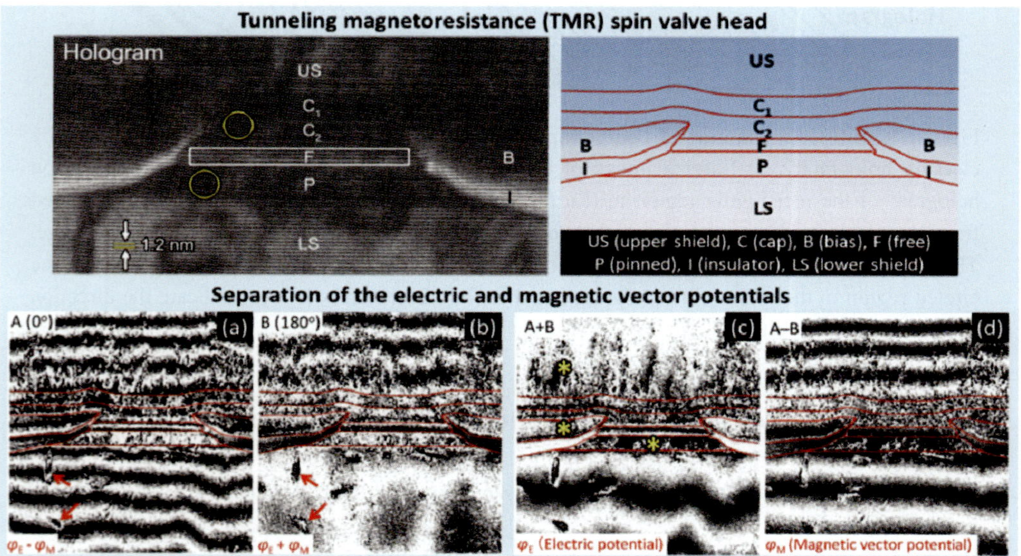

Fig. 2. Hologram of TMR spin valve head, which was cut from the original hologram (4096 × 4096 pixels). The labels in the image represent each layers depending on their functional properties, for example, US (upper shield), C_1 and C_2 (cap), F (free), P (pinned), LS (lower shield), B (bias), and I (insulator). The fringe spacing of 1.2 nm is indicated. (a) and (b) were obtained before and after flipping the specimen 180°, respectively. (c) Equipotential contour map derived by summation (A+B). (d) Magnetic flux map obtained by subtraction (A−B). The phase amplification for Figs. 2a - 2d is 2. The arrows indicate nanocrystallines, satisfying the Bragg condition along the incident electron beam. The yellow asterisks in Fig. 2c indicate the positions for thickness measurement.

$$\Delta\phi = C_E V_0(x,y)t(x,y) - \frac{2\pi e}{h}\iint B(x,y)dxdz,$$

where C_E is an interaction constant (0.00652 radV^{-1}nm^{-1} for 300 kV electrons), $V_0(x, y)$ is the projected mean inner potential, $t(x, y)$ is the projected thickness, h is the Planck's constant, e is the elementary electric charge, and $B(x, y)$ is the in-plane component of the magnetic flux density integrated through t.

In the hologram in Fig. 2, the complex hetero-nanostructures of the spin valve head were observed, which consist of various materials with magnetic or nonmagnetic properties. Depending on their functional properties, they are designated as cap (C_1, C_2), insulator (I), upper shield (US), lower shield (LS), bias (B), free (F), and pinned (P) layers, so there are many interfaces in this specimen (see also the schematic illustration in Fig. 2). For precisely characterizing the magnetic properties in each layer with electron holography, three factors should be considered: i.e., thickness, mean inner potential of materials, and dynamical diffraction effect. Our methodology of choice was therefore flipping the specimen from $0°$ to $180°$ and recording two holograms where only the sign of the phase shift due to the magnetic vector potential was reversed.

Figures 2a and 2b depict the reconstructed phase images of two holograms taken at $0°$ and $180°$, in which the phase shift $\varphi(x, y)$ is represented by $\cos\varphi(x, y)$. Care was taken to ensure the same diffraction contrast between the two holograms, as indicated by the red arrows, and to align them. The phase shifts due to the electric and magnetic vector potentials are hereafter referred to φ_E and φ_M, respectively. In Figs. 2a and 2b, the phase shifts were affected by the superimposed electric and magnetic vector potentials, i.e., $\varphi_E - \varphi_M$ and $\varphi_E + \varphi_M$. Note that in Figs. 2a and 2b, the LS layer does not show a similar density of black/white contour lines despite the identical field of view. This was due to the difference in the sign of the slope between φ_E and φ_M, which appeared to cancel each other out in the LS layer of Fig. 2b.

By simple summation and subtraction of those two phase images, φ_E and φ_M can be separated, as shown in Figs. 2c and 2d. Figure 2c displays the equipotential contour map that represents the projected thickness and the mean inner potential of materials, whereas Fig. 2d reveals the in-plane component of the magnetic flux in this viewing field. In Fig. 2d, the phase shifts of the US, B, and F layers at some regions were measured to be 0.162 rad/nm, 0.125 rad/nm, and 0.105 rad/nm, respectively. Given measured thicknesses, the magnetic flux densities of the US, B, and F layers were estimated to be 1.0 T, 0.9 T, and 1.0 T, respectively. This is in good agreement with the saturation magnetization of the materials in bulk, *i.e.*, NiFe (1.0 T) and CoPt (0.8 T). Considering the uncertainty of about 10% due to the mean inner potential and/or dynamical diffraction effect, it is reasonable that the magnetic flux density measured here should have an error of ± 0.1 T.

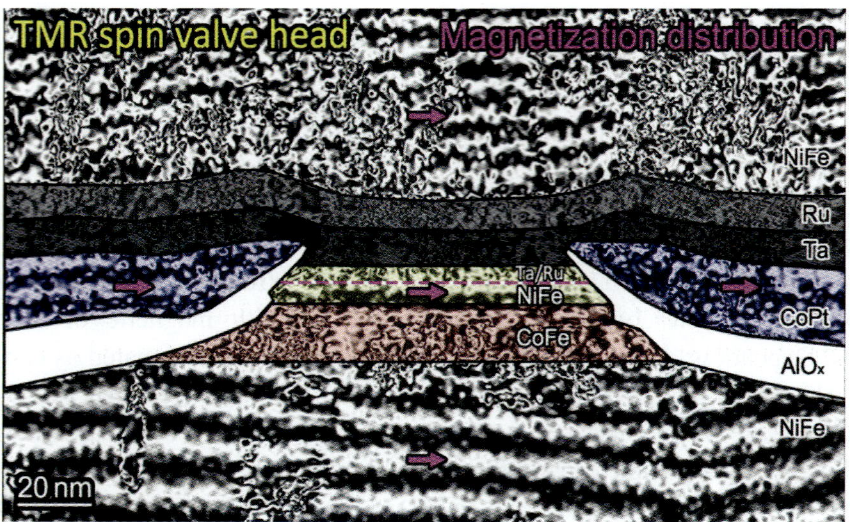

Fig. 3. Magnetization distribution of TMR spin valve head. The reconstructed phase images were amplified by 6. Noisy and relatively thinner lines in the image are attributed to diffraction contrast and phase amplification. Arrows represent the directions of the in-plane component of the magnetization.

Directly visualized in Fig. 3 is the magnetization distribution in the remanent state by separating the electric potential and averaging phase images reconstructed from ten holograms consecutively taken under the same conditions, which provides improved phase resolution. In the magnetic flux map, the direction and the density of black (or white) contour lines correspond to those of the in-plane component of the magnetic flux density projected along the electron beam. The thick contour lines were clearly observed in both the NiFe (not colored) and CoPt layers (colored in blue). Furthermore the contour line was visible inside the 5 nm NiFe (free layer) below the dotted line in yellow colored region. In the CoFe layer (colored in pink), the in-plane component of the magnetic flux seems to be zero (or extremely small), meaning that its magnetization direction is normal to the film plane. Further electron holography studies on the magnetization process are expected with high voltage electron microscope (for example, 1,000 or 1,200 kV) in the near future, although it is very challenging because of the needs for phase and spatial resolutions better than that reported here. The methodology reported here is very promising for improving TMR head design and solving problems caused by complexity and miniaturization in spin-electronic devices.

3.2. *Real-space observation of skyrmion lattice in a helimagnet MnSi thin sample*

Chiral spin textures with different length scales have been attracting increasing interest for the study of quantum magneto-transport and possible application to magnetic data storage and spin-electronic devices.[8-10,21,24] The topological spin texture, a "skyrmion", in MnSi magnets is particularly attractive for carrying information in devices because spin transfer torque emerges at low current densities ($\sim 10^6$ A/m^2).[25] Recent experimental[26,27] and theoretical studies[28-30] have demonstrated that skyrmion crystals appear *via* the magnetic phase transition from a helical spin structure as a function of the applied field or ambient pressure combined with temperature. The length scale (period) for the helical structure, depending on both ferromagnetic exchange interaction J_{ex} and Dzyaloshinsky-Moriya interaction D, generally ranges from several to tens of nanometers.

Skyrmions in helimagnets are made up of downward core spins and upward peripheral spins swirling up with a unique spin chirality that is determined by the underlying chiral crystal structure. A skyrmion carrying a topological quantum number acts as an effective magnetic flux. As the charge carriers flow over the skyrmion crystal, they are deflected by the emergent electromagnetic force (Lorentz force analog) induced by such magnetic fluxes, resulting in the topological Hall effect.[27,29] High-density skyrmions or short-period skyrmion crystals enhance the Hall effect by means of the fictitious field (real-space Berry phase). Since the gigantic topological Hall effect should have a major scientific effect on emerging spintronics, it is desirable to achieve a short-period skyrmion crystal targeted at spintronic applications. Recent *in situ* Lorentz microscopy studies have revealed the magnetic configuration of skyrmion lattices and the magnetic phase diagrams for FeGe and Fe$_{0.5}$Co$_{0.5}$Si thin samples, in which the skyrmion lattice constant is 70 nm for FeGe and 90 nm for Fe$_{0.5}$Co$_{0.5}$Si.[31, 32] This is particularly noteworthy in the fields of basic and applied nanomagnetism given the topologically non-trivial and stable spin configuration on a nm-length scale.

Lorentz micrographs in real space (Fig. 4) show the helical structure, skyrmion lattice and spin distributions of skyrmion. Stripe magnetic domains (helical structure), which were imaged as bright and/or dark lines, were clearly visible at zero applied magnetic field. The magnetization with a helical spin spiral along <111> was reversed 180° in alternate domains. The period was estimated to be 18 nm, as shown in helical structure. When a magnetic field of 0.12 T was applied normal to the thin-film (image) plane, the stripe characteristic of the domains was suppressed.

Particularly interesting is the appearance of a peculiar domain pattern with six-fold symmetry (see the skyrmion lattice in Fig. 4), like an atomic lattice, when the field reached 0.18 T. The lattice constant agreed well with the measured period of the stripe domain, as shown in the helical structure. The reconstructed images in the insets in the upper panels present the characteristic of each magnetic structure more clearly. The

     H. S. Park

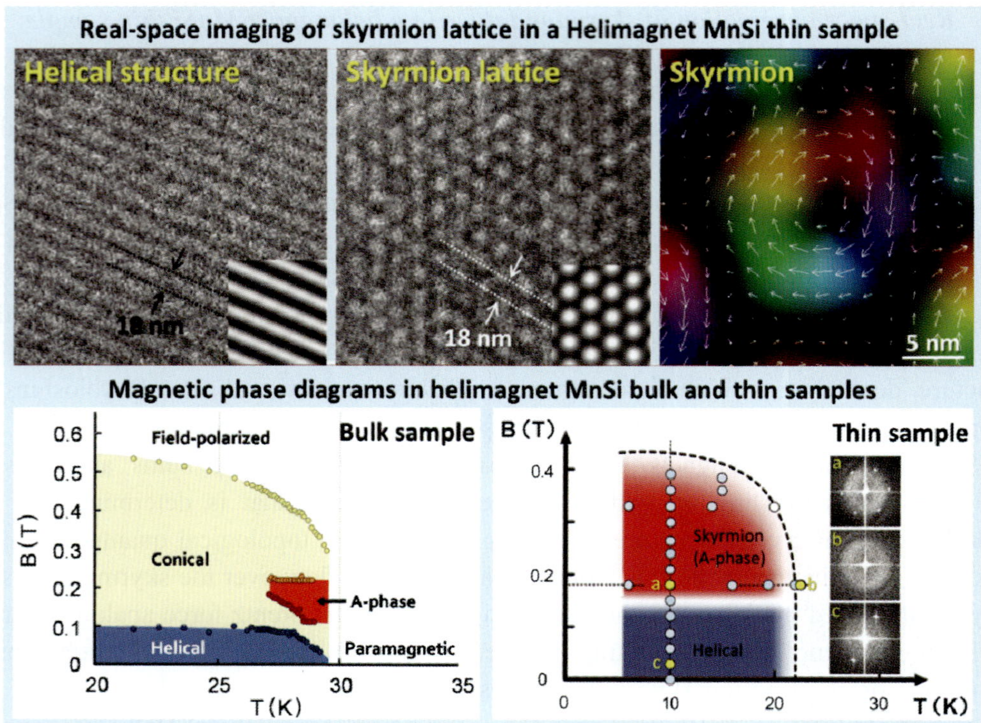

Fig. 4. **Upper:** Lorentz micrographs showing helical structure and skyrmion lattice. **Lower left:** Magnetic phase diagram of helimagnet MnSi bulk in temperature-magnetic field plane, reproduced using reported data points.[26] **Lower right:** Magnetic phase diagram for MnSi thin sample produced by referring to patterns in diffractograms. Diffractograms a, b, and c represent skyrmion, disordered, and helical phases, respectively. Circles indicate measured points.

topological spin textures are evident in the lateral magnetization distribution map indicated as "Skyrmion". The white arrows indicate the size and direction of the magnetic component at each point. The areas colored in black (the cores and outsides of the particles) represent the zero in-plane component of the magnetization.

To obtain a deeper understanding of the magnetic phase transition, we carried out a series of experiments focusing on the applied field and temperature. The lower right panel in Fig. 4 shows the magnetic phase diagram we obtained for MnSi thin sample.[18] The circles indicate the measured points, and the color map represents the magnetic phases, *i.e.*, the helical and skyrmion-lattice structures, which were respectively derived from the existence of the two-fold and six-fold symmetries in the diffractograms. Diffractograms representing the skyrmion (six-fold), disordered (broad halo ring), and helical (two-fold) phases are displayed at the three points indicated by yellow circles. This phase diagram differs from that for bulk MnSi (lower left panel in Fig. 4), *e.g.*, the suppression of the conical phase, the field dependence of the magnetic structures, and the slight decrease in T_c. We ascribe these differences to the nanoscale confinement, *i.e.*, the dimension difference between thin sample and bulk: the perpendicular magnetic anisotropy arises in thin sample more effectively due to surface effects and/or lattice strain. This elucidates not only the development of a skyrmion structure even below 10 K (suppressing the conical phase) but also the field dependence of the magnetic structures. These observations unambiguously demonstrate that a skyrmion lattice can be stabilized even at the lowest temperature in thin sample, in contrast to the case for bulk. The method reported here, *in situ* observation in real and Fourier spaces, provides a more thorough understanding of magnetic phase transition on the nanometer scale, which may suggest new designs for spin-electronic devices using a skyrmion. The detailed spin configurations in the vicinity of a skyrmion lattice using electron holography are in progress and reported elsewhere.

4. Concluding remarks

We have shown here that, with electron holography and Lorentz microscopy, magnetization distributions in a TMR spin valve head and a helimagnet MnSi thin sample can be successfully studied with high spatial and phase resolutions. By averaging and separating the electric and magnetic potentials, we quantitatively measured the local variations in thickness and magnetic flux densities of the nanostructured layers. *In situ* observations of magnetic microstructures with applied magnetic fields and/or temperature variations directly provide deeper understandings of magnetism in matters. The electron microscopy techniques reported here are very promising not only for solving problems caused by complexity and miniaturization in future spin-electronic devices, but also for investigating the magnetic orders in the vicinity of an interface. Lastly, I anticipate further exploration into the emergent quantum world using a dream microscope of the late Dr. Akira Tonomura, *i.e.*, an atomic-resolution holography electron microscope with an accelerating voltage of 1,200 kV that is being developed by his colleagues at the Central Research Laboratory, Hitachi, Ltd.

Acknowledgments

This research was supported by the grant from the Japan Society for the Promotion of Science (JSPS) through the "Funding Program for World-Leading Innovative R&D on Science and Technology (FIRST Program)," initiated by the Council for Science and Technology Policy (CSTP). I would like to express my sincere gratitude to the late Dr. A. Tonomura, Prof. D. Shindo, Prof. Y. Tokura, Prof. Y. Murakami, Dr. T. Matsuda, Dr. K. Hirata, Dr. X. Z. Yu, Dr. Y. A. Ono, Dr. N. Osakabe, Dr. H. Shinada, Dr. T. Tanigaki, Mr. S. Aizawa, Mr. K. Yanagisawa, Ms. Y. Tamura, and Ms. K. Matsuyama, past and present, and other members of FIRST Tonomura Project, who contributed significantly to the research discussed here; their contributions are acknowledged in the publications referenced here.

References

1. A. Tonomura, *Electron Holography.*2nd edition (Springer-Verlag, Berlin, Heidelberg, 1999).
2. A. Tonomura, T. Matsuda, J. Endo, H. Todokoro, and T. Komoda, Development of a field-emission electron microscope, *J. Electron Microsc.* **28**(1), 1-11, (1979).
3. A. Tonomura, N. Osakabe, T. Matsuda, T. Kawasaki, J. Endo, S. Yano, and H. Yamada, Evidence for Aharonov-Bohm effect with magnetic field completely shielded from electron wave, *Phys. Rev. Lett.* **56**(8), 792–795, (1986).
4. A. Tonomura, J. Endo, T. Matsuda, T. Kawasaki, and H. Ezawa, Demonstration of single-electron buildup of an interference pattern, *Amer. J. Phys.* **57**(2), 117-120, (1989).
5. T. Matsuda, S. Hasegawa, M. Igarashi, T. Kobayashi, M. Naito, H. Kajiyama, J. Endo, N. Osakabe, A. Tonomura, and R. Aoki, Magnetic field observation of a single flux quantum by electron-holographic interferometry, *Phys. Rev. Lett.* **62**(21), 2519-2522, (1989).
6. K. Harada, T. Matsuda, J. Bonevich, M. Igarashi, S. Kondo, G. Pozzi, U. Kawabe, and A. Tonomura, Real-time observation of vortex lattices in a superconductor by electron microscopy, *Nature* **360**, 51-53, (1992).
7. http://www.first-tonomura-pj.net/e/index.html
8. A. Fert and P. Grünberg, Les Prix Nobel, The Noble prizes 2007, K. Grandin, Edition., The origin, development and future of spintronics, Nobel Foundation, Stockholm (2008).
9. Y. Tokura, Multiferroics as quantum electromagnets, *Science* **312**(5779), 1481-1482, (2006).
10. T. Shinjo, *Nanomagnetism and Spintronics.* (Elsevier, Amsterdam, 2009).
11. S. Chikazumi, *Physics of Ferromagnetism.* (Oxford Science Publications, Oxford, 1997).
12. A. Hubert and R. Schafer, *Magnetic Domains: the Analysis of Magnetic Microstructures.* (Springer-Verlag, Berlin, 1998).
13. M. R. Freeman and B. C. Choi, Advances in magnetic microscopy, *Science* **294**(5546), 1484-1488, (2001).
14. M. D. Graef and Y. Zhu, *Magnetic Imaging and its Applications to Materials.* (Academic Press, London, 2001).
15. H. S. Park, J. S. Baskin, and A. H. Zewail, 4D Lorentz electron microscopy imaging; magnetic domain wall nucleation, reversal, and wave velocity, *Nano Lett.* **10**(9), 3796-3803, (2010).
16. H. S. Park, K. Hirata, K. Yanagisawa, Y. Ishida, T. Matsuda, D. Shindo, and A. Tonomura, Nanoscale magnetic characterization of tunneling magnetoresistance spin valve head by electron holography, *Small* **8**(23), 3640-3646, (2012).
17. H. S. Park, Y. Murakami, K. Yanagisawa, T. Matsuda, R. Kainuma D. Shindo, and A. Tonomura, Electron holography studies on narrow magnetic domain walls observed in a Heusler alloy $Ni_{50}Mn_{25}Al_{12.5}Ga_{12.5}$, *Adv. Funct. Mater.* **22**(16), 3434-3437, (2012).
18. A. Tonomura, X. Z. Yu, K. Yanagisawa, T. Matsuda, Y. Onose, N. Kanazawa, H. S. Park, and Y. Tokura, Real-space observation of skyrmion lattice in helimagnet MnSi thin samples, *Nano Lett.* **12**(3), 1673-1677, (2012).
19. K. Harada, J. Endo, N. Osakabe, and A. Tonomura, Direction-free magnetic field application system, *e-J. Surf. Sci. Nanotech.* **6**, 29-34, (2008).

20. K. Harada, A. Tonomura, Y. Togawa, T. Akashi, and T. Matsuda, Double-biprism electron interferometry, *Appl. Phys. Lett.* **84**(17), 3229-3231, (2004).
21. A. Chappert, A. Fert, and F. N. Van Dau, The emergence of spin electronics in data storage, *Nature Mater.* **6**, 813-823, (2007).
22. B. Dieny, V. S. Speriosu, S. Metin, S. S. P. Parkin, B. A. Gurney, P. Baumgart, and D. R. Wilhoit, Magnetotransport properties of magnetically soft spin-valve structures, *J. Appl. Phys.* **69**(8), 4774-4779, (1991).
23. J. S. Moodera, L. R. Kinder, T. M. Wong, and R. Meservey, Large magnetoresistance at room temperature in ferromagnetic thin film tunnel junctions, *Phys. Rev. Lett.* **74**(16), 3273-3276, (1995).
24. Y. Tokura and N. Nagaosa, Orbital physics in transition-metal oxides, *Science* **288**, 462-468, (2000).
25. F. Jonietz, S. Muhlbauer, C. Pfleiderer, W. Munzer, A. Bauer, T. Adams, R. Georgii, P. Boni, R. A. Duine, K. Everschor, M. Garst, and A. Rosch, Spin transfer torques in MnSi at ultralow current densities, *Science* **330**(6011), 1648-1651, (2010).
26. S. Muhlbauer, B. Binz, F. Jonietz, C. Pfleiderer, A. Rosch, A. Neubauer, R. Georgii, and P. Boni, Skyrmion lattice in a chiral magnet, *Science* **323**(5916), 915-919, (2009).
27. A. Neubauer, C. Pfleiderer, B. Binz, A. Rosch, R. Ritz, P. G. Niklowitz, and P. Boni, Topological Hall effect in the A phase of MnSi, *Phys. Rev. Lett.* **102**(18), 186602, (2009).
28. A. N. Bogdanov and D. A. Yablonskiî, Thermodynamically stable "vortices" in magnetically ordered crystals, *Sov. Phys. JETP.* **68**(1), 101-103, (1989).
29. M. Onoda, G. Tatara, and N. Nagaosa, Anomalous Hall effect and skyrmion number in real and momentum spaces, *J. Phys. Soc. Jpn.* **73**, 2624-2627, (2004).
30. U. K. Rößler, A. N. Bogdanov, and C. Pfleiderer, Spontaneous skyrmion ground states in magnetic metals, *Nature* **442**(7104), 797-801, (2006).
31. X. Z. Yu, N. Kanazawa, Y. Onose, K. Kimoto, W. Z. Zhang, S. Ishiwata, Y. Matsui, and Y. Tokura, Near room-temperature formation of a skyrmion crystal in thin-films of the helimagnet FeGe, *Nature Mater.* **10**, 106-109, (2011).
32. X. Z. Yu, Y. Onose, N. Kanazawa, J. H. Han, Y. Matsui, N. Nagaosa, and Y. Tokura, Real-space observation of a two-dimensional skyrmion crystal, *Nature* **465**, 901-904, (2010).

Probing the Proton with Electron Microscopy

Jerome I. Friedman

Department of Physics, Massachusetts Institute of Technology,
77 Massachusetts Avenue, Cambridge, MA 02139, USA
E-mail: jif@mit.edu

This article is written as a tribute and memorial to Dr. Akira Tonomura who was an outstanding experimental physicist and a friend. Early in his career, he opened a new era in electron microscopy by demonstrating in 1968 that electron holography, proposed by Gabor in 1949, was possible; and later he developed Lorentz "phase" microscopy, which allows one to generate real-space, real-time images. All through his career, he perfected these designs into superb instruments that he employed to investigate fundamental questions in physics. Dr. Tonomura set world standards for electron microscopy

In this article, I will describe an experiment involving another form of electron microscopy that was utilized to probe the insides of the proton and neutron. Although this experiment utilized instruments that were very different from those used by Dr. Tonomura, the same basic principles of electron microscopy were used to investigate the inner structure of these particles. The experimental configuration that served as a high-powered electron microscope consisted of a two-mile long, 20 GeV electron linear accelerator with associated spectrometers and particle detectors located at the Stanford Linear Accelerator Center (SLAC).

When SLAC was being constructed, physicists from Cal Tech, MIT, and SLAC formed a collaboration to investigate the structure of the proton at this new accelerator because it would provide the highest electron energies in the world. Higher energies meant that the proton could be probed at shorter distances than had been studied in the past. The plan was to carry out a program of elastic electron-proton scattering, following Robert Hofstadter's pioneering approach to studying the proton. The electron was the ideal probe particle because its structure was known, its interaction was well understood; and of course, in the 1950s, Hofstadter had used elastic electron-proton scattering to measure the proton's form factor and its root mean squared radius.[1,2]

There were two models of the proton at that time. The one that was overwhelmingly favored was the bootstrap model. It was based on the concept of nuclear democracy. The idea was that there was no hadron more fundamental than any other hadron, and the hadrons were considered to be composites of one another. (For a review of strong interaction physics of the period, see Ref. 3.).

In applying the bootstrap model, the proton would often be discussed as a bare neutron that extended out to about two tenths of a fermi, surrounded by a positive pion cloud. There were other pictures of course; but the general point of view was that hadrons did not have elementary constituents, namely point-like constituents that were described by a field theory. A consequence of this picture was that hadrons were thought to have diffuse substructures and no elementary building blocks.

The other model was the quark model of Murray Gell-Mann[4] and George Zweig.[5] The reason quarks were proposed was that SU(3) was so successful in providing a Periodic Table for the known hadrons. This classification scheme was both descriptive and predictive, and quarks provided the dynamics to account for its success. The quark model was proposed because the mathematical triplets that quarks represented provided a natural way to account for the structure of the SU(3) families.

When the quark model was first proposed, it included three kinds of quarks. There was the *up*, the *down* and the *strange* quark. They all had spin 1/2, and they had the peculiar characteristic of fractional charges. The idea of fractional charges was an unpopular idea, and most physicists thought that this was one of the reasons why the quark model could not be correct. In constructing hadrons out of quarks, the model employed three quarks to make a baryon and a quark—antiquark pair to make a meson.

In response to this model, physicists searched for quarks in every possible way: looking at accelerator production, cosmic rays, the terrestrial environment, sea water, meteorites, air, etc. No quarks were found in any of these searches.[6] And of course, that was what was expected, because the quark model was thought to be totally unrealistic. The general point of view in 1966 was that quarks were most likely just mathematical representations: useful but not real. The real picture of particles at that time was that they have diffuse substructures and no elementary building blocks.

Experimental program

The experiment at SLAC was initially focused on measurements of elastic e-p scattering to measure the proton's magnetic form factor. The results obtained[7] are shown in Fig. 1. These results were just a continuation of the so-called dipole form factor, which had been obtained in previous measurements at the Cambridge Electron Accelerator and the Deutsches Electronen Synchrotron (DESY).

Because nothing new had been discovered, the collaboration concluded the elastic scattering program. However, the MIT and SLAC Groups decided to continue electron scattering, but to concentrate on inclusive inelastic scattering.

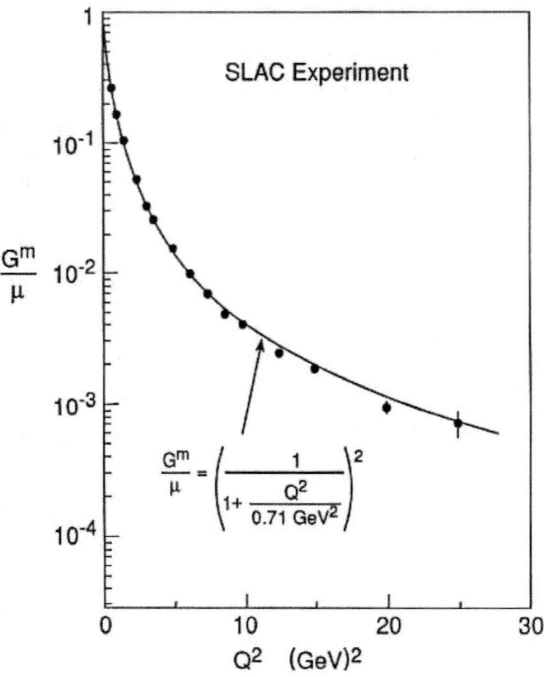

Fig. 1. Magnetic Form Factor of the proton from elastic scattering (Ref. 7).

The initial measurements in the inelastic scattering program studied the electro-production of resonances as a function of four-momentum transfer. It was thought that higher mass resonances might become more prominent when excited by virtual photons, and a search was carried out at the very highest masses that could be reached. For completeness the inelastic continuum was also investigated, since this was a new energy region that had not been previously explored. The proton resonances that were measured[8] showed no unexpected kinematic behavior. Their transition form factors fell about as rapidly as the elastic proton form factor with increasing values of the four momentum transfer, q.

However, some surprising features were found when the continuum region (now commonly called the deep inelastic region) was investigated. This investigation consisted of measuring the spectra of inelastically scattered electrons, with only the scattered electrons detected, over a wide range of incident energies, scattering angles and energy losses, extending to energy losses as large as could be reliably measured.

Inelastic scattering has a special advantage in studying a particle's internal structure. Elastic scattering is interesting because the target particle's form factor can be measured, giving information about the time average of the charge distribution and the magnetic moment distribution; and of course that is useful. But if there are constituents in the target particle, they cannot be seen in that way because they are moving around very rapidly and a time average is being measured. Because of this, inelastic scattering has to be employed to observe constituents.

This can be seen in terms of a rough calculation using the uncertainty principle. What is really required is to take a snapshot in time. So the interval of time during which the energy is exchanged must be very short; and for that to happen, there must be a large energy exchange ΔE. For example, if there is a ΔE of 2 GeV, the snapshot is taken in about 3×10^{-25} seconds. And if the constituents are moving at approximately the velocity of light, they have moved about 10^{-14} cm in that time. Because the size of the proton is about 10^{-13} cm, there is the possibility of seeing constituents inside the proton. Observing constituents requires large energy exchanges, and deep inelastic scattering is required for obtaining such large values of ΔE.

To carry out these measurements, a monochromatic beam from the linear accelerator was passed through a liquid hydrogen target, and then through a series of monitors. The scattered electrons were momentum analyzed by one of three magnetic spectrometers installed in End Station A. In separate experiments, the SLAC 20 GeV, 8 GeV, and 1.6 GeV spectrometers were used to cover different kinematic regions; however, most of the measurements were made with the two larger devices. Downstream of the magnetic elements of these spectrometers were placed scintillation counter hodoscopes, which registered the momentum and scattering angle of each scattered electron. In conjunction with the hodoscopes there were particle identification counters that were employed to identify electrons amid a background of negative pions. These consisted of a gas Cerenkov counter, a total absorption counter for electromagnetic cascades, and a few counters used to sample early shower development in the total absorption counter. Initially, we employed a liquid hydrogen target and later used liquid deuterium to study the neutron.

The first unexpected feature of these early results[9,10] was that the deep inelastic cross sections showed only a weak falloff with increasing q^2. When the experiment was planned, there was no clear theoretical picture of what to expect. The observations of Robert Hofstadter and his coworkers in their pioneering studies of elastic electron scattering from the proton showed that the proton had a size of about 10^{-13} cm and a smooth charge distribution. This result plus the theoretical framework that was most widely accepted at the time suggested, when the experiment was planned, that the deep inelastic electron proton cross-sections would fall rapidly with increasing q^2. The experimental yields that were observed were between a factor of 10 and 1000 times greater than were expected on the basis of a model of off-mass-shell photoproduction with the inclusion of the proton form factor. We were surprised, and made extensive checks of our radiative corrections routine before we were convinced that our results were correct. The weak momentum transfer dependence of the inelastic cross-sections for excitations well beyond the resonance region is illustrated in Fig. 2.

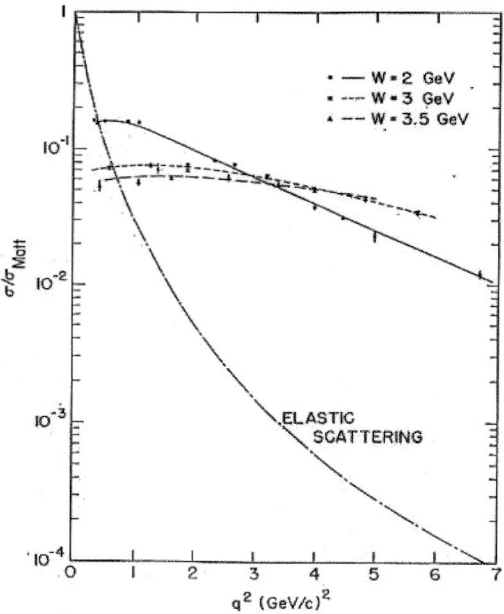

Fig. 2. $(d^2\sigma/dE^2)/\sigma$Mott in GeV^{-1} vs. q^2 for $W = 2$, 3 and 3.5 GeV at $\theta = 10°$. The lines drawn though the data are meant to guide the eye. Also shown is the cross section for elastic e-p scattering divided by σMott, $(d\sigma/d\Omega)/\sigma$Mott, calculated for $\theta = 10°$, using the dipole form factor. The relatively slow variation with q^2 of the inelastic cross section compared with the elastic cross section is clearly shown.

The differential cross-section divided by the Mott cross-section,

$$\sigma\text{Mott} = [\, e^4/(4E^2)\,][(\cos^2(\theta/2))/\sin^4(\theta/2)],$$

is plotted as a function of the square of the four-momentum transfer, $q^2 = 2EE'\,(1 - \cos\theta)$, for constant values of the invariant mass of the recoiling target system, W, where

$$W^2 = 2M(E-E') + M^2 - q^2.$$

The quantity E is the energy of the incident electron, E' is the energy of the final electron, and θ is the scattering angle, all defined in the laboratory system; and M is the mass of the proton. The cross section is divided by the Mott cross section in order to remove the major part of the well-known four-momentum transfer dependence arising from the photon propagator. The q^2 dependence that remains is related to the properties of the target system. Results for $\theta = 10°$ are shown in the figure for each value of W. As W increases, the q^2 dependence appears to decrease. The striking difference between the behavior of the deep inelastic and elastic cross-sections is also illustrated in this figure, where the elastic cross section, divided by the Mott cross-section for $\theta = 10°$, is shown.

The second surprising feature in the data, scaling, was found by following a suggestion of James Bjorken.[11] To describe the concept of scaling, one has to introduce the general expression for the differential cross section for unpolarized electrons scattering from unpolarized nucleons, with only the scattered electrons detected[12]:

$$\frac{d^2\sigma}{d\Omega dE'} = \sigma_{\text{Mott}}\left[W_2 + 2W_1 \tan^2\frac{\theta}{2}\right].$$

The functions W_1 and W_2 are called structure functions, and depend on the properties of the target system. As there are two polarization states of the virtual photon, transverse and longitudinal, two such functions are required to describe this process. In general, W_1 and W_2 are expected to be functions of both q^2 and v, where v is the energy loss of the scattered electron. However, on the basis of models that satisfy current algebra, Bjorken conjectured that in the limit of q^2 and v approaching infinity, the two quantities vW_2 and W_1 become functions only of the ratio $\omega = 2Mv/q^2$; that is

$$2MW_1 (v, q^2) \rightarrow F1 (\omega)$$

$$vW_2 (v, q^2) \rightarrow F2 (\omega)$$

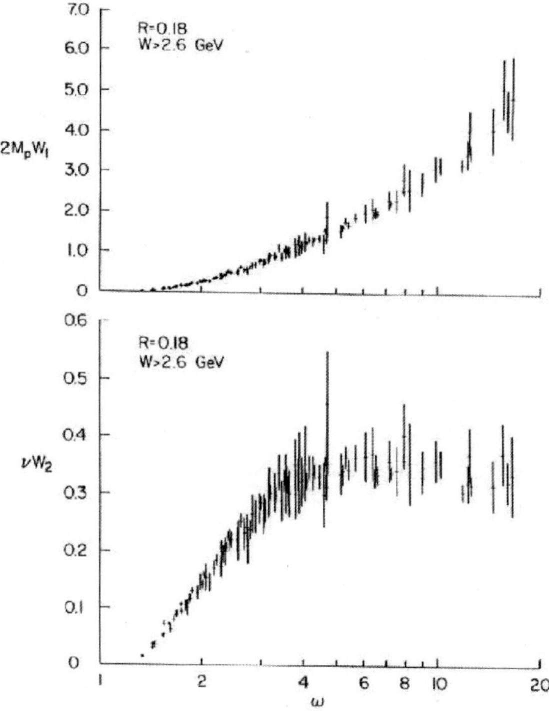

Fig. 3. Results for proton structure functions to test Bjorken scaling (from Ref. 13).

The observed scaling behavior[13] of the structure functions is shown in Fig. 3, where experimental values of νW_2 and $2MW_1$ are plotted as a function of ω for values of q^2 ranging from 2 to 20 GeV^2. The data demonstrated scaling within experimental errors for $q^2 > 2$ GeV^2 and W > 2.6 GeV. The dynamical origin of scaling was not clear at that time, and a number of models were proposed to account for this behavior and for the weak q^2 dependence of the inelastic cross section. While most of these models were firmly embedded in S-matrix and Regge pole formalism, the experimental results caused some speculation regarding the existence of a possible point-like structure in the proton. In his plenary talk at the XIV International Conference on High Energy Physics held in Vienna in 1968, where preliminary results on the weak q^2 dependence and scaling were presented, Wolfgang Panofsky[8] reported "... theoretical speculations are focused on the possibility that these data might give evidence on the behavior of point-like charged structures in the nucleon." However, this was not the prevailing point of view at the time. This picture challenged the views of most of the physics community, and only a small number of physicists took such a possibility seriously. One of these was Bjorken[14] who had proposed in his 1967 Varenna lectures that deep inelastic electron scattering might provide evidence of elementary constituents.

The MIT-SLAC results indicated the existence of point-like constituents within the proton. But were they quarks? To demonstrate that these constituents were quarks, it would have to be shown that they were spin 1/2 particles, and they must have fractional charges consistent with the quark model. Curtis Callan and David Gross[15] showed that the ratio of the structure functions depended on the spins of the constituents. For spin 1/2 constituents, they predicted F2 / F1 = 2x, where x = 1/ω; and this result was tested experimentally. The experimental results[16,17] clearly indicated that the spin was 1/2. The first requirement for identifying the constituents as quarks was satisfied.

What about the charges of the constituents? Deep inelastic neutrino nucleon scattering measurements made between 1971 and 1974 with Gargamelle, the large heavy liquid bubble chamber at CERN, played an essential role in answering this question. The Gargamelle measurements demonstrated two important features. One was the linearly increasing cross-section with energy, which indicated pointlike structure in the nucleon, confirming the electron scattering results. And secondly, their measurements of F2, when combined with those from the electron scattering results, demonstrated that the constituents in the nucleon have the fractional charges of quarks. The prediction for the ratio of the nucleon structure functions from neutrino scattering and electron scattering on the basis of the quark model gives a value of 3.6:

$$\int[F2^{\nu n}(x)+ F2^{\nu p}(x)]dx / \int[F2^{en}(x)+ F2^{ep}(x)]dx = 2/(Q_u^2 + Qd^2) = 2 / [(2/3)^2 + (1/3)^2] = 3.6$$

The experimental value from the Gargamelle and MIT-SLAC experiments[18] was 3.4±0.7. This clearly demonstrated that the constituents of the nucleon have the fractional charges of quarks. When Donald Perkins presented this result at the International High Energy Conference at Fermilab in 1972, he called it an "astonishing verification of the quark model". And astonishing it was, because of the strong negative feelings about quarks in the physics community.

After the London Conference[19] in 1974, with its strong confirmation of the quark model, a general change of view developed with regard to the structure of hadrons. The bootstrap approach and the concept of nuclear democracy were in decline, and by the end of the 1970's, the quark structure of hadrons became the dominant view for developing theory and planning experiments. A crucial element in this change was the general acceptance of Quantum Chromodynamics, QCD,[20,21] which eliminated the last paradox, namely, why are there no free quarks? The infrared slavery mechanism of QCD provided a reason to accept quarks as physical constituents without demanding the existence of free quarks. The asymptotic freedom property of QCD also readily provided an explanation of scaling, but logarithmic deviations from scaling were required in this theory. These deviations were later con-firmed in higher energy muon and neutrino scattering experiments at FNAL and CERN and later shown with high precision at the electron- proton collider at DESY. The quark model, with quark interactions described by QCD, became the accepted view of the structure of hadrons. This picture, which is one of the foundations of the Standard Model, has not been contradicted by any experimental evidence.

References

1. R. Hofstadter and R. W. McAllister, Electron scattering from the proton, *Phys. Rev.* **98**(1), 217-218, (1955).
2. R. W. McAllister and R. Hofstadter, Elastic scattering of 188-MeV electrons from the proton and the alpha particle, *Phys. Rev.* **102**(3), 851-856, (1956).
3. S. C. Frautschi, *Regge Poles and S Matrix Theory.* (W. A. Benjamin, New York, 1963).
4. M. Gell-Mann, A schematic model of baryons and mesons, *Phys. Lett.* **8**(3), 214-215, (1964).
5. G. Zweig, An SU(3) model for strong interaction symmetry and its breaking, in *Developments in the Quark Theory of Hadrons*, Vol. 1, Eds. D. B. Lichtenberg and S. P. Rosen, pp. 22-101 (1964), and CERN Geneva-TH. 401 (Rec, Jan. 64), 24p, http://cdsweb.cern.ch/record/352337?In=en.
6. W. Jones, A review of quark search experiments, *Rev. Mod. Phys.* **49**(4), 717-752, (1977).
7. D. H. Coward, H. DeStaebler, R. A. Early, J. Litt, A. Minten, L. W. Mo, W. K. H. Panofsky, R. E. Taylor, M. Breidenbach, J. I. Friedman, H. W. Kendall, P. N. Kirk, B. C. Barish, J. Mar, and J. Pine, Electron-proton elastic scattering at high momentum transfers, *Phys, Rev. Lett.* **20**(6), 292-295, (1968).
8. W. K. H. Panofsky, Electromagnetic interactions: Low-q2 electrodynamics, elastic and inelastic electron and muon scattering, in *Proceedings of 14th International Conference on High Energy Physics*, Eds. J. Prentki and J. Steinberger, pp. 23-29, Vienna, Austria (1968). The experimental report, presented by the present author, is not published in the Conference Proceedings. It was, however, produced as a SLAC preprint.

9. E. D. Bloom, D. H. Coward, H. DeStaebler, J. Drees, G. Miller, L. W. Mo, R. E. Taylor, M. Breidenbach, J. I. Friedman, G. C. Hartmann, and H. W. Kendall, High-energy inelastic e-p scattering at 6° and 10°, *Phys. Rev. Lett.* **23**(16), 930-934, (1969).

10. M. Breidenbach, J. I. Friedman, H. W. Kendall, E. D. Bloom, D. H. Coward, H. DeStaebler, J. Drees, L. W. Mo, and R. E. Taylor, Observed behavior of highly inelastic electron-proton scattering, *Phys. Rev. Lett.* **23**(16), 935-939, (1969).

11. J. D. Bjorken, Asymptotic sum rules at infinite momentum, *Phys. Rev.* **179**(5), 1547-1553, (1969).

12. S. D. Drell and J. D. Walecka, Electrodynamic processes with nuclear targets, *Ann. Physics (NY)* **28**, 18-33, (1964).

13. G. Miller, E. D. Bloom, G. Buschhorn, D. H. Coward, H. DeStaebler, J. Drees, C. L. Jordan, L. W. Mo, R. E. Taylor, J. I. Friedman, G. C. Hartmann, H. W. Kendall, and R. Verdier, Inelastic electron-proton scattering at large momentum transfers and the inelastic structure functions of the proton, *Phys. Rev. D* **5**(3), 528-544, (1972).

14. J. D. Bjorken, Current algebra at small distances, in *Proceedings of the International School of Physics "Enrico Fermi," Course XLI: Selected Topics in Particle Physics*, Ed. J. Steinberger, pp. 55-81 (Academic Press, New York, 1968).

15. C. G. Callan, Jr. and D. J. Gross, High-energy electroproduction and the constitution of the electric current, *Phys. Rev. Lett.* **22**(4), 156-159, (1969).

16. E. M. Riordan, A. Bodek, M. Breidenbach, D. L. Dubin, J. E. Elias, J. I. Friedman, H. W. Kendall, J. S. Poucher, M.R. Sogard, and D. H. Coward, Extraction of R=σL/σT from deep inelastic *e-p* and *e-d* cross sections, *Phys. Rev. Lett.* **33**(9), 561-564, (1974).

17. E. M. Riordan, A. Bodek, M. Breidenbach, D. L. Dubin, J. E. Elias, J. I. Friedman, H. W. Kendall, J. S. Poucher, M. R. Sogard, and D. H. Coward, Test of scaling of the proton electromagnetic structure functions, *Phys. Lett. B* **52**(2), 249-252, (1974).

18. D. H. Perkins, Neutrino interactions, in *Proceedings of the 16th International Conference on High Energy Physics*, J. D. Jackson and A. Roberts (Eds.), Vol. 4, pp. 189-248, Batavia, Illinois, USA (Sept., 1972).

19. D. C. Cundy, in *Proceedings of the 17th International Conference on High Energy Physics*, J. R. Smith (Ed.), p. IV-131, Rutherford Laboratory, Chilton, Didcot, Oxon, UK (1974).

20. D. J. Gross and F. Wilczek, Ultraviolet behavior of non-Abelian gauge theories, *Phys. Rev. Lett.* **30**(26), 1343-1346, (1973).

21. H. D. Politzer, Reliable perturbative results for strong interactions, *Phys. Rev. Lett.* **30**(26), 1346-1349, (1973).

Hanbury Brown–Twiss Interferometry with Electrons: Coulomb vs. Quantum Statistics

Gordon Baym[*] and Kan Shen[†]

*Department of Physics, University of Illinois at Urbana-Champaign,
1110 W. Green St., Urbana, Illinois 61801, USA.*

gbaym@illinois.edu

A longstanding goal of Akira Tonomura was to observe Hanbury Brown–Twiss anti-correlations between electrons in a field-emission free electron beam. The experimental results were reported in his 2011 paper[1] with Tetsuji Kodama and Nobuyuki Osakabe. An open issue in such a measurement is whether the observed anti-correlations arise from quantum statistics, or are simply produced by Coulomb repulsion between electrons. In this paper we describe a simple classical model of Coulomb effects to estimate their effects in electron beam interferometry experiments, and conclude that the experiment did indeed observe quantum correlations in the electron arrrival times.

1. Introduction

Since the pioneering detection by Hanbury Brown and Twiss (HBT) of "bunching" of photons in a light beam,[3] HBT experiments with massive bosons, e.g., atomic beams[4,5] and π and K mesons in high energy nuclear collisions[6,7] have shown similar two-particle correlations. Seeing anti-correlation – or anti-bunching – effects in experiments with identical fermions where the two-particle intensity (I) correlation function

$$C(\mathbf{r}_2 - \mathbf{r}_1, t_2 - t_1) = \frac{\langle I_1 I_2 \rangle}{\langle I_1 \rangle \langle I_2 \rangle}. \tag{1}$$

---

[*]This paper is dedicated to the memory of the late Akira Tonomura. He was at the same time a remarkable scientist and a warm friend with whom author GB spent many a happy moment, from Tokyo to Urbana and Washington to Scandinavia; his constant interest in HBT correlations in a free electron beam was a valuable source of inspiration.

The research discussed here, based in good measure on the Ph.D. dissertation[2] of author KS, has been supported in part by the U.S. National Science Foundation over the years, most recently by NSF Grants PHY07-01611 and PHY09-69790. GB is grateful to the Aspen Center for Physics, supported in part by NSF Grant PHY10-66293, where parts of this research were carried out.

[†]Currently at Quantitative Strategies, Credit Suisse, 11 Madison Ave, New York, NY 10010, USA.

should fall, at small separations (either in position or momentum space), to zero for particles of the same spin (or to 1/2 for unpolarized particles) has proven more elusive. Experiments with neutrons are complicated by a low energy nuclear resonance, while experiments with protons are complicated in addition by Coulomb repulsion.[8,9] On the other hand, anti-bunching of neutral cold fermionic ^{40}K atoms,[10] and the corresponding bunching of neutral cold bosonic ^{87}Rb atoms[11] emitted from optical lattices has been successfully observed. In adddition, anti-bunching with neutral atomic ^{3}He beams, as well as bunching with neutral atomic ^{4}He atomic beams, was clearly demonstrated in the experiment of Jeltes *et al.*[5]

Detecting anti-bunching in a beam of electrons has been a major experimental challenge over the years, owing to the low degeneracy as well as the short coherence time of the beams. Starting in the 1990s Akira Tonomura and his group focussed on seeing this striking effect of quantum statistics with electrons in a field-emission electron beam. Following his group's theoretical feasibility analysis,[12,13] electron HBT experiments have been realized in free space;[1,14] such experiments with electrons show a reduction in the correlation function for small space-time separation, generally attributed to anti-bunching. On the other hand, repulsive Coulomb interactions between electrons also reduce the probability of two electrons being close in space. Whether the observed anti-bunching effect is due to electron quantum statistics or rather Coulomb repulsion is the issue we deal with in this paper. We conclude that the recently reported experiment of Kodama, Osakabe, and Tonomura[1] very cleanly sees HBT in the arrival time correlations of electrons pairs; In the experiment of Ref. 14, at a significantly lower beam energy, Coulomb effects account for several percent of the HBT signal.

HBT interferometry has in fact been seen for electrons in semiconductor devices,[15,16] where, owing to screening, Coulomb effects are less important. For example, in the HBT experiment of Ref. 16, the screening length ($\sim$ 5nm) is typically much smaller than the Fermi wavelength ($\sim$ 40nm).[17] Interestingly, two dimensional mesoscopic semiconductors open the possibility of seeing HBT correlations for fractional statistics [18,19] as well as Aharonov-Bohm physics.[20]

The importance of correcting for Coulomb interactions has long been recognized in interpreting high energy nuclear experiments.[21] For example, the raw correlation function of distinguishable pions of opposite charge ($\pi^+\pi^-$), produced in the E877 ultra-relativistic heavy ion collision experiment,[22] shows a very similar buildup at small momentum differences to those of identical charged pions ($\pi^+\pi^+$ and $\pi^-\pi^-$). The Coulomb interaction between opposite charges tends to increase the probability of a pair of bosons being close in momentum, while reducing that for like charges. Only after the effects of Coulomb interactions are extracted, does one see the expected effects of quantum statistics (see, e.g., Refs. 7 and 23).

Our aim in this note is to present a simple schematic discussion of Coulomb effects in interferometry experiments with electrons, based on the classical behavior of electrons taken pairwise. We do not attempt to explain the detailed results of Tonomura's group on HBT with electrons, but rather aim to estimate the role of Coulomb interactions in their search for quantum correlations.

The Coulomb problem for a pair of electrons is characterized by four length scales, 1) the electron Bohr radius $a_0 \equiv \hbar^2/me^2$, with m the electron mass; 2) the size r_0 of the emitting region transverse to the beam; 3) the typical separation z_0 of the particles along the beam direction; and importantly, 4) the classical turning point of the pair, r_{tp}, defined by

$$e^2/r_{tp} = q^2/2m_{red}, \tag{2}$$

where q is magnitude of the final relative momentum of the two particles and $m_{red} = m/2$.

The traditional method of correcting for Coulomb interactions is to employ the Gamow correction, which assumes that the characteristic separation of the pair of particles is much smaller than their classical turning point, namely, that the particles are produced well within the classically forbidden regime bounded by r_{tp}. The actual rate observed in an experiment is taken to be that in the absence of Coulomb interactions times the Gamow correction, $|\psi_c(0)|^2$, which is the absolute

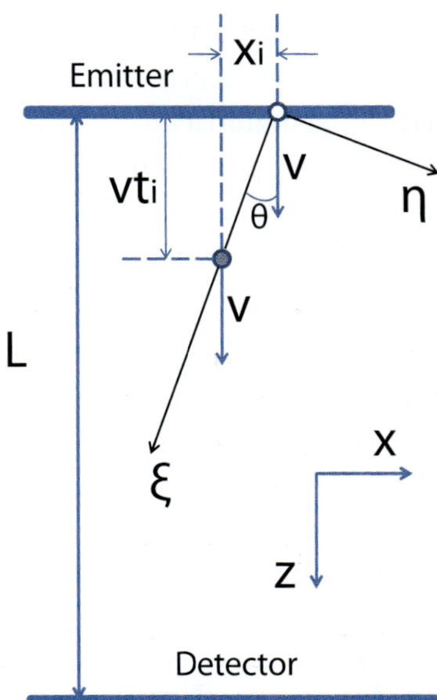

Fig. 1.   Schematic of emission of two electrons. The first (closed circle) and second (open circle) travel downward to the detector plate at velocity $\simeq v$. Coulomb acts along the ξ direction.

value squared of the relative Coulomb wave function at the origin,

$$\psi_c(0) = \left(\frac{2\pi\eta}{e^{2\pi\eta} - 1} \right)^{1/2}, \tag{3}$$

where the dimensionless parameter η equals $zz'e^2/\hbar v_{rel}$, with v_{rel} the relative velocity of the two particles of charges ze and $z'e$. On the other hand, in a field emission source, the relative energy of a pair is $\simeq \Delta E$, where ΔE is the initial electron energy spread in the beam. Thus $r_{tp} \simeq e^2/\Delta E \simeq (1.46/\Delta E_{eV})$ nm, where ΔE_{eV} is the energy spread measured in electron volts. Since r_{tp} is generally tens of nm, $r_{tp} \ll r_0$, indicating that a pair of electrons in such an experiment is typically emitted *outside* the pair's classical turning point. Coulomb effects are dominated by classical physics rather than by a quantum Gamow correction.

2. Classical model

To bring out the effect of Coulomb interactions, we model the experiment as independent emissions of electrons from a tip, followed by acceleration to final velocity v in the beam direction (z) and energy $E_f = mv^2/2$. We first neglect quantum statistics, and focus on the Coulomb effects in a single pair of particles, since the major contribution to the correlation function is from particles nearby in space and time, a configuration in which we can, to a first approximation, neglect many-body effects. We assume that the emission points of the pair are separated by x_i in space and t_i in time. Thus the initial spatial separation of the pair is $s_i = \left(x_i^2 + (vt_i)^2 \right)^{1/2}$.

The Coulomb repulsion between the electrons increases their relative separation in space and time, as is illustrated schematically in Fig. 2 for two electrons emitted at the same point in the tip at times t_A and then t_B. The electron separation at later

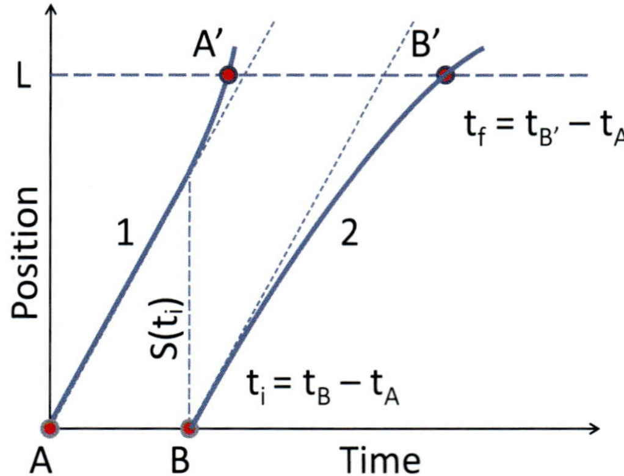

Fig. 2. Schematic picture of the effect of Coulomb repulsion on the trajectories of two electrons emitted successively from the same point in the tip.

time is readily calculated from the conservation of energy of the relative motion of the two electrons,

$$E_{rel} = m\dot{s}^2/4 + e^2/s. \tag{4}$$

A lower bound on the size of the Coulomb hole can derived by neglecting the initial relative kinetic energy of the pair; then after integration of $\dot{s}$, one finds that the final separation of the pair, s_f, is determined implicitly by

$$\sqrt{\frac{4e^2}{m}}\,\Delta t = s_i^{3/2}\left\{\sqrt{\sigma(\sigma-1)} + \ln\left[\sqrt{\sigma}+\sqrt{\sigma-1}\right]\right\}, \tag{5}$$

where $\sigma = s_f/s_i$, and $\Delta t \simeq L/v$ is the time elapsed between emission and detection of the pair. Approximately,

$$s_f = \begin{cases} s_c \left(s_c/s_i\right)^{1/2}, & s_i \leqslant s_c \\ s_i, & s_i > s_c, \end{cases} \tag{6}$$

where

$$s_c \equiv v\tau_c \equiv \left(\frac{2e^2 L^2}{E_f}\right)^{1/3} \tag{7}$$

defines a critical Coulomb distance and time τ_c; numerically, $s_c \simeq 6.5 \times 10^{-4} L_{cm}^{2/3}/E_{f,KeV}^{1/3}$ cm.

As seen in the left panel of Fig. 3, the relation between the final separation s_f when the electrons reach the detector plate and s_i is not monotonic, but rather is decreasing at small s_i and increasing at large s_i. For very small initial separation, the Coulomb interaction significantly accelerates the two electrons away from each other, making the final separation large, while for very large initial separation, the Coulomb interaction is negligible, and the final separation is essentially equal to that initially. No matter how small s_i is, the final spatial separation is finite, i.e., there is a *Coulomb hole* in the distribution of final separations. Coulomb forces increase the spatial separation between a pair of electrons, $s(t)$, with time, so that $s_f > s_i$.

Since the angle θ that the relative position vector of the electrons makes with respect to the beam axis (see Fig. 1) is conserved in the motion, the final separations at detection are related by

$$x_f/x_i = t_f/t_i = s_f/s_i; \tag{8}$$

thus the final separation in time of the two electrons is given by $t_f = (s_f/s_i)t_i$. At small t_i/τ_c, the final t_f has the structure shown in the right of Fig. 3, dependent on x_i. However, for $t_i/\tau_c \gtrsim 0.1$, the curves on the right of Fig. 3 converge to that in the

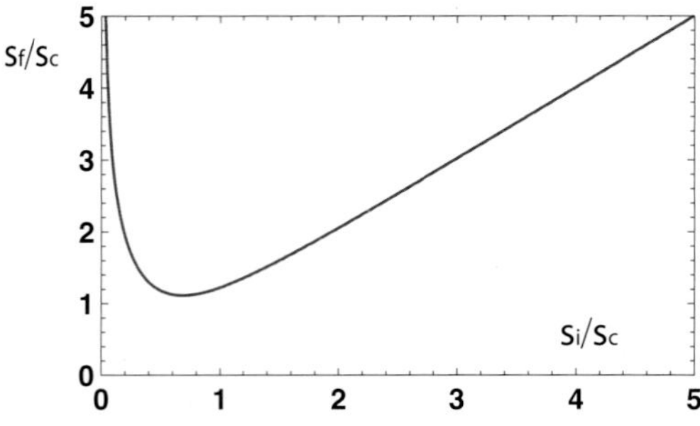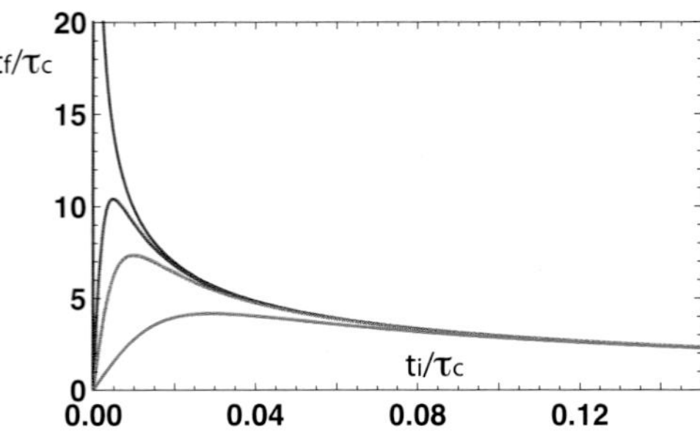

Fig. 3. a) Final vs. initial spatial separations, with Coulomb interactions included classically, and b) final vs. initial time separations for different initial transverse separations $x_i = 0$, 5, 10, and 30 nm (top to bottom). For large t_i/τ_c, the curves all converge to that in the left panel.

left panel of Fig. 3, which has a minimum at $s_f \approx s_c$. To a first approximation, the minimum $t_f \approx s_c/v = \tau_c$ is independent of the initial spatial separation.

Experimentally one measures the correlation function $C(t_f)$ in terms of the subsequent time intervals t_f between detection of particles, averaged over the distribution of initial emission intervals t_i. For independent emissions, the distribution of times between adjacent emissions from the tip is Poissonian

$$P_0(t_i) = \frac{1}{\bar{t}_i} e^{-t_i/\bar{t}_i}, \tag{9}$$

where $\bar{t}_i$ is the average time separation between two emissions. In the absence of Coulomb corrections and quantum statistics, $C_0(t_i) = 1$. The final $C(t_f)$ and $P(t_f)$ are given in terms of the map (5) between s_f and s_i; since the map is not simply one-to-one, one needs to sum over the two branches. The resulting $P(t_f)$ and $C(t_f)$, calculated with the approximate solution (6) are shown as thin lines in Fig. 4.

In experiments, the finite time resolution of the detectors would smooth out the sharp Coulomb holes in Fig. 4. Measurement of an observable $\tilde{f}(t)$ at time t averages the actual $f(t')$ over t', weighted by the detector time resolution function $R(t - t')$:

$$\tilde{f}(t) = \int_{-\infty}^{+\infty} R(t - t')f(t')dt', \tag{10}$$

with $f(t) = f(-t)$ for $t < 0$. In Fig. 4, effects of time resolution are indicated by the thick curves, where we take a Gaussian time resolution function

$$R(t - t') = \frac{1}{\sqrt{2\pi}t_r} e^{-(t-t')^2/2t_r^2} , \tag{11}$$

with a characteristic time scale t_r chosen here for illustration to equal τ_c. Typically, $t_r \gg \tau_c$. The dip at low t_f in right panel illustrates clearly how Coulomb correlations can mimic quantum correlations.

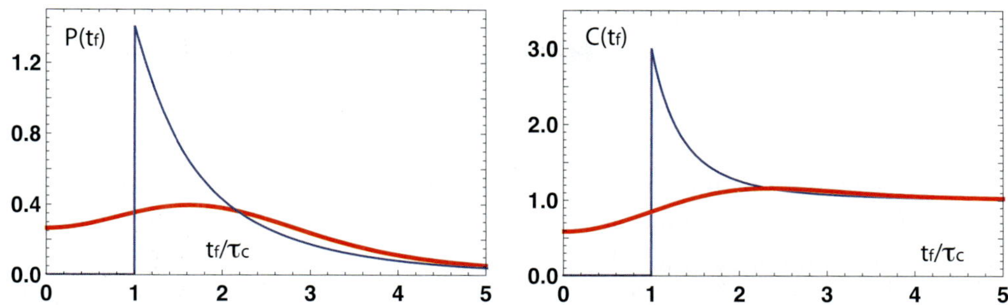

Fig. 4. Effects of Coulomb repulsion as seen in a) the final time distribution $P(t_f)$ vs. t_f, with $\bar{t}_i = 0.2$ ns; the vertical axis is in units of $1/\bar{t}_i$, and b) the normalized second-order final correlation function $C(t_f)$ vs. t_f. The thin lines are calculated directly from Eq. (6); the thick lines include finite time resoluton via Eqs. (10) and (11), with $t_r = \tau_c$.

3. HBT anti-correlations

To assess the importance of the Coulomb hole it is necessary to compare it with the regime of suppression of the correlation function from quantum statistics. The Pauli principle suppresses the correlation function between same spin electrons emitted from nearby points on a time scale

$$t_{HBT} = \hbar/T_f, \tag{12}$$

where T_f is the effective longitudinal temperature of the electron gas after acceleration. Since the acceleration of the beam is essentially adiabatic, the entropy per electron is conserved, and the temperature of the gas falls with expansion. Owing to acceleration the density n of the gas drops, since in a steady state the current nv remains constant from emission through acceleration. To estimate the final gas temperature, we note that the entropy per particle of a gas with an anisotropic temperature depends on $T^{1/2}T_\perp/n$, where T is the longitudinal and $T_\perp$ the transverse temperature. Thus in anisotropic expansion, $T^{1/2}T_\perp/n$ remains constant, and for $T_\perp$ constant, Tv^2 is invariant. Initially $v = \sqrt{T_i/m}$ and after acceleration to velocity much greater than that of the thermal motion, $v = \sqrt{2E_f/m}$. Thus the final longitudinal temperature of the beam is given by

$$T_f\sqrt{2E_f/m} \simeq T_i\sqrt{T_i/m}, \tag{13}$$

so that with Eq. (12) and $T_i \simeq 2\Delta E$, we have

$$T_f = 2(\Delta E)^2/E_f \tag{14}$$

(essentially the result derived heuristically in Ref. 1) and

$$t_{HBT} = \frac{\hbar E_f}{2(\Delta E)^2}. \tag{15}$$

The ratio of the Coulomb to HBT suppressions is thus

$$\frac{t_{HBT}}{\tau_c} \simeq 2^{-5/6} \left(\frac{a_0}{L}\right)^{2/3} \frac{E_f^{11/6} \mathrm{Ry}^{1/6}}{(\Delta E)^2} = 0.9 \frac{E_{f,\mathrm{KeV}}^{11/6}}{\Delta E_{\mathrm{eV}}^2 L_{\mathrm{cm}}^{2/3}}. \tag{16}$$

This simple calculation indicates the importance, in an HBT measurement of correlations in arrival times, of accelerating the electrons to a large final beam energy in order to overcome Coulomb effects.

In addition to time anti-correlations in the beam, HBT correlations should appear in the electron spatial separations, analogous to the spatial correlations seen in the original Hanbury Brown–Twiss measurement of the angular diameter of the star Sirius.[3] When two electrons are emitted at the same time, but spatially separated, the correlation function is suppressed in space on a scale of order the particle wavelength divided by the angular size of the source, or

$$s_{HBT} \simeq \frac{\hbar L}{mvr_0} = 6 \times 10^{-4} \frac{L_{cm}}{E_{f,KeV}^{1/2} r_{10nm}} \ \mathrm{cm} \tag{17}$$

where r_{10nm} is the transverse size of the emission region in units of 10 nm. Comparing with the minimal Coulomb hole, s_c, we find

$$\frac{s_{HBT}}{s_c} \simeq \frac{L_{cm}^{1/3}}{E_{f,KeV}^{1/6} r_{10nm}}. \tag{18}$$

To reach this ratio, the initial spatial separation must be of order s_c; however, a more realistic bound on s_i is the transverse size of the emission tip, which leads to a considerably larger Coulomb hole, as one can infer from Fig. 1.

In the experiment of Kodama, Osakabe, and Tonomura[1] $E_f \sim$ 50-100 KeV, and $\Delta E \simeq 0.17$ eV; with $L \sim 100$ cm one estimates that $t_{HBT}/\tau_c \sim (2-6) \times 10^3$, sufficiently high that one does not need to worry about Coulomb effects in measuring pure time-of-arrival correlations. On the other hand, in the opposite regime, when measuring spatial HBT correlations, one has $s_{HBT}/s_c \sim 0.5$, indicating that Coulomb effects must be taken into account in analyzing the experiment.

In contrast, in the lower energy experiment of Kiesel *et al.*,[14] where $E_f \sim 0.9$ KeV, $\Delta E \simeq 0.13$ eV, and $L \sim 1$ cm, one has $t_{HBT}/\tau_c \sim 44$, and thus Coulomb effects while small are not entirely negligible. For spatial correlations, however, $s_{HBT}/s_c \sim 0.25$. and thus Coulomb effects are dominant.

In conclusion, a full analysis of the HBT experiment of Kodama, Osakabe, and Tonomura requires correcting for Coulomb effects. The simple model presented here forms a useful and simple basis for including Coulomb interactions among the electrons. A detailed analysis requires taking into account the distribution of initial spatial separations and velocities of the electrons in addition to the time separations, the effects of realistic finite time resolution, as well as the effects of the quadrupole magnets which give one the freedom to adjust the angular size of the beam. While Coulomb effects are present independent of the relative spin of the electron pair, Pauli quantum correlations occur only between same spin electrons; thus to distinguish optimally Coulomb repulsions from quantum correlations one would ideally like to repeat the experiments with spin polarized electron beams.

References

1. T. Kodama, N. Osakabe, and A. Tonomura, Correlation in a coherent electron beam, *Phys. Rev. A* **83**(6), 063616, (2011).
2. K. Shen, Coulomb interactions in Hanbury Brown–Twiss experiments with electrons, Ph.D. dissertation, University of Illinois at Urbana-Champaign, Urbana, Illinois (2009).
3. R. Hanbury Brown and R. Q. Twiss, Correlation between photons in two coherent beams of light, *Nature* **177**, 27-29, (1956); R. Hanbury Brown and R. Q. Twiss, A test of a new type of stellar interferometer on Sirius, *Nature* **178**, 1046-1048, (1956).
4. M. Yasuda and F. Shimizu, Observation of two-atom correlation of an ultracold neon atomic beam, *Phys. Rev. Lett.* **77**(15), 3090-3093, (1996).
5. T. Jeltes, J. M. McNamara, W. Hogervorst, W. Vassen, V. Krachmalnicoff, M. Schellekens, A. Perrin, H. Chang, D. Boiron, A. Aspect, and C. I. Westbrook, Comparisonof the Hanbury Brown–Twiss effect for bosons and fermions, *Nature* **445**, 402-405, (2007).
6. D. H. Boal, C.-K. Gelbke, and B. K. Jennings, Intensity interferometry in subatomic physics, *Rev. Mod. Phys.* **62**(3), 553-602, (1990).
7. G. Baym, The physics of Hanbury Brown–Twiss intensity interferometry: from stars to nuclear collisions, *Acta Phys. Polonica B* **29**, 1839-1884, (1998).
8. W. Bauer, C.-K. Gelbke, and S. Pratt, Hadronic interferometry in heavy-ion collisions, *Annu. Rev. Nucl. Part. Sci.* **42**, 77-100, (1992).
9. R. Ghetti, L. Carlén, M. Cronqvist, B. Jakobsson, F. Merchez, B. Norén, D. Rebreyend, M. Rydehell, Ö. Skeppstedt, and L. Westerberg, Simultaneous neutron-nuetron proton-neutron and proton-proton interferometry measurements, *Nucl. Inst. and Meth.* **A335**, 156-164, (1993).
10. T. Rom, Th. Best, D. van Oosten, U. Schneider, S. Fölling, B. Paredes, and I. Bloch, Free fermion antibunching in a degenerate atomic Fermi gas released from an optical lattice, *Nature* **444**, 733-736, (2006).
11. S. Fölling, F. Gerbier, A. Widera, O. Mandel, T. Gericke, and I. Bloch, Spatial quantum noise interferometry in expanding ultracold atom clouds, *Nature* **434**, 481-484, (2005).
12. S. Saito, J. Endo, T. Kodama, A. Tonomura, A. Fukuhara, and K. Ohbayashi, Electron counting theory, *Phys. Lett. A* **162**(6), 442-448, (1992).
13. T. Kodama, N. Osakabe, J. Endo, A. Tonomura, K. Ohbayashi, T. Urakami, S. Ohsuka, H. Tsuchiya, Y. Tsuchiya, and Y. Uchikawa, *Phys. Rev. A* **57**(4), 2781-2785, (1998).
14. H. Kiesel, A. Renz, and F. Hasselbach, Observation of Hanbury Brown–Twiss anticorrelations for free electrons, *Nature* **418**, 392-394, (2002).
15. M. Henny, S. Oberholzer, C. Strunk, T. Heinzel, K. Ensslin, M. Holland, and C. Schönenberger, The fermionic Hanbury Brown and Twiss experiment, *Science* **284**(5412), 296-298, (1999).
16. W. D. Oliver, J. Kim, R. C. Liu, and Y. Yamamoto, Hanbury Brown and Twiss-type experiment with electrons, *Science* **284**(5412), 299-301, (1999).
17. W. D. Oliver, The generation and detection of electron entanglement, Ph.D. dissertation, Stanford University (2003).

18. S. Vishveshwara, Revisiting the Hanbury Brown–Twiss setup for fractional statistics, *Phys. Rev. Lett.* **91**(19), 196803, (2003).

19. G. Campagnano, O. Zilberberg, I. V. Gornyi, D. E. Feldman, A. C. Potter, and Y Gefen, Hanbury Brown–Twiss interference of anyons, *Phys. Rev. Lett.* **109**(10), 106802, (2012).

20. S. Vishveshwara, M. Stone, and D. Sen, Correlators and fractional statistics in the quntum Hall bulk, *Phys. Rev. Lett.* **99**(19), 190401, (2007); D. Sen, M. Stone, and S. Vishveshwara, Quasiparticle propagation in quantum Hall systems, *Phys. Rev. B* **77**(11), 115442, (2008).

21. S. Pratt, Coherence and Coulomb effects on pion interferometry, *Phys. Rev. D* **33**(1), 72-79, (1986).

22. D. Miskowiec, E877 collaboration, Pion-pion correlations in Au+Au collisions at AGS energy, *Nucl. Phys.* **A590**, 473c, (1995).

23. G. Baym and P. Braun-Munzinger, Physics of Coulomb corrections in Hanbury Brown–Twiss interferometry in ultrarelativistic heavy ion collisions, *Nucl. Phys.* **A610**, 286c-296c, (1996).

Vortex Molecules in Thin Films of Layered Superconductors

Alexander I. Buzdin

Condensed Matter Theory Group, LOMA, UMR 5798, University of Bordeaux,
F-33405 Talence, France
E-mail: a.bouzdine@loma.u-bordeaux1.fr [*]

In bulk layered superconductors the vortices tilted with respect to the anisotropy axes attract each other at long distance, which leads to the vortex chains structures. In thin film the intervortex interaction is modified by an extremely slow decay of the supercurrent induced by a single vortex line (Pearl's effect). The interplay between these interactions in thin films is responsible for a formation of a minimum of the interaction potential vs. the intervortex distance. This minimum exists only for relatively strong tilting. Depending on the strength and the tilt of the magnetic field we may expect the formation of the vortex molecules rearranging in multiquanta flux lattices. The increase in the field tilting should be accompanied by the series of the phase transitions between the vortex lattices with different number of vortices per unit cell. The Lorentz microscopy technique seems to be an ideal tool to observe such effects.

1. Introduction

For the first time I heard about the *vortices in superconductors* during my studies at the Department of Low Temperature Physics in Moscow State University. I was really fascinated by these objects and the story of their theoretical discovery by Alexei Abrikosov. However, in spite the fact that I have started myself working on the theory of vortices and even predicted some new phenomena such as the attraction between the tilted vortices in layered superconductors, the perception of vortices as physical objects remained pretty formal for me.

The situation changed after the lecture of Akira Tonomura at the Materials & Mechanisms of Superconductivity Conference (M2S) in 1994 in Grenoble on which I assisted. He presented the Lorentz microscopy measurements of vortex structures.[1] For the first time I saw the moving vortices which formed a regular vortex structure, emerging, collapsing, and vanishing. Tonomura's investigations of vortices gave life to

---

[*] Also at Institut Universitaire de France, Paris.

these objects by inspiring their beauty and reality. I keep memories of this talk of Tonomura as the best experimental presentation I have ever assisted.

The understanding of the properties of type II superconductors is ultimately related to the paradigm of "vortex matter." Thermal excitations, vortex pinning, crystal anisotropy, and spatial and time varying magnetic field, all these reveal panoply of different transitions in this vortex matter, which makes its physics very rich.[2] Both the equilibrium and transport properties of the vortex matter are essentially affected by the behavior of the intervortex interaction potential. The direct observation of vortices and their dynamics by the technique developed by Tonomura was really the outstanding breakthrough in the physics of superconductivity.

Tonomura and I had a first long scientific discussion in Prague at the Conference on Vortex Matter in Superconductor in 2002. Tonomura was interested in the observation of the vortex chains in high T_c superconductors at tilted fields using his technique of electron holography. The questions we discussed were related to the role of the demagnetization factor of the sample and the degree of the electron anisotropy. Soon Tonomura[3,4] managed to observe these vortex chains as well as the combined Abrikosov and Josephson lattices. Nevertheless, during our subsequent meetings he constantly came back to the question of the demagnetization factor in his experiment. It was also the main subject of our email correspondence.

When I visited Tonomura at Hitachi in 2008, he showed me very beautiful pictures of vortices in $YBa_2Cu_3O_{7-\delta}$ (hereafter abbreviated as YBaCuO), which clearly demonstrated that at weak tilting the vortex chains were absent. This finding was quite puzzling because according to my previous works, the tilted vortices always should attract each other at large distances and then form chains. After our two-hour discussion we finally realized that the thickness of his YBaCuO samples was comparable with the London penetration depth and then the very concept of the demagnetization factor was not applicable in this case. Immediately we recalled that in such a case a contribution from Pearl's interaction should be included. This interaction is always repulsive and could mask the subtle attraction of the tilted vortices. The subsequent calculations confirmed this idea and it became clear that the interplay between the surface Pearl's interaction and the volume attraction should lead to formation of the vortex molecules. Tonomura's sense of observation and his strong intention to understand in depth the observed phenomena were crucial for understanding the mechanism of the vortex molecules formation.

In this paper I will analyze the physics of such vortex molecule formation and the role of electron anisotropy.

2. Interaction between tilted vortices in thin-film superconductors

In isotropic bulk superconductors the potential of the interaction between vortices is well known to be repulsive and screened at intervortex distance R greater than the London penetration depth λ. As a result, in perfect crystals quantized Abrikosov vortices form a triangular lattice. In thin films of anisotropic superconductors, however, this standard interaction potential behavior appears to be strongly modified because of the interplay between the long-range repulsion predicted in the pioneering work by J. Pearl[5] and the attraction caused by the tilt of the vortex lines with respect to the anisotropy axes.[6,7]

The long-range repulsion force between vortices in thin superconducting films is caused by an extremely slow decay of the supercurrent j_s induced by a single vortex line [5] expressed by $j_s \propto 1/r^2$, where r is the distance from the vortex center. The intervortex interaction occurs mainly through the magnetic field outside the film, and the repulsive force between two vortices decaying as $\sim 1/r^2$ in contrast to the exponential decay at $r > \lambda$ in bulk superconductors. However, this dramatic difference in the behavior of the interaction potential has not been verified experimentally yet, because most of the measurable physical quantities do not experience qualitative changes related to Pearl's prediction. In particular, the triangular lattice remains as an energetically favorable vortex configuration for any sample thickness. We demonstrated that the situation changes dramatically for thin films of layered superconductors where Pearl's effect appears to be responsible for a strong modification of the vortex configurations formed in a tilted external magnetic field.[8] This possibility of probing Pearl's potential is closely related to the phenomenon of the long-range attraction between tilted vortex lines in anisotropic systems.[6,7] In bulk anisotropic superconductors this phenomenon is known to result in the formation of vortex chains in the regime of low magnetic fields. Indeed, the attraction between two vortices leads to formation of a vortex pair. Then a third vortex will be attracted by this pair, etc. The interaction between any two vortices in the chain (except the nearest neighbors) is attractive, which is sufficient to stabilize the chain. These vortex chains have been observed experimentally by the decoration technique[9] in $YBa_2Cu_3O_{7-\delta-\delta}$ scanning-tuneling microscopy[10] in $NbSe_2$, and Lorentz microscopy measurements[4] in $YBa_2Cu_3O_{7-\delta}$.

If we consider a thin film sample we obtain the interplay between two different long-range potentials: (i) attraction of the tilted vortices ($U_{att} \sim -1/R^2$) and (ii) Pearl's repulsion ($U_{rep} \sim 1/R$). By varying either the film thickness or the tilting angle we can modify the balance between these interactions, which should determine energetically favorable vortex configurations in samples with thickness d comparable to the London penetration depth λ. This interplay results in a very rich vortex physics that, besides the fundamental interest, is important for understanding the electrodynamics of superconducting films.[8,11,12]

In general case the anisotropy axis **c** is oriented perpendicular to the film plane. Our quantitative description of the behavior of the interaction potential is based on the standard London theory. Within this model it is possible to find an exact solution describing the field and current distributions for an arbitrary configuration of vortex lines in a film of finite thickness[13,14] by an appropriate modification of the image method. An anisotropic superconductor in the London limit is known to be characterized by two penetration depths $\lambda_\perp$ and $\lambda_\parallel$ which are, in fact, the lengths of magnetic field screening by currents flowing in directions perpendicular and parallel to the layers, respectively. In the case of large anisotropy $\lambda_\perp \gg \lambda_\parallel$ we can neglect the currents flowing perpendicularly to the layers when considering rather small distances from the vortex core (i.e., r $\ll \lambda_\perp$). This conclusion can be supported by the analyses of the field distribution and vortex interaction in a bulk superconductor carried out within the framework of the anisotropic London theory[6] and the model of non-interacting atomic superconducting layers.[15] All the relevant expressions in these two approaches appear to coincide in the limit r $\ll \lambda_\perp$. It is natural to expect that in this regime the model of non-interacting superconducting layers should be adequate for thin film samples as well.

Let us consider a finite stack of superconducting layers as shown in Fig. 1.

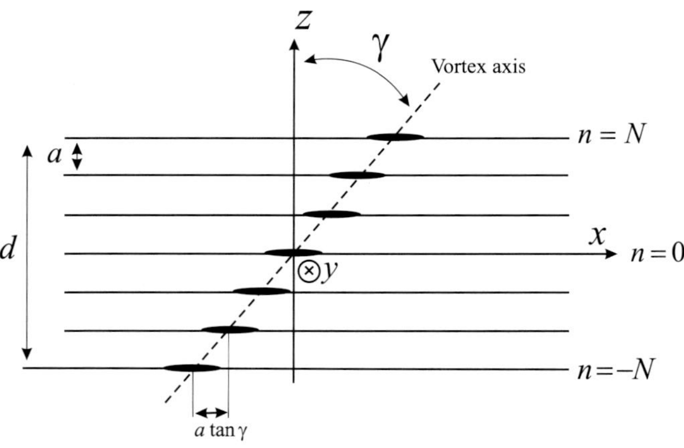

Fig. 1. Schematic representation of a layered superconducting structure with a tilted vortex line: *a* is the distance between layers, n is the layer number, and (*x,y,z*) is the coordinate system.

For simplicity, we restrict ourselves to the case of straight tilted vortex lines, which is a good approximation for a thin superconducting film.

We assume here two vortex lines to be parallel and shifted by a certain vector R in the plane of the layers. Analogously to the bulk limit, the most energetically favorable configuration corresponds to the case $R_y = 0$ which is a consequence of the angular dependence of the attraction force.[6,7,16] Hence, we set $R_y = 0$ and present some typical plots[8] of the interaction energy ε_{int} vs. distance $R_x = R$ in Fig. 2.

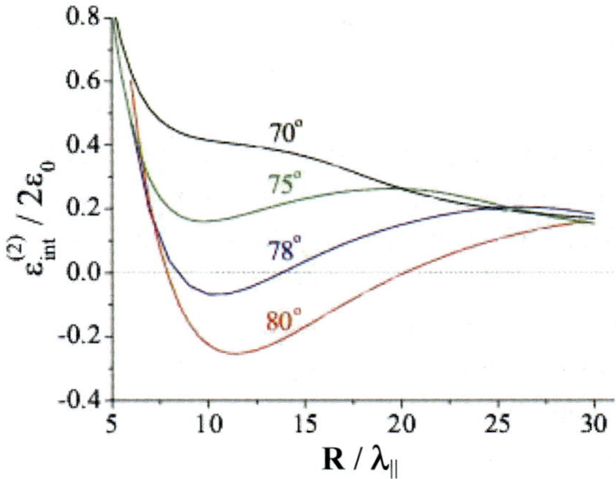

Fig. 2. Typical plots of the interaction energy per vortex vs, the distance R between two vortices for a film of thickness d = $\lambda_\parallel$ and different tilting angles $\gamma = 70°, 75°, 78°, 80°$ ($\varepsilon_0 = \Phi_0^2/16\pi^3\lambda_\parallel$).

Analyzing the dependence ε_{int} (R), one can separate three contributions to the energy of vortex interaction: (a) a short-range repulsion that decays exponentially with increasing intervortex distance R (for $R > \lambda_\parallel$); (b) an intervortex attraction which is known to be specific for tilted vortices in bulk anisotropic systems.[6,7] This attractive energy term decays as R^{-2} and strongly depends on the angle γ between the vortex axis and the c direction; (c) a long-range (Pearl) repulsion which decays as R^{-1} and results from the surface contribution to the energy. Note that the third term does exist even for a large sample with thickness d (see Ref. 5) although in the limit $d \gg \lambda_\parallel$ it is certainly masked by the dominant bulk contribution. At $R \gg \lambda_\parallel$ the short-range interaction term vanishes and one can write an approximate expression describing the behavior of the interaction energy as

$$\varepsilon_{int} \cong \frac{\Phi_0^2}{8\pi^2}\left(-\frac{d_{eff}}{R^2}tan^2\gamma + \frac{2}{R}\right), \tag{1}$$

where the effective film thickness is given by the relation $d_{eff} = d - 2\lambda_\parallel \tanh(d/2\lambda_\parallel)$.

The formula (1) describes interplay between the long-range attractive force (first term) and the repulsive force (second term). Note that the London penetration depth increases with an increase in temperature; thus, the effective thickness decreases and the long range attraction force appears to be suppressed with increasing temperature. For large R, the energy is always positive corresponding to the vortex repulsion similar to the one between the pancakes in a single layer system. With a decrease in the distance R, the attractive force comes into play resulting in the change of the sign of the energy at

$$R^* \sim 0.5 d_{eff} tan^2(\gamma).$$

Naturally this is possible if the tilting angle is not too small, because our analysis is only valid for $R \gg \lambda_\parallel$. This restriction provides us the condition for the attraction existence:

$$tan^2\gamma > tan^2\gamma_c = \frac{2\lambda_\parallel}{d_{eff}}. \tag{2}$$

Therefore, the formation of vortex chains can be energetically favorable only at the tilting $\gamma > \gamma_c$. Note that for relatively thick films it simply means that its thickness d should be larger than the London penetration depth $\lambda_\parallel$.

Modern vortex imaging techniques elaborated in large part by Tonomura provide a possibility to experimentally observe the cross-over between different intervortex interaction regimes in thin film samples. In particular, the magnetic field distributions induced by vortices in thin films can be probed by the penetrating electron beam used in Lorentz microscopy measurements.[1,17] This technique, owing to the low penetration power of the existing 300-kV field-emission beam, permits us to work with films of thicknesses smaller than $0.5 - 1$ μm. It is, therefore, "par excellence" an ideal tool to study the peculiarities of the vortex structures in thin films. In Ref. 8 it was demonstrated how the Lorentz microscopy technique permits to discover the special characters of the intervortex interaction in YBaCuO films of thickness $d \simeq 0.5$ μm, placed in a tilted magnetic field. The vortex structure changes qualitatively for a fixed external magnetic field direction by increasing absolute value of the field B_0. The increase in the field value B_0 causes an increase in the tilting angles of vortex lines according to the estimate. Therefore, the vortex attraction prevails at higher B_0 values, while at low fields the attraction force is overcome by the repulsive force due to Pearl's effect, and the vortex chains are expected to disappear. Indeed, one can see that such a qualitative change in the vortex structure is confirmed[8] by the experimental data: at low fields, which correspond to small γ values, the vortex chains are completely absent, while at rather high fields $B_0 > B^*$, where B^* is the critical field, the formation of vortex chains appears to be energetically favorable. Taking $\lambda_\parallel \sim 0.2$ μm for YBaCuO, we can use the formula (2) to estimate the critical angle γ_c for our films: $tan\gamma_c \sim 1$. This gives a rough estimate[8] of B^* to be ~ 10-100 G, which appears to be in good agreement with the experimental data.[8]

3. Vortex molecules

Even in the regime when the intervortex attraction exists, the formation of infinite chains can be questioned for rather thin films. The point is that, despite the fact that two vortices attract each other, further increase in the number of vortices arranged in a chain can be energetically unfavorable because of the slower decay of the repulsive force compared to the attractive one. Therefore, for rather thin samples, there appears an intriguing possibility to observe vortex chains of finite length, i.e., vortex molecules. The calculations[8,11] confirm this and indeed the number of vortices energetically favorable in a molecule grows as we increase the film thickness and/or the tilting angle because of the increasing attraction term in the pair potential ε_{int}. In Ref. 11 also the equilibrium form of the tilted vortices was calculated and analyzed how it influences the intervortex interaction.

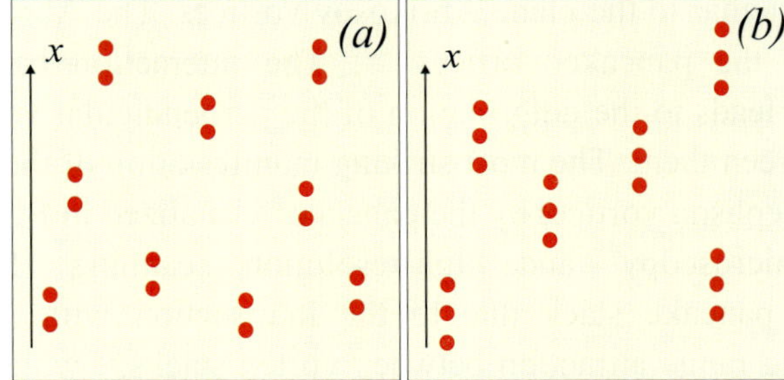

Fig. 3. Schematic pictures of vortex matter consisting of (a) dimeric molecules and (b) trimeric molecules. Vortex positions are denoted by filled circles.

Figure 4 shows schematic pictures of vortex matter consisting of dimeric (Fig. 4a) and trimeric (Fig. 4b) molecules, indicating that the vortex lattices contain more than one vortex per unit cell. The transitions between different multiquanta vortex lattices should occur with the change of the tilting angle and field strength.[11] Finally, for rather thick samples with $d \gg \lambda_\parallel$ we get a standard infinite chain structure typical of bulk systems. Note that the formation of an infinite vortex chain may be considered in some sense as a polymerization of the vortex molecules. Certainly, the crossover from the vortex molecule state to the infinite chain structure is strongly influenced by the increase in the vortex concentration governed by the component of the external magnetic field perpendicular to the film. Indeed, one can expect such a cross-over to occur when the mean intervortex spacing approaches the molecular size.

The experiments[8] were performed at constant orientation of the applied magnetic field. Therefore, by varying the magnetic field, changes of both the tilting angle and vortex concentration were realized. As the vortex concentration was relatively high, we could not expect to observe the molecules (in this regime the average distance between vortices must be much larger than the size of the molecule). To observe the vortex molecules (or multiquanta vortex lattices) it would be preferable to change only the parallel component of the magnetic field, by varying the vortex tilting angle while not affecting the vortex concentration, which must be very low to avoid the inter-molecule interaction.

The preceding analysis is applicable to the moderately anisotropic layered superconductors such as $NbSe_2$ and YBaCuO, when the magnetic field penetrates in the form of the tilted vortices. For strongly anisotropic superconductors such as $Bi_2Sr_2CaCu_2O_{8+\delta}$ (BSCCO), the situation is qualitatively different and the tilted magnetic field penetrates in the form of the crossing lattice comprising the in-plane Josephson vortices and perpendicular to the plane Abrikosov vortices. The Abrikosov vortex, in this case, is a stack of the pancake vortices.[18,19] The interaction between pancake and Josephson vortices[20] leads to the deformation of the perpendicular vortex line and results in the attraction between them. The most striking manifestation of this phenomenon is the decoration of the Josephson vortices by the pancakes visualized in BSCCO single crystals by the Lorentz microscopy[4] and high-resolution scanning Hall probe.[16,21] The deformation of the pancake stack due to the intersection with Josephson vortex is responsible for a long range attraction between vortex stacks,[22] which is quite similar to the vortex attraction in the case of moderately anisotropic superconductors in the tilted field.[6,7] The competition against Pearl's repulsion is quite important for the thin films and may completely overcome the attraction or lead to the formation of the clusters of Abrikosov vortices – vortex molecules,[12] similar to the case of the moderate anisotropy. Our analysis in Ref. 12 shows that for BSCCO the optimal thickness of the film to observe the exotic (multiquanta) vortex lattices is around $d \sim (10\text{-}100)\, \lambda_\parallel$.

4. Conclusion

The Lorentz microscopy technique is ideally adapted to study the vortices in thin film superconductors and my last email correspondence with Akira Tonomura was just devoted to the discussion of the conditions of the experimental observation of these vortex molecules. Passing away of Tonomura-Sensei, who was a great person of exceptional wisdom, is really irremediable loss for science. Visiting him at Hitachi Advanced Research Laboratory, I was always impressed by his original, very deep philosophical approach to the science, his prophetic vision of the role of physics in promoting and discovering the laws of nature. We had also very interesting conversations when I accompanied Abrikosov during his visit to Tonomura's Laboratory at Hitachi Advanced Research Laboratory in 2005 (See Fig. 4). Abrikosov strongly appreciated the works of Tonomura on the Aharonov-Bohm effect and vortex physics and considered them as Nobel Prize level achievements.

Fig. 4. A. Abrikosov and A. Tonomura together with A. Buzdin and K. Kadowaki at Tsukuba University in October 2005.

Acknowledgments

I am very grateful to Akira Tonomura for many discussions which stimulated my works and formed the basis of the research presented in this article. I am indebted to my colleagues A. Mel'nikov and A. Samokhvalov for their collaboration in the works reviewed in the present article.

References

1. A. Tonomura, Electronic Holography, 2nd Ed. (Springer Series in Optical Sciences 70) (Springer, Berlin, Heidelberg, 1999).
2. G. Blatter, G. Blatter, M. V. Feigel'man, V. B. Geshkenbein, A. I. Larkin, and V. M. Vinokur, Vortices in high-temperature superconductors, *Rev. Mod. Phys.* **66**(4), 1125-1388, (1994).
3. T. Matsuda, O. Kamimura, H. Kasai, K. Harada, T. Yoshida, T. Akashi, A. Tonomura, Y. Nakayama, J. Shimoyama, K. Kishio, T. Hanaguri and K. Kitazawa, Oscillating rows of vortices in superconductors, *Science* **294**, 2136-2138, (2001).
4. A. Tonomura, H. Kasai, O. Kamimura, T. Matsuda, K. Harada, T. Yoshida, T. Akashi, J. Shimoyama, K. Kishio, T. Hanaguri, K. Kitazawa, T. Masui, S. Tajima, N. Koshizuka, P. L. Gammel, D. Bishop, M. Sasase, and S. Okayasu, Observation of structures of chain vortices inside anisotropic high-T_c superconductors, *Phys. Rev. Lett.* **88**(23), 237001, (2002).

5. J. Pearl, Current distribution in superconducting films carrying quantized fluxoids, *Appl. Phys. Lett.* **5**(4), 65-66, (1964).

6. A. I. Buzdin, and A. Yu. Simonov, Penetration of inclined vortices into layered superconductors, *JETP Lett.* **51**, 191-195, (1990).

7. A. M. Grishin, A. Yu. Martynovich, and S. V. Yampolskii, Magnetic field inversion and vortex chains in anisotropic superconductors, *Sov. Phys. JETP* **70**, 1089-1098, (1990).

8. A. I. Buzdin, A. S. Mel'nikov, A. V. Samokhvalov, T. Akashi, T. Masui, T. Matsuda, S. Tajima, H. Tadatomo, and A. Tonomura, Crossover between magnetic vortex attraction and repulsion in thin films of layered superconductors, *Phys. Rev. B* **79**(9), 094510, (2009).

9. P. L. Gammel, D. J. Bishop, J. P. Rice, and D. M. Ginsberg, Images of the vortex chain state in untwinned $YBa_2Cu_3O_{7-\delta}$ crystals, *Phys. Rev. Lett.* **68**(22), 3343-3346, (1992).

10. H.F. Hess, C. A. Murray, and J. V. Waszczak, Scanning-tunneling-microscopy study of distortion and instability of inclined flux-line-lattice structures in the anisotropic superconductor $2H\text{-}NbSe_2$, *Phys. Rev. Lett.* **69**(14), 2138-2141, (1992).

11. A. V. Samokhvalov, D. A. Savinov, A. S. Mel'nikov, and A. I. Buzdin, Vortex clusters and multiquanta flux lattices in thin films of anisotropic superconductors, *Phys. Rev. B* **82**(10), 104511, (2010).

12. A. V. Samokhvalov, A. S. Mel'nikov, and A. I. Buzdin, Attraction between pancake vortices and vortex molecule formation in the crossing lattices in thin films of layered superconductors, *Phys. Rev. B* **85**(18), 184509, (2012).

13. V. Pudikov, Peculiarity of vortex interaction near the surface in highly layered superconductors, *Physica C* **212**(1-2), 155-163, (1993).

14. G. Carneiro and E. H. Brandt, Vortex lines in films: Fields and interactions, *Phys. Rev. B* **61**(9), 6370-6376, (2000).

15. A. I. Buzdin, Vortex structure in the presence of tilted columnar defects, *JETP Lett.* **68**, 544-548, (1998).

16. S. J. Bending and M. J. W. Dodgson, Vortex chains in anisotropic superconductors, *J. Phys.: Condens. Matter.* **17**(35), R955, (2005).

17. K. Harada, T. Matsuda, J. Bonevich, M. Igarashi, S. Kondo, G. Pozzi, U. Kawabe, and A. Tonomura, Real-time observation of vortex lattices in a superconductor by electron microscopy, *Nature* **360**, 51-53, (1992).

18. A. I. Buzdin and D. Feinberg, Electromagnetic interaction of vortices in layered superconducting structures, *J. Phys. (Paris)* **51**, 1971 (1990).

19. J. R. Clem, Two-dimensional vortices in a stack of thin superconducting films: A model for high-temperature superconducting multilayers, *Phys. Rev. B* **43**(10), 7837-7846, (1991).

20. A. Koshelev, Crossing lattices, vortex chains, and angular dependence of melting line in layered superconductors, *Phys. Rev. Lett.* **83**(1), 187-190, (1999)

21. A. Grigorenko, S. Bending, T. Tamegai, S. Ooi, and M. Henini, A one-dimensional chain state of vortex matter, *Nature* b, 728 (2001).

22. A. Buzdin and I. Baladié, Attraction between pancake vortices in the crossing lattices of layered superconductors, *Phys. Rev. Lett.* **88**(14), 147002, (2002).

Coherent Quantum Phase Slip

Jaw-Shen Tsai

RIKEN Center for Emergent Matter Science
2-1, Wako, Saitama 351-0198, Japan and
NEC Smart Energy Research Laboratories
34 Miyukigaoka, Tsukuba 305-8501, Japan
E-mail: tsai@cj.jp.nec.com

Observation of motion of individual magnetic vortex in trapped in superconductor was one of the many beautiful achievements that Akira Tonomura shown to us. Here we report the recently found quantum tunneling of such magnetic flux through a thin superconducting wire, the quantum phase slip.

At around the year 2001, Dr. Akira Tonomura was trying to form a new research group in RIKEN, under the auspices of the Frontier Research Project, and he kindly approached our research group at NEC, asking us to join his research group. We were very much exited for his invitation and offer. The research group included his team in Hitachi, Franco Nori's team, Yoshichika Otani's team, and mine (See Fig. 1). The first thing we thought about at that time, was to promptly come up with a suitable name for the group, for the name would naturally dictate the direction of our research. We tried to pick a name that represented the greatest denominator of the research topics among the four teams and the name "Single Quantum Dynamics" was thus chosen.

At Tonomura's team, observation of isolated quantized magnetic vortices in superconductors using the electron holography technique have been realized. At our team, we had demonstrated the quantum state control in Josephson-junction-based superconducting circuits. We naturally considered carrying out collaborative research between two teams from the beginning. However, it turned out to be a difficult feat. In two sets of experiments, the typical energy scale involved had a huge difference in magnitudes. Nevertheless, I enormously enjoyed working within this research group under his directorship.

After about 12 years from its initiation, the Single Quantum Dynamics Research Group was finally dissolved in March 2013, just about one year after the death of Akira Tonomura. I cherish the precious period during which I was able to work under his magnificent leadership. He was such an extraordinary scientist, remarkable mentor, as well as a very dear friend of mine.

Fig. 1. The author (Jaw-Shen Tsai) with the other team leaders of the Single Quantum Dynamics Research Group of RIKEN. From left to right: Yoshichika Otani, Franco Nori, Akira Tonomura (Group Director), and Tsai (at RIKEN, May 13, 2010).

In this article, I would like to present the latest notable result of our team, the first demonstration of the quantum phase slip. When I told him about our result over the telephone, Tonomura was very much excited about our achievement and immediately appreciated the importance of it, in spite of his already degrading bodily condition.

Superconductivity describes a phenomenon in which electrons pass through certain types of materials without any resistance when cooled below a given temperature. Among the most important applications of superconductivity is the Josephson junction, named after physicist Brian Josephson, who in 1962 predicted[1] that a superconducting current could tunnel between superconductors separated by a thin insulating layer. This phenomenon, the Josephson effect, has been applied in a variety of areas including magnetometer design, voltage standardization, and quantum computing.

Researchers have long known of an intriguing theoretical parallel[2,3] to the Josephson effect in which insulator and superconductor are reversed: rather than electric charges (Cooper pairs) jumping from one superconducting layer to another across an insulating layer, magnetic flux quanta jumps from one insulator too another across a superconducting layer (Fig. 2). Quantum tunneling of electrons in the Josephson junction is replaced in this parallel by the coherent "slip" of the phase, a quantum variable that, in superconducting circuits, plays a dual role to that of electric charge. This fundamental transport phenomenon in superconducting system is known as "coherent quantum phase slip (CQPS)" and has been limited to theory until recently.

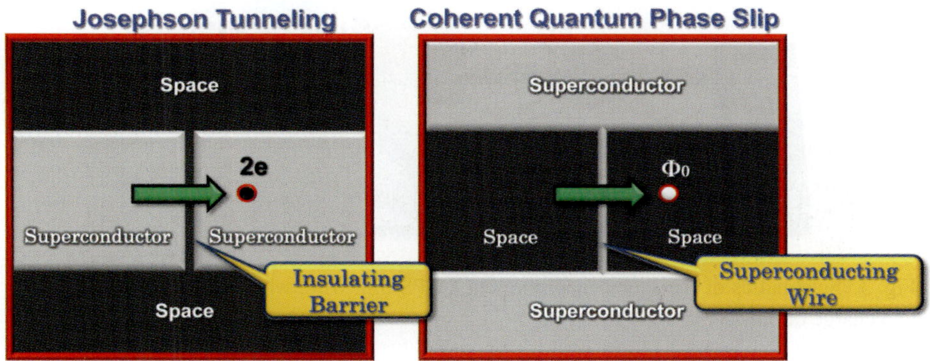

Fig. 2. Schematics of Josephson tunneling (left) and coherent quantum phase slip (right). Left figure: Two superconductors separated by thin insulator (in this case space). Current is produced by superconducting electron pairs tunneling across the insulator. Right figure: Two insulators separated by superconductor. Current is produced by magnetic flux tunneling across the superconductor.

In 2012 we succeeded in the first direct observation[4] of CQPS in a narrow superconducting wire of indium-oxide (InOx). The wire is inserted into a larger superconducting loop to form a new device called a phase-slip qubit, with the superconducting layer (the thin wire) sandwiched between insulating layers of empty space.

We studied the effect in qubit spectroscopy experiments. Since the tunneling probability of the quantum phase slip is exponentially proportional to the cross-sectional area as well as the conductivity (just above the superconducting transition) of the thin wire, one should use a very thin wire that is made of material such as disordered superconductor where high normal resistivity is usually found.

To demonstrate the CQPS effect unambiguously, we imbedded a thin wire in a flux qubit configuration[2] as in Fig. 3.

In such a device, energy gap should appear in the energy dispersion relation, because of the coherent transfer of flux through the thin wire. Such energy gap is readily observed in a spectroscopy experiment that has been developed for the characterization of the flux qubit.

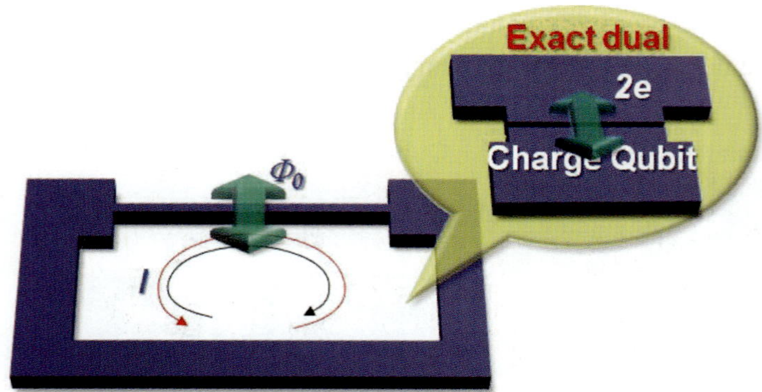

Fig. 3. Schematics of flux qubit incorporating CQPS thin wire, which is an exact conjugate to a charge qubit.

Figure 4 shows the actual picture of the device. It is a flux qubit with CQPS thin wire, implanted in a coplanar microwave resonator. The thin wire, flux qubit, as well as the center transmission line of the resonator is made of indium oxide (InO_x). The material was chosen because it is a well-studied disordered superconductor and relatively easy to prepare. The film we used had a superconducting transition temperature of about 2.7K with the film thickness of 35 nm and the normal sheet resistance of 1.7 kΩ per square.

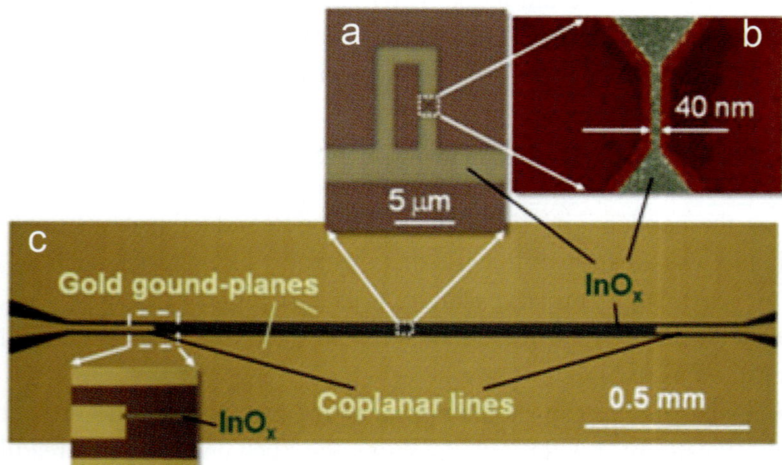

Fig. 4. The sample picture. **a**: InO_x loop with a narrow wire segment on the right side is attached to the resonator (horizontal line) at the bottom. **b**: False-color scanning-electron micrograph of the narrow InO_x segment. **c**: Step-impedance resonator, comprising a 3-μm-wide InO_x strip galvanically coupled to a gold coplanar line.

To detect the superposition of flux states we measure transmission t through the resonator at frequency versus external magnetic field B_{ext}. The fourth mode peak of the transmission exhibits well pronounced periodic structure: sharp negative dips in the amplitude $|t|$, and in the phase rotation. The period corresponds with high accuracy to one flux quantum through the area of 32 mm^2 of the loop shown in Fig. 4, with a 40-nm-wide wire.

We were able to perform spectroscopy measurements by monitoring resonator transmission while tuning the magnetic flux penetrating this loop by scanning B_{ext} and scanning the probe microwave frequencies. We detected a band gap in the energy curves from the transmission phase plot, showing the resonance excitation of the two-level system as shown in Fig. 5. The green-blue line corresponds to the expected energy splitting, which is well fit by the theoretical model. The obtained energy gap from Fig. 4 is about 4.9 GHz. This clearly demonstrates coherent coupling between the flux states in the loop. This gap is a result of quantum mechanics, which prevents the two states from occupying the same energy level, forcing them to tunnel across the superconducting layer—and through a quantum phase-slip in the narrow wire—to avoid it.

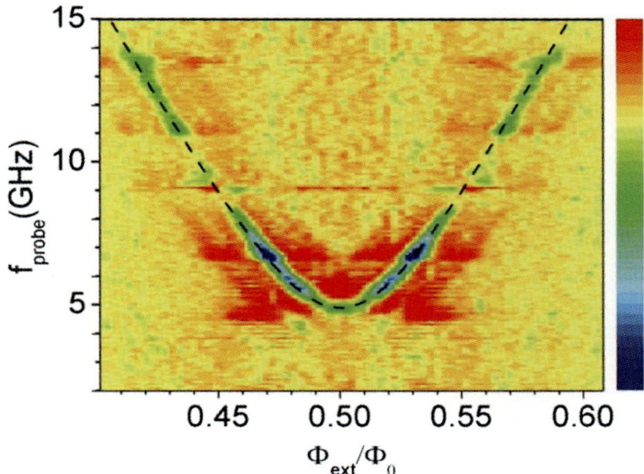

Fig. 5. Energy band gap obtained using energy spectroscopy, showing an occurrence of coherent tunneling.

The spectroscopy experiment showed a clear coherent nature of the flux tunneling. Thus we have demonstrated the CQPS in thin wire. The demonstration was carried out just 50 years after the initial proposal of the Josephson effect, which is the quantum conjugate of CQPS. The successful experiment also ushers in a novel class of devices that exploit the unique functionality of quantum phase-slip to gorge a new path in superconducting electronics.

                                  J.-S. Tsai

Acknowledgments

The original authors of Ref. 1, O. V. Astafiev, L. B. Ioffe, S. Kafanov, Yu. A. Pashkin, K. Yu. Arutyunov, D. Shahar, and O. Cohen, are acknowledged. This work was supported by the grant from the Japan Society for the Promotion of Science (JSPS) through the "Funding Program of World-Leading Innovative R&D on Science and Technology (FIRST Program)," initiated by the Council for Science and Technology Policy (CSTP). This work was also supported by Grants-in-Aid for Scientific Research, MEXT Kakenhi, on "Quantum Cybernetics".

References

1. B. D. Josephson, Possible new effects in superconductive tunnelling, *Phys. Lett.* **1**(7), 251-253, (1962).
2. J. E. Mooij and C. J. P. M. Harmans, Phase-slip flux qubits, *New J. Phys.* **7**, 219, (2005).
3. J. E. Mooij and Yu. V. Nazarov, Superconducting nanowires as quantum phase-slip junctions, *Nature Physics* **2**, 169-172, (2006).
4. O. V. Astafiev, L. B. Ioffe, S. Kafanov, Yu. A. Pashkin, K. Yu. Arutyunov, D. Shahar, O. Cohen, and J. S. Tsai, Coherent quantum phase slip, *Nature* **484**, 355-358, (April 2012).

Coherency of Spin Precession in Metallic Lateral Spin Valves

YoshiChika Otani[*], Hiroshi Idzuchi[†] and Yasuhiro Fukuma[‡]

RIKEN Center for Emergent Matter Science, Wako, Saitama 351-0198, Japan
E-mail: yotani@issp.u-tokyo.ac.jp

Diffusive pure spin currents in lateral spin valves lose phase coherency in precession while undergoing scattering events, leading to a broad distribution of the dwell time in a transport channel from the injector to the detector. Here we demonstrate the lateral spin-valves with dual injectors enable us to detect a genuine precession signal from the Hanle effect, demonstrating that the phase coherency in precession is improved with an increase of the channel length. The coherency in the spin precession shows a universal behavior as a function of the normalized separation between the injector and the detector in material-independent fashion for metals and semiconductors including graphene.

Commemoration of Dr. Akira Tonomura

First of all, I (Otani) would like to express my sincere condolence on the passing of Dr. Akira Tonomura on 2nd May, 2012 who had been the director of Single Quantum Dynamics Research Group at the Advanced Science Institute of RIKEN since 2001, which my Quantum Nanoscale Magnetics research team belonged to. I have known him since I was a postdoc researcher at Laboratoire Louis Néel CNRS Grenoble, France in 1992. At that time Dr. Tonomura was the man of the times after his famous demonstrative electron holography experiment of the Aharonov-Bohm effect. We were seeking a possibility to invite Dr. Tonomura as an honor speaker for the institute seminar through my Ph.D. supervisor, Prof. Chikazumi, who had been serving as an advisor at Hitachi Central Research Laboratory. Our invitation was sudden; however, he kindly took time from his busy schedule to visit our institute to give an invited lecture on his experiments. His beautiful experiments deeply impressed us and made me decide to prepare samples to be observed by his electron holography. Since then I have been collaborating with his research group. During our collaboration he was always

---

[*] *Institute for Solid State Physics, University of Tokyo, Kashiwa, 277-8581, Japan*
[†] *Institute for Solid State Physics, University of Tokyo, Kashiwa, 277-8581, Japan*
[‡] *Frontier Research Academy, Kyushu Institute of Technology, Iizuka 820-8502, Japan.*

encouraging us and promoting opportunities for challenging experiments in the new research field such as spintronics. The result shown here is one of the examples which came to fruition through his encouragement.

1. Introduction

Recent advance in nano-scale fabrication technology has opened up a possibility for studying a diffusive transport of accumulated spins in a nonmagnet by means of nonlocal spin injection.[1-4] The mean free path for consecutive spin flip events is called "spin diffusion length", and is much longer than that of electrons collisions.[5] The diffusive transport of spins, i.e., the pure spin currents, thus provides not only a variety of scientific interests but also an additional data transfer functionality for future spintronic device applications.[6,7]

Hanle effect measurement is one of the most effective means to characterize the spin transport and relaxation in nonmagnets.[8,9] When the magnetic field is applied perpendicular to the spin orientation, a collective spin precession is induced by its torque. In ballistic spin transport, spins can coherently rotate at a frequency proportional to the applied magnetic field. This enables us to control the direction of the spins in the channel and to manipulate the output signal of lateral spin valves (LSVs) by adjusting an effective external parameter such as the Rashba field tunable via a gate voltage.[10] This scheme realizes an active spin device such as a spin-transistor.[11] However, in a diffusive pure spin current in nonmagnets, the incoherent precession of collective spins causes drastic decreases of the spin accumulation.[12-21] We here demonstrate the dual spin injectors for LSVs are effective to examine coherency in the collective precession of the diffusive spins, which enhance the spin accumulation in the channel and also suppress the spurious signal in the Hanle effect measurements. In this way we can detect a genuine in-plane precession signal in a 10 μm long Ag wire. The coherent characteristics are investigated by the spin precession signal, revealing that the dwell time distribution narrows when the spins diffuse longer distance in the channel between the injector and the detector of LSVs.

2. Experimental procedure

Lateral spin valves with Permalloy/MgO/Ag junctions were prepared on a Si/SiO_2 substrate by means of suspended shadow mask evaporation. All the layers were e-beam deposited in an ultra-high vacuum (10^{-6} Pa).

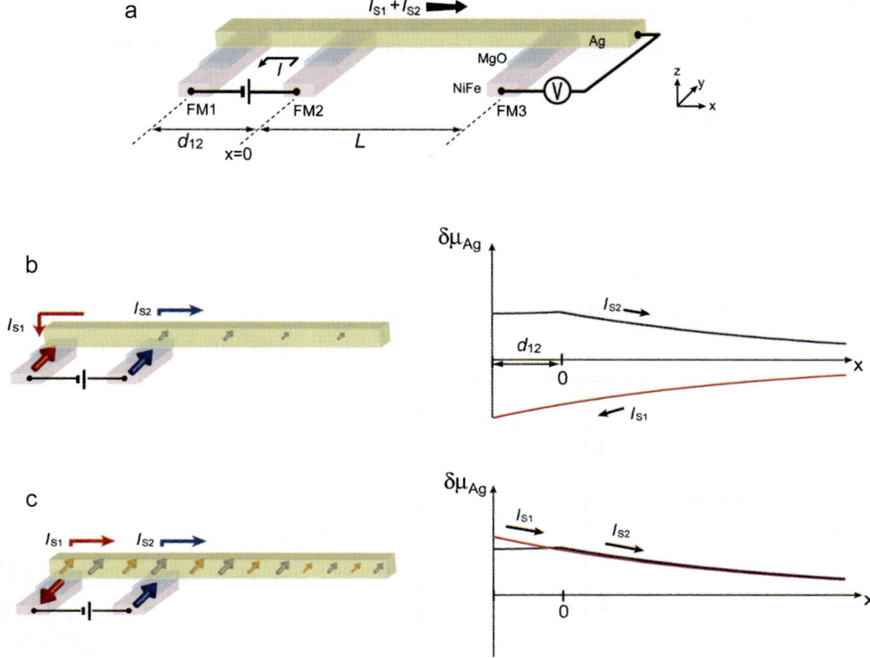

Fig. 1. Schematic diagram of LSV with dual injectors in non-local measurement configuration. **(a)** The Permalloy wires are 140 nm in width and 20 nm in thickness. The Ag wire is 120 nm in width and 100 nm in thickness. The separation between FM1 and FM2, d_{12}, is 350 nm. **(b)** and **(c)** Illustrations for parallel or anti-parallel configuration on the left and the spatial variations of spin accumulation in silver $\delta\mu_{Ag}$ on the right. The red and blue curves show $\delta\mu_{Ag}$ induced by spin injectors of FM1 and FM2, respectively. In the parallel configuration **b**, the spin currents I_{S1} and I_{S2} flow in the opposite directions. In the anti-parallel configuration **c**, the I_{S1} and I_{S2} flow in the same direction.

Firstly, a 20-nm-thick Permalloy layer was obliquely deposited at 45° tilted from the substrate normal, followed by the second deposition of an interface MgO layer at the same tilting angle of 45°. Then, a 100-nm-thick Ag layer was obliquely deposited normal to the Si substrate. Finally, a 3-nm-thick capping MgO layer was deposited to prevent surface contamination of the devices. All the prepared devices were annealed at 400°C for 40 min in a N_2 (97%) + H_2 (3%) mixture atmosphere.

The non-local measurements were carried out by using a dc current source and nano-voltmeter. The bias current in the range between 200 and 400 µA was applied to the injector. The magnetic field was applied parallel to the Permalloy wires for the spin valve measurements. For the Hanle effect measurements, the magnetic field was applied perpendicular to the Si substrate. The field direction was carefully aligned to prevent in-plane field component that switches the Permalloy magnetization during the measurements.

A conventional LSV consists of a pair of injector and detector ferromagnetic wires that are bridged by a nonmagnetic wire. The spins are injected by applying a bias voltage across the ferromagnetic/nonmagnetic interface and accumulate in its vicinity. Their density decays exponentially with a factor of $\exp(-d/\lambda_S)$ where d is the distance from the injector and λ_S is the spin diffusion length. Unlike the above LSV, our structure consists of three $Fe_{20}Ni_{80}$ (Permalloy) wires bridged by a Ag wire as shown in Fig. 1a. The current I is applied between FM1 and FM2 for the spin injection into the Ag wire, and the spin accumulation in voltage V is detected by using the third Permalloy wire FM3. To avoid the back flow of the induced spin currents into FM wires, we employ the Permalloy/MgO/Ag junction in the present study.[21,22] Our scheme shown in Fig. 1b confines the spin current solely to the detector since the unnecessary side edge, i.e., relaxation volume, on the left of FM1 is removed. Figure 1b shows $\delta\mu_{Ag}$ from each injector with parallel magnetic configuration and the magnitude of each spin current from FM1 and FM2 can be twice the conventional I_S due to the confinement effect. Because of the opposing spin current across the FM1/MgO/Ag and FM2/MgO/Ag interfaces, the currents flowing across the interfaces cancel out for the parallel magnetic configuration of dual injectors, whereas for the anti-parallel configuration, the spin currents induced by FM1 and FM2 are constructive as depicted in Fig. 1c. As a result, this scheme enhances the total spin current by up to a factor of four compared to the conventional one.

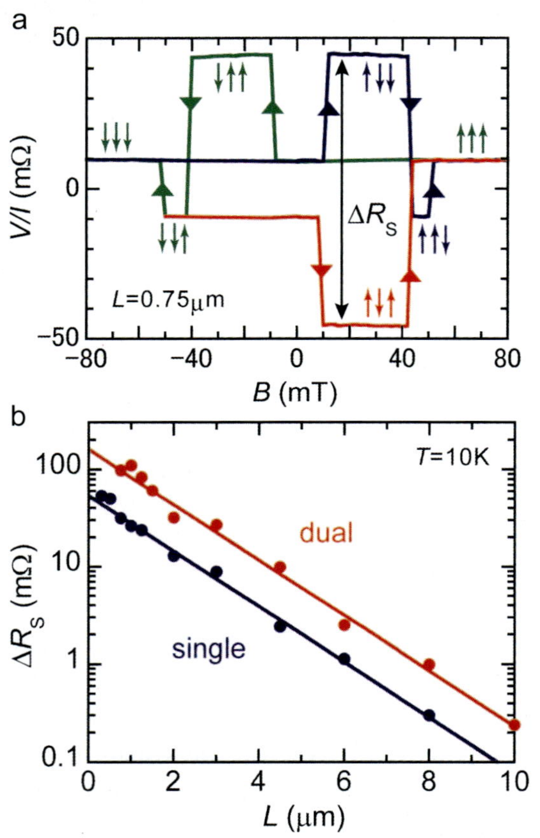

Fig. 2. **(a)** Spin signal as a function of magnetic field for double injector LSV (DLSV) at 10 K. **(b)** Spin valve signal ΔR_S as a function of injector-detector separation at $T = 10$ K. The solid lines were obtained by fitting the data points to equation (1).

The non-local spin valve (SV) signal of the conventional single injector LSV (SLSV) show high and low values which correspond respectively to the parallel and antiparallel configurations of the injector and detector FMs, of which overall change ΔR_S amounted to 31.5 mΩ at 10 K in the present study. In contrast, as in Fig. 2a, the non-local SV signals for the dual injector LSV (DLSV) exhibit 3 distinct switching fields at around 10 mT, 40 mT and 50 mT, corresponding to the switching fields of FM1, FM2 and FM3, respectively. The hysteresis loop shows three-level signals associated with the magnetic configurations among three FM electrodes. Since the antiparallel configuration of FM1 and FM2 maximizes the spin accumulation in DLSV, the overall change ΔR_S can therefore be evaluated as a difference between the magnetic configurations between FM2 and FM3 by keeping the antiparallel configuration of FM1 and FM2. The center-to-center injector and detector separation L dependence of ΔR_S in Fig. 2 shows that ΔR_S of DLSV is almost 3 times larger than that of SLSV and decreases exponentially with increasing L due to Elliot-Yafet spin relaxation mechanism in the Ag wire.[23,24]

When the interface resistance is larger than the spin resistance of the nonmagnetic wire R_N, the analytical expression of ΔR_S for DLSV is approximated by using the solution of the one-dimensional spin diffusion equation

$$\Delta R_S = \alpha P_I^2 R_N e^{-L/\lambda_N} , \tag{1}$$

where $\alpha = 1 + \exp(-2d_{12} / \lambda_N) + 2\exp(-d_{12} / \lambda_N)$, P_I is interfacial polarization, λ_N is the spin diffusion length of the nonmagnet, and d_{12} is the separation between FM1 and FM2. The ΔR_S for DLSV is enhanced remarkably by a factor of α compared to that of SLSV, corresponding to the reduced equation (1) of ΔR_S for SLSV[25] with $d_{12} \gg \lambda_N$. The first and second terms in α represent the spin current injected from FM2 and the third term represents the spin current injected from FM1. The obtained experimental results in Fig. 2 were fitted to equation (1) by adjusting parameters P_I and λ_N, yielding $P_I = 0.36$, $\lambda_N = 1500$ nm and $\alpha = 3.2$. The results are consistent with our previous data in LSVs with permalloy/MgO/Ag junctions.[21]

The Hanle effect measurements were performed on LSVs by applying perpendicular magnetic fields. Figure 3a shows the modulated non-local spin signal for SLSV and DLSV. A parabolic background signal is observed for the SLSV, the origin of which is the magnetization process of FMs. When the applied magnetic field is increased above the demagnetizing field of the FM wires, the magnetizations for the injector and the detector are tilted up along the field direction, pushing the background signal up towards the value of parallel configuration for FMs. To describe the both contributions of spin precession and magnetization process, we decompose them into that of spin precession in x-y plane and that of the z component reflecting the magnetization process. The non-local spin signal V/I in the presence of B_Z is thus given by the sum of the above two contributions;

$$\frac{V}{I} = R_{\mathrm{S}}^{\mathrm{Hanle}}\left(\omega_{\mathrm{L}}, \mathbf{e}^{\mathrm{FM1}}\cdot\mathbf{e}_y, \mathbf{e}^{\mathrm{FM2}}\cdot\mathbf{e}_y, \mathbf{e}^{\mathrm{FM3}}\cdot\mathbf{e}_y\right) + R_{\mathrm{S}}^{\mathrm{Hanle}}\left(0, \mathbf{e}^{\mathrm{FM1}}\cdot\mathbf{e}_z, \mathbf{e}^{\mathrm{FM2}}\cdot\mathbf{e}_z, \mathbf{e}^{\mathrm{FM3}}\cdot\mathbf{e}_z\right), \tag{2}$$

with

$$R_{\mathrm{S}}^{\mathrm{Hanle}}\left(\omega_{\mathrm{L}}, a^{\mathrm{FM1}}, a^{\mathrm{FM2}}, a^{\mathrm{FM3}}\right) = \frac{1}{2}P_{\mathrm{I}}^2 R_{\mathrm{N}}\,\mathrm{Re}\left[\alpha_\omega\left(\lambda_\omega/\lambda_{\mathrm{N}}\right)\exp(-L/\lambda_\omega)\right], \tag{3}$$

where $\alpha_\omega = a^{\mathrm{FM2}}a^{\mathrm{FM3}}\left\{1+\exp(-2d_{12}/\lambda_\omega)\right\} - 2a^{\mathrm{FM1}}a^{\mathrm{FM3}}\exp(-d_{12}/\lambda_\omega)$, $\lambda_\omega = \lambda_{\mathrm{N}}/\sqrt{1+i\omega_{\mathrm{L}}\tau_{\mathrm{sf}}}$, $\tau_{\mathrm{sf}} = \lambda_{\mathrm{N}}^2/D_{\mathrm{N}}$ is the spin relaxation time, D_{N} is the diffusion constant, $\omega_{\mathrm{L}} = \gamma_e B_z$ is the Larmor frequency, $\gamma_e = g\mu_{\mathrm{B}}/\hbar$ is the gyromagnetic ratio, g is the g-factor, μ_{B} is the Bohr magneton and $a^{\mathrm{FM}i}$ is the projection of the unit vector of the magnetization of FMi $\mathbf{e}^{\mathrm{FM}i}$ on y or z-axis. Note that $R_{\mathrm{S}}^{\mathrm{Hanle}}\left(\omega_{\mathrm{L}}, \mathbf{e}^{\mathrm{FM1}}\cdot\mathbf{e}_y, \mathbf{e}^{\mathrm{FM2}}\cdot\mathbf{e}_y, \mathbf{e}^{\mathrm{FM3}}\cdot\mathbf{e}_y\right)$ represents the nonlocal resistance for the precessional frequency ω_{L} and $R_{\mathrm{S}}^{\mathrm{Hanle}}\left(0, \mathbf{e}^{\mathrm{FM1}}\cdot\mathbf{e}_z, \mathbf{e}^{\mathrm{FM2}}\cdot\mathbf{e}_z, \mathbf{e}^{\mathrm{FM3}}\cdot\mathbf{e}_z\right)$ represents the nonlocal resistance without spin precession.

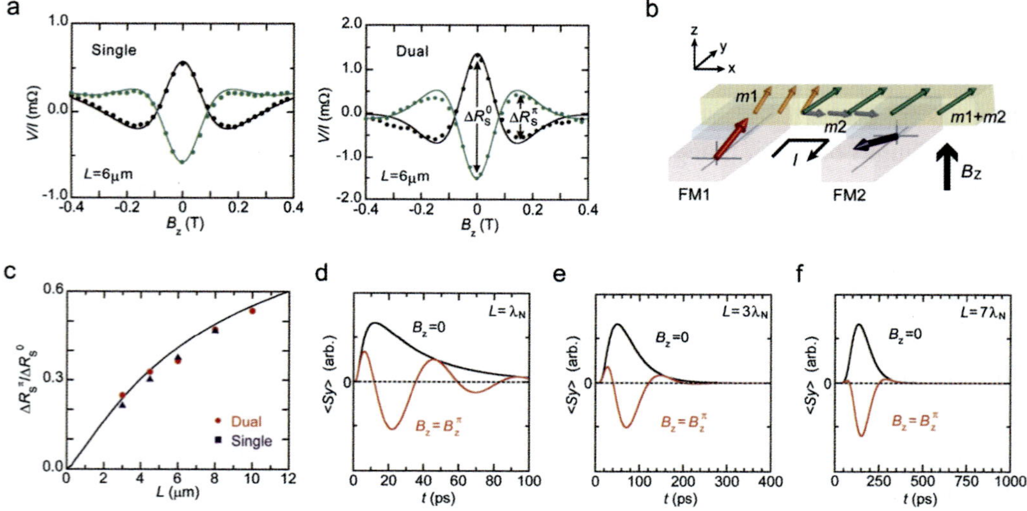

Fig. 3. **(a)** Non-local spin signal modulated by spin precession as a function of perpendicular field for single and dual injector LSVs with $L = 6$ μm at 10 K. The solid lines are the fitting curves using equation (2). **(b)** Schematic diagram of magnetizations of injectors and non-equilibrium magnetization m in the Ag wire in the presence of high B_Z. In Ag, y-components of $m1$ and $m2$ are constructive but their z-components are canceled out each other out for both anti-parallel and parallel magnetization configurations of the injectors. **(c)** Coherent parameter of the spin precession $\Delta R_{\mathrm{S}}^\pi/\Delta R_{\mathrm{S}}^0$ as a function of L. The solid lines are the fitting curves derived from equation (2) using the same parameters as used in **a**. Data are corrected by taking into account of the influence of magnetization process. **(d)-(f)** Density of y-directional spin arrived at the detector as a function of the dwell time for LSVs with different L. The black and red lines, respectively, represent the distributions of the dwell time in the channel with $B_z = 0$ and with $B_z = B_z^\pi$ that causes π rotation after the spin injection.

For SLSV, we obtain $P_{\mathrm{I}} = 0.37$ and $\lambda_{\mathrm{N}} = 1420$ nm by fitting equation (2) to the experimental data as shown in Fig. 1a. These values are consistent with those obtained alternatively from the L dependence of ΔR_{S} in the previous section. For DLSV, the z component of the injectors is canceled out because of the opposite direction of the applied current to the junctions as depicted in Fig. 3b. This allows us to detect the genuine precession signal and to investigate the dynamic properties of the spin current as will be discussed below.

In the diffusive pure-spin transport, the collective spin precession decoheres due to broadening of the dwell time distribution in the channel between the injector and the detector.[10] The amplitude of the spin valve signal at $B_Z = 0$ decreases after the π rotation at $B_Z^{\pi} = 0.16$ T, as can be seen in Fig. 3a. In order to better quantify the coherency in the collective spin precession, we define the coherent parameter as the ratio $\Delta R_{\mathrm{S}}^{\pi} / \Delta R_{\mathrm{S}}^{0}$, where $\Delta R_{\mathrm{S}}^{\pi}$ and $\Delta R_{\mathrm{S}}^{0}$ are respectively the amplitude of the spin valve signal right after the π rotation and that in zero field right before the rotation begins. The $\Delta R_{\mathrm{S}}^{\pi} / \Delta R_{\mathrm{S}}^{0}$ increases with increasing L, and the experimental trend is well reproduced by equation (2) as shown in Fig. 3c. To understand the observed trend in more detail, we employ the one-dimensional diffusion model which gives the y-component of net spin density at the detector, as a function of the dwell time t in the presence of B_Z.[10]

$$\langle S_y \rangle \propto 1 / \sqrt{4\pi D_{\mathrm{N}} t}\, \exp\left(-\left(\frac{L^2}{4 D_{\mathrm{N}} t}\right) - \frac{t}{\tau_{\mathrm{sf}}}\right) \cos(\omega_{\mathrm{L}} t), \tag{4}$$

The $\langle S_y \rangle$ versus t curves for $L = \lambda$ with $B_Z = 0$ and $B_Z = B_Z^{\pi}$ are shown in Fig. 3d. When $B_Z = 0$, $\langle S_y \rangle$ takes a broad peak structure followed by a long exponential tail. The detected spin signals in LSVs are proportional to the $\langle S_y \rangle$ integrated over time. The distribution of $< S_y >|_{B_Z=0}$ becomes narrower as the channel length gets longer, of which evolution is depicted in three distribution curves under $B_Z = 0$ of Figs. 3d-f. The long exponential tail observed in Fig. 3d diminishes in proportion to $1/\sqrt{t}\, \exp(-t/\tau_{\mathrm{sf}})$. When $B_Z = B_Z^{\pi}$ is applied, the integrated value of $\langle S_y \rangle$ over time cancels out for a short separation $L \sim \lambda$ (Fig. 3d) whereas it does not cancel for a long separation $L \gg \lambda$ (Figs. 3e and f), indicating that the coherence of collective spin precession is well preserved for long spin transport. This trend is experimentally observed as an increase of $\Delta R_{\mathrm{S}}^{\pi} / \Delta R_{\mathrm{S}}^{0}$ from 0.21 to 0.53 with the increase in L from 3 to 10 μm as shown in Fig. 3c.

3. Discussion

To better understand the coherence in collective spin precession, $t = \tau_{\mathrm{sf}} T$ is substituted into the distribution function at $B_Z = 0$. We then obtain $< S_y > \propto 1/\sqrt{T}\, \exp\left(-(L/2\lambda_{\mathrm{N}})^2/T - T\right)$, where T is dimensionless time. This implies that the distribution of the dwell time, i.e., the coherency, is characterized only by the effective length L/λ_N and more importantly it does not depend on the kind of materials as long as their transport is diffusive. To check this idea the L/λ_N dependence of $\Delta R_S^\pi / \Delta R_S^0$ is summarized in Fig. 4 by using the data so far reported for metals, semiconductors and graphene. Interestingly the relation between the coherent parameter and the effective length shows a universal behavior and the experimental data are well reproduced by equation (2). We should note here that the effective length L/λ_N, not the spin lifetime, is an important parameter to manipulate the spin precession coherently in the diffusive pure-spin transport while the spin accumulation is relaxed during the diffusive transport in the channel. Therefore, the dual spin injector with Permalloy/MgO/Ag junctions can offer advantages for

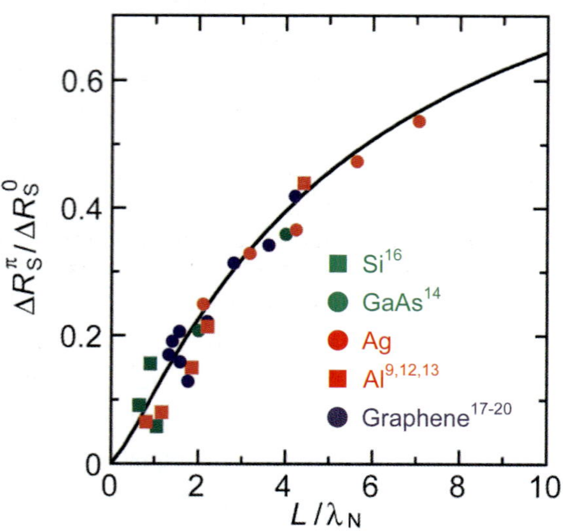

Fig. 4. Coherent parameter of the spin precession $\Delta R_S^\pi / \Delta R_S^0$ as a function of L/λ_N. The solid line is a universal curve obtained from equation (2).

realizing giant spin accumulation as well as coherent spin precession along a 10 μm-long Ag wire which is much longer than the spin diffusion length. In the Hanle effect measurement of spin-polarized electron transport in a 350 μm-thick undoped single-crystal Si wafer, coherent spin precession up to 13π is reported.[26] However, the diffusion constant of the spin current is lower than that in Ag and the collector current, which is proportional to the amount of spins in the channel, is two orders of the magnitude smaller than the detected spin current in DLSV with $L = 10$ μm, even though the spin-polarized current in Si is accelerated by means of electric fields. Therefore, the dual spin injection scheme and the long diffusive pure-spin transport in our study can be useful in developing a new class of spintronic devices. The material-independent perspective for the spin precession will be beneficial for us to design pure-spin-current-based memory and transistor by using a variety of metallic and semi-conductive materials including graphene.

References

1. M. Johnson and R. H. Silsbee, Interfacial charge-spin coupling: injection and detection of spin magnetization in metals, *Phys. Rev. Lett.* **55**(17), 1790-1793, (1985).

2. M. Johnson and R. H. Silsbee, Thermodynamic analysis of interfacial transport and of the thermomagnetoelectric system, *Phys. Rev. B* **35**(10), 4959-4972, (1987).

3. P. C. Van Son, H. van Kempen, and P. Wyder, Boundary resistance of ferromagnetic-nonferromagnetic metal interface, *Phys. Rev. Lett.* **58**(21), 2271-2273, (1987).

4. F. J. Jedema, A. T. Filip, and B. J. van Wees, Electrical spin injection and accumulation at room temperature in an all-metal mesoscopic spin valve, *Nature* **410**, 345-348, (2001).

5. J. Bass and W. P. Pratt, Jr., Spin-diffusion lengths in metals and alloys, and spin-flipping at metal/metal interfaces: an experimentalist's critical review, *J. Phys.: Condens. Matter.* **19**(18), 183201, (2007).

6. I. Žutić, J. Fabian, and S. D. Sarma, Spintronics: Fundamentals and applications, *Rev. Mod. Phys.* **76**(2), 323-410, (2004).

7. C. Chappert, A. Fert, and F. N. Van Dau, The emergence of spin electronics in data storage, *Nature Mater.* **6**, 813-823, (2007).

8. M. Johnson and R. H. Silsbee, Coupling of electronic charge and spin at a ferromagnetic-paramagnetic metal interface, *Phys. Rev. B* **37**(10), 5312-5325, (1988).

9. F. J. Jedema, H. B. Heershe, A. T. Filip, J. J. A. Baselmans, and B. J. van Wees, Electrical detection of spin precession in a metallic mesoscopic spin valve, *Nature* **416**, 713-716, (2002).

10. Y. A. Bychkov and E. I. Rashba, Oscillatory effects and the magnetic susceptibility of carriers in inversion layers, *J. Phys. C: Solid State Phys.* **17**(33), 6039, (1984).

11. S. Datta and B. Das, Electronic analog of the electro-optic modulator, *Appl. Phys. Lett.* **56**(7), 665-667, (1990).

12. S. O. Valenzuela and M. Tinkham, Direct electronic measurement of the spin Hall effect, *Nature* **442**, 176-179, (2006).

13. A. Van Staa, J. Wulfhorst, A. Vogel, U. Merkt, and G. Meier, Spin precession in lateral all-metal spin valves: experimental observation and theoretical description, *Phys. Rev. B* **77**(21), 214416, (2008).

14. X. Lou, C. Adelmann, S. A. Crooker, E. S. Garlid, J. Zhang, K. S. M Reddy, S. D. Flexner, C. J. Palmstrom, and P. A. Crowell, Electrical detection of spin transport in lateral ferromagnet-semiconductor devices, *Nature Phys.* **3**, 197-202, (2007).

15. O. M. J. Vant Erve, C. A. Affouda, A. T. Hanbicki, C. H. Li, P. E. Thompson, and B. T. Jonker, Information processing with pure spin currents in silicon: spin injection, extraction, manipulation, and Detection, *IEEE Trans. Electron Devices* **56**(10), 2343-2347, (2009).

16. T. Sasaki, T. Oikawa, T. Suzuki, M. Shiraishi, Y. Suzuki, and K. Noguchi, Temperatrue dependence of spin diffusion length in silicon by Hanle-type spin precession, *Appl. Phys. Lett.* **96**(12), 122101, (2010).

17. N. Tombros, C. Józsa, M. Popinciuc, H. T. Jonkman, and B. J. Van Wees, Electronic spin transport and spin precession in single graphene layers at room temperature, *Nature* **448**, 571-574, (2007).

18. M. Popinciuc, C. Józsa, P. J. Zomer, N. Tombros, A. Veligura, H. T. Jonkman, and B. J. Van Wees, Electronic spin transport in graphene field-effect transistors, *Phys. Rev. B* **80**(21), 214427, (2009).

19. T. Maassen, F. K. Dejene, M. H. Guimarães, C. Józsa, and B. J. Van Wees, Comparison between charge and spin transport in few-layer graphene, *Phys. Rev. B* **83**(11), 115410, (2011).

20. W. Han, K. Pi, K. M. McCreary, Y. Lin, J. J. I. Wong, A. G. Swartz, and R. K. Kawakami, Tunneling spin injection into single layer graphene, *Phys. Rev. Lett.* **105**(16), 167202, (2010).

21. Y. Fukuma, L. Wang, H. Idzuchi, S. Takahashi, S. Maekawa, and Y. Otani, Giant enhancement of spin accumulation and long-distance spin precession in metallic lateral spin valves, *Nature Mater.* **10**, 527-531, (2011).
22. Y. Fukuma, L. Wang, H. Idzuchi, and Y. Otani, Enhanced spin accumulation obtained by inserting low-resistance MgO interface in metallic lateral spin valves, *Appl. Phys. Lett.* **97**(1), 012507, (2010).
23. R. J. Elliott, Theory of the effect of spin-orbit coupling on magnetic resonance in some semiconductors, *Phys. Rev.* **96**(2), 266-279, (1954).
24. Y. Yafet, g factors and spin-lattice relaxation of conduction electrons, in *Solid State Physics* Vol. 14, Eds. F. Seitz and D. Turnbull, pp. 1-98 (Academic Press, New York, 1963).
25. S. Takahashi and S. Maekawa, Spin injection and detection in magnetic nanostructures, *Phys. Rev. B* **67**(5), 052409, (2003).
26. B. Huang, D. J. Monsma, and I. Appelbaum, Coherent spin transport through a 350 micron thick silicon, Wafer, *Phys. Rev. Lett.* **99**(17), 177209, (2007).

Transverse Relativistic Effects in Paraxial Wave Interference

Konstantin Y. Bliokh,[1,2] Yana V. Izdebskaya,[3] and Franco Nori[1,4*]

[1]*RIKEN Center for Emergent Matter Science, Wako, Saitama 351-0198, Japan*
[2]*A. Usikov Institute of Radiophysics and Electronics, NASU, Kharkov 61085, Ukraine*
[3]*Nonlinear Physics Center, Research School of Physics and Engineering,*
The Australian National University, Canberra ACT 0200, Australia
[4]*Physics Department, University of Michigan, Ann Arbor, Michigan 48109-1040, USA*
**E-mail: fnori@riken.jp*

This article reviews some of our recent results, to be published elsewhere, on wave interference and vortices. We consider relativistic deformations of interfering paraxial waves moving in the transverse direction. Owing to *superluminal* transverse phase velocities, noticeable deformations of the interference patterns arise when the waves move with respect to each other with *non-relativistic* velocities. Similar distortions also appear on a mutual tilt of the interfering waves, which causes a phase delay analogous to the relativistic time delay. We illustrate these observations by the interference between a vortex wave beam and a plane wave, which exhibits a pronounced deformation of the radial fringes into a fork-like pattern (relativistic Hall effect).

This article is dedicated to the memory of Dr. Akira Tonomura, a great scientist and a wonderful person, who did epochal and ground-breaking experiments on wave interference and vortices, as well as in other areas of physics. He is greatly missed by the many people who were blessed to know him. One of the authors (F.N.) was his collaborator, friend, and a member of his research group.

1. Relativistic deformations: Lorentz contraction and velocity addition

Special relativity is based on the Lorentz transformations of space-time, which describe transitions from a 'laboratory' reference frame to a frame moving with velocity $\mathbf{v}$:

$$t' = \gamma\left(t - \mathbf{v}\cdot\mathbf{r}/c^2\right), \quad \mathbf{r}' = \gamma\left(\mathbf{r} - \mathbf{v}t\right). \tag{1}$$

Here $\gamma = 1/\sqrt{1 - v^2/c^2}$ is the Lorentz factor, and quantities in the moving frame are indicated by primes. Importantly, the $\mathbf{r}$-dependent time delay in Eq. (1), $-\mathbf{v}\cdot\mathbf{r}/c^2$, revises the concept of simultaneity and causes interesting distortions of objects when observed in a moving reference frame.[1,2] Such distortions are absent in non-relativistic physics based on Galilean transformations and the invariance of time. As we argue below, one can distinguish two types of relativistic deformations:

(1) Lorentz length contraction of motionless objects;

(2) Shape distortions of moving objects, related to the relativistic velocity addition.

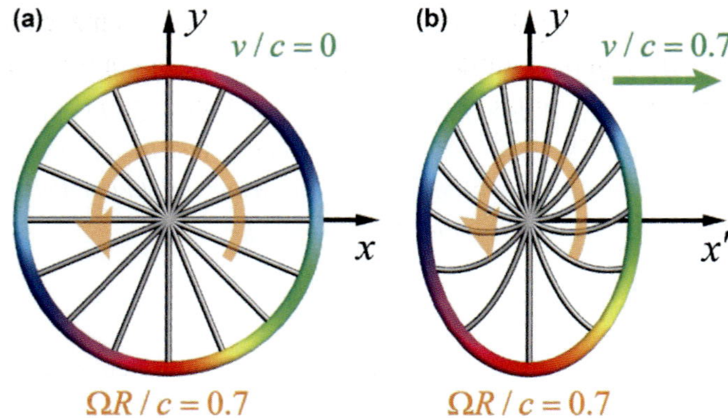

Fig. 1. Virtual snapshots of a relativistic flywheel of radius R rotating with angular velocity Ω : **(a)** in the laboratory frame and **(b)** in the frame moving with velocity v in the x-direction. Two relativistic deformations are seen in **(b)**: (i) the Lorentz x-contraction squeezing the circle into an ellipse (becomes noticeable at $v \sim c$) and (ii) a characteristic distortion of the radial spokes along the orthogonal y-direction (relativistic Hall effect).[3,4] The latter effect is caused by the relativistic addition of the rotational velocity $u = \Omega R$ and frame velocity v , so that it becomes noticeable at $\Omega R v \sim c^2$.

Throughout this paper we consider reference frames moving with respect to each other with velocity v in the x-direction. Let a material point move with velocity u along the x-axis of the laboratory frame: $x(t) = x_0 + ut$. Then, applying the Lorentz transformation (1), one can find that its coordinate in the moving frame becomes

$$x'(t') = \frac{\gamma^{-1}x_0 + (u - v)t'}{1 - uv/c^2}.$$ (2)

For a motionless point, $u = 0$, the coordinate $x'(0) = \gamma^{-1}x_0$ indicates the *Lorentz contraction*, whereas for moving point the velocity $u' \equiv dx'/dt' = (u - v)/(1 - uv/c^2)$ yields the relativistic *velocity addition* formula. In the general case $u \neq 0$, the coordinate $x'(0)$ in Eq. (2) indicates the transformation of the x-scale of the object and includes both the Lorentz contraction, γ^{-1} , in the numerator and the velocity addition effect, $1 - uv/c^2$, in the denominator. Although both of these distortions originate from the same Lorentz transformation of time (1), below we show that they can occur independently in various situations. Note that the velocity-addition deformation is a first-order effect in v/c and also depends on u/c, whereas the Lorentz contraction is a second-order effect $\sim v^2/c^2$.

A nice illustration of the above relativistic deformations appears when observing a spinning body (flywheel) in a moving reference frame,[3] as shown in Fig. 1. First, the circular flywheel experiences the Lorentz x-contraction with the factor of γ^{-1}, and becomes elliptical. Second, the rotating radial spokes in the wheel become distorted and asymmetrically redistributed along the orthogonal y-axis because of opposite velocity additions on the $y>0$ and $y<0$ sides of the wheel. The Lorentz contraction depends only on the frame motion and becomes noticeable at $v\sim c$. In contrast, the y-deformation is essentially related to the rotational velocity of the wheel, $u=\Omega R$ (R is the radius and Ω is the angular velocity of the wheel); this deformation becomes noticeable at $uv\sim c^2$. We note that the y-distortion of a spinning body on a Lorentz boost in the x-direction is intimately related to the Lorentz transformation of the angular momentum, and can be regarded as the relativistic Hall effect.[4]

2. Deformations of wave intensity and phase: Superluminal wavefronts

Recently, we found[4] that "spinning" waves carrying angular momentum, the so-called *optical vortices* or *vortex beams*, also experience relativistic deformations resembling those in Fig. 1. The vortex beams are well known and widely used in optics.[5, 6] A few years ago they were also described for quantum electrons[7] and first generated experimentally by Tonomura's group[8] (see also Refs. 9 and 10).

Let us consider a scalar monochromatic vortex beam propagating along the z-axis. In what follows we are interested in the wave distributions in the transverse (x,y)-plane and we omit all z-dependences. Then, the vortex beam is described by the wave function

$$\psi(x,y,t)\propto A(r)\exp(i\ell\varphi-i\omega t),\tag{3}$$

where (r,φ) are the polar coordinates in the (x,y)-plane, $\ell=0,\pm1,\pm2,...$ is the vortex charge (the quantum number of the angular momentum along the z-axis), ω is the frequency, and $A(r)$ is the radial amplitude distribution. Hereafter, we assume the Laguerre-Gaussian beams[5] with the zero radial index and $A(r)\propto(\kappa r)^{|\ell|}\exp\left[-(\kappa r)^2\right]$, where $\kappa\ll k$ is the characteristic radial wave number. Such beams have an annular intensity distribution with characteristic radius $R\sim|\ell|/\kappa$. The phase fronts of the vortex (3) represent $|\ell|$ radial lines which rotate in the (x,y)-plane with the angular velocity $\Omega_{\text{ph}}=\omega/\ell$. Upon transition to the moving reference frame, the annular intensity distribution $I=|\psi|^2=|A(r)|^2$ experiences the Lorentz x-contraction with the factor γ^{-1}. At the same time, the rotating radial phase fronts of the vortex undergo relativistic Hall-effect y-deformations entirely similar to the spokes of a spinning flywheel in Fig. 1 (see also Fig. 2b).[4]

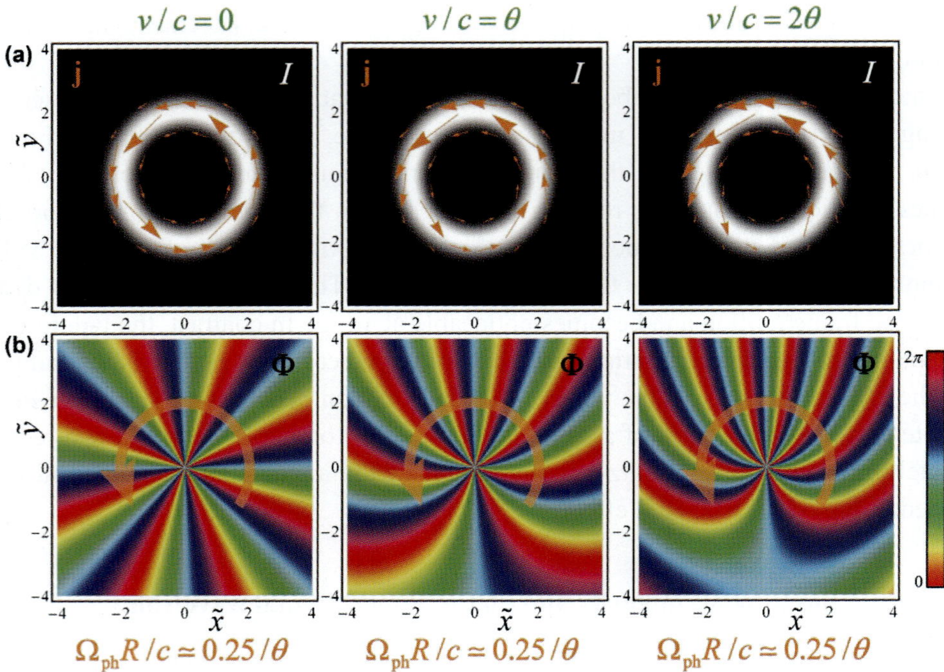

Fig. 2. Instant transverse distributions of: **(a)** Intensity $I = |\psi|^2$, current $\mathbf{j} = \mathrm{Im}\left(\psi^*\nabla\psi\right) = I\nabla\Phi$, and **(b)** phase $\Phi = \arg\psi$ in the paraxial vortex beam (with $\ell = 8$ and paraxiality parameter $\theta = \kappa/k \ll 1$) propagating along the z-axis. The dimensionless coordinates $\tilde{x} = \kappa x$ and $\tilde{y} = \kappa y$ are used. The distributions are shown in the frames moving in the x-direction with velocities $v/c = 0, \theta, 2\theta \ll 1$. Such non-relativistic velocities make the Lorentz x-contraction of the intensity distribution negligible, but nonetheless drastically deform wavefronts, similar to the Hall-effect y-distortion of the flywheel spokes in Fig. 1. This is explained by the relativistic velocity addition with the superluminal motion of the radial wavefronts (4): $u_{\mathrm{ph}}/c = \Omega_{\mathrm{ph}}R/c \simeq 0.25/\theta \gg 1$ ($R \simeq 2/\kappa$ is the beam radius and $\Omega_{\mathrm{ph}} = \omega/8$ is the angular velocity of the wavefronts rotation).

Thus, there is a correspondence between the relativistic deformations of mechanical bodies and waves, but there is also a remarkable difference. Namely, the transverse velocity of the wavefront motion, i.e., the *phase velocity*, can be *superluminal*. Indeed, consider a paraxial wave propagating mostly along the z-axis, with the longitudinal wave number $k_z \simeq k$ and a characteristic transverse wave number $k_\perp \sim \kappa \ll k$. To quantify the paraxiality, we will use the small parameter $\theta = \kappa/k \ll 1$. Then, the transverse phase velocity in the (x, y) plane is estimated as

$$u_{\mathrm{ph}} \sim \frac{\omega}{\kappa} = \frac{c}{\theta} \gg c \tag{4}$$

(for the sake of simplicity we assume waves with $\omega = ck$). For instance, in the above optical-vortex example, the rotational velocity of the wavefronts at the beam radius is

$u_{ph} = \Omega_{ph} R \sim c / \theta \gg c$. Therefore, the velocity-addition deformations of the wavefronts become noticeable at

$$v \sim \frac{c^2}{u_{ph}} \sim \theta c \ll c, \tag{5}$$

i.e., at essentially *non-relativistic* velocities of the frame motion. This is demonstrated in Fig. 2, which displays the transverse intensity, current, and phase distributions for the paraxial vortex beam observed in reference frames moving with small velocities $v \sim \theta c$. One can see no Lorentz contraction in the vortex intensity distribution but a pronounced y-distortion of the wavefronts due to the relativistic velocity addition with $u_{ph} \sim c / \theta$.

Of course, superluminal motion of the wavefronts is non-observable *per se*. But the shape of the phase fronts plays a crucial role in the wave interference. Then the question arises: *Can one observe relativistic deformations of the wave interference patterns at non-relativistic velocities?* We address this question below.

3. Transformations of the wave interference patterns

Let us consider a generic wave interference of two complex scalar fields:

$$\psi(\mathbf{r},t) = \psi_1(\mathbf{r},t) + \psi_2(\mathbf{r},t). \tag{6}$$

The interference pattern is described by the resulting intensity distribution, $I(\mathbf{r},t) = |\psi(\mathbf{r},t)|^2$. Note that interference fringes can move with arbitrarily large velocities, but, of course, can be observed only in the case of subluminal motion.

To find transformations of the interference pattern on transition to the moving reference frame, one should substitute the Lorentz transformation (1) in the wave function (6). Here we argue that one can distinguish two basic types of relativistic effects in wave interference:

    (1) The observer moves with respect to both waves and observes the same interference picture but in the moving frame;

    (2) The observer and the second wave move with respect to the first wave. In other words, the second wave is used as a probe attached to the observer and sensing the first wave in the moving frame.

For these two cases, the wave functions in the moving frame can be written, respectively, as

$$\psi'(\mathbf{r}',t') = \psi_1\left[\mathbf{r}(\mathbf{r}',t'),t(\mathbf{r}',t')\right] + \psi_2\left[\mathbf{r}(\mathbf{r}',t'),t(\mathbf{r}',t')\right], \tag{7a}$$

$$\psi'(\mathbf{r}',t') = \psi_1\left[\mathbf{r}(\mathbf{r}',t'),t(\mathbf{r}',t')\right] + \psi_2(\mathbf{r}',t'). \tag{7b}$$

Here $\mathbf{r}(\mathbf{r}',t')$ and $t(\mathbf{r}',t')$ denote the Lorentz transformation given by Eq. (1). In the following Subsections 3.1 and 3.2 we analyze the relativistic deformations in the interference patterns $I'(\mathbf{r}',t') = |\psi'(\mathbf{r}',t')|^2$ for the two cases (7a) and (7b), respectively.

3.1. *Moving interference patterns*

First, we examine the frame moving with respect to both waves. In this case, Eq. (7a) shows that the interference intensity pattern is transformed as any material object, i.e., via the Lorentz transformations (1): $I'(\mathbf{r}',t') = I\left[\mathbf{r}(\mathbf{r}',t'),t(\mathbf{r}',t')\right]$. This is quite natural since in quantum mechanics any matter distribution is associated with the intensity of the wave function.

As the simplest example, let us consider the interference of two plane waves propagating along the z- and x-axes, with transverse wave numbers $k_{x1,2} = \kappa_{1,2}$. As before, we are only interested in the distributions in the $z = 0$ plane and omit all z-dependences. Thus, the two interfering wave functions are:

$$\psi_1(x,t) = \exp(i\kappa_1 x - i\omega_1 t), \quad \psi_2(x,t) = \exp(i\kappa_2 x - i\omega_2 t). \tag{8}$$

The interference pattern for these waves, $I(x,t)$, represents an array of fringes with period $\Delta = 2\pi / |\kappa_1 - \kappa_2|$ and moving with velocity $u_f = (\omega_1 - \omega_2)/(\kappa_1 - \kappa_2)$ along the x-direction. Let us now choose one fringe in this interference pattern, which has a coordinate $x(t) = x_0 + u_f t$ in the laboratory frame. Then, performing the Lorentz transformation (1) and (7a), one can readily ascertain that the coordinate of this fringe in the moving frame is given by Eq. (2) with $u = u_f$ and the corresponding Lorentz contraction and velocity addition. The only difference is that the fringe motion can be superluminal, $u_f \gg c$, and then the relativistic velocity-addition effects formally occur at $v \sim c^2 / u_f \ll c$. However, superluminal fringes and, hence, their deformations remain fundamentally unobservable.

It is worth noticing that the Lorentz transformation (1), when applied to a plane wave $\exp(i\mathbf{k} \cdot \mathbf{r} - i\omega t)$, results in the following transformation of the wave parameters:

$$\omega' = \gamma(\omega - \mathbf{k} \cdot \mathbf{v}), \quad \mathbf{k}' = \gamma(\mathbf{k} - \omega \mathbf{v} / c^2), \tag{9}$$

so that the wave function becomes $\exp(i\mathbf{k}' \cdot \mathbf{r}' - i\omega' t')$ in the moving frame. The shift of the wave vector in Eq. (9), $-\omega \mathbf{v} / c^2$, originates from the time delay in Eq. (1), and it is this shift that causes deformations of the interference patterns. In the paraxial geometry $k_z \simeq k$, $v_x = v$, the transformation (9) represents a *tilt* of the wave vector in the (z,x)-plane by the angle $\alpha \simeq v / c$. This will be used in what follows.

For comparison with the examples in Sections 1 and 2, let us consider now an interference pattern which mimics a spinning flywheel in Fig. 1. Such pattern appears when interfering the co-propagating vortex beam (3) and plane wave (the z-dependences are omitted):[11]

$$\psi_1(x,y,t) = A(r)\exp(i\ell\varphi - i\omega_1 t), \quad \psi_2(t) = \exp(-i\omega_2 t). \tag{10}$$

The intensity distribution of the superposition (10) represents a circular array of $|\ell|$ radial "spokes" with vortex radius $R \sim |\ell|/\kappa$, rotating with angular velocity $\Omega_f = (\omega_1 - \omega_2)/\ell$ (Fig. 3a). Thus, the velocity of the circular motion of the radial fringes is $u_f \sim (\omega_1 - \omega_2)/\kappa$, and it can take on any values depending on the frequency difference. Figure 3 shows deformations of the interference pattern of waves (10) in the moving frame.

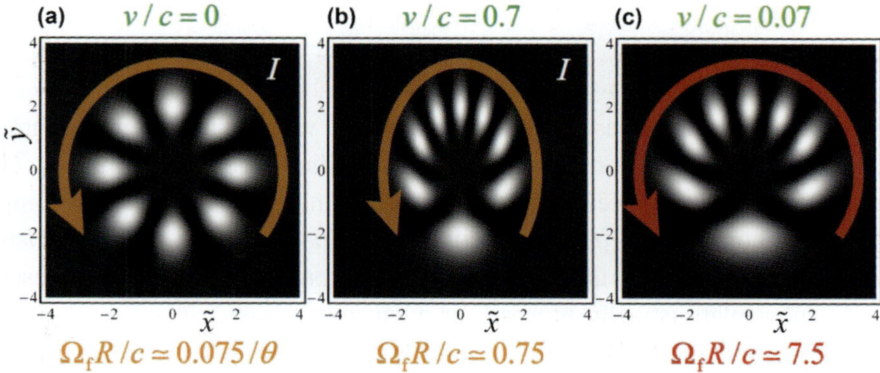

Fig. 3. Transverse intensity pattern for interference[10] of co-propagating paraxial vortex beam ($\ell = 8$) and plane wave. The dimensionless coordinates $\tilde{x} = \kappa x$ and $\tilde{y} = \kappa y$ are used. **(a)** In the laboratory frame, the radial fringes rotate with angular velocity $\Omega_f = (\omega_1 - \omega_2)/8$ and linear velocity $u_f \sim \Omega_f R$ ($R \simeq 2\kappa^{-1}$ being the vortex radius), which can take on arbitrary values depending on the wave parameters. **(b)** In the moving frame with $v \sim c$ and with the fringe velocity $\Omega_f R \sim c$, the pattern shows both the Lorentz x-contraction and velocity-addition y-distortion of the fringes, entirely similar to the mechanical flywheel in Fig. 1. **(c)** Choosing parameters corresponding to a superluminal fringe velocity $\Omega_f R \gg c$, the velocity-addition distortion (but no Lorentz contraction) occurs for non-relativistic frame motion, $v \ll c$, but cannot be observed.

When $u_f \sim c$ and $v \sim c$, both the Lorentz x-contraction of the circle and velocity-addition y-distortion of the spokes appear (Fig. 3b), entirely similar to those in Fig. 1. At the same time, when $u_f \gg c$ and $v \sim c^2/u_f \ll c$, the Lorentz contraction is negligible, while the velocity-addition deformation of the radial fringes is present (Fig. 3c), akin to the distortion of the vortex wavefronts in Fig. 2b. Still, as we mentioned before, this effect cannot be detected.

3.2. *Waves moving with respect to each other*

Finally, we examine the second type of relativistic interference, when the two waves move with respect to each other. Assuming that the observer is attached to the second wave, the transformation to the moving frame is described by Eq. (7b). From here on, we consider only *non-relativistic* velocities of the frame motion, $v \ll c$, $\gamma \simeq 1$.

Performing the transformation (7b) with (1) in the simplest case of two interfering plane waves (8), we find that the interference fringe with the coordinate $x(t) = x_0 + u_{\mathrm{f}} t$ in the laboratory frame will have the following coordinate in the moving frame:

$$
x'(t') \simeq \frac{x_0 + \left(u_{\mathrm{f}} - v \dfrac{\kappa_1}{\kappa_1 - \kappa_2} \right) t'}{1 - \dfrac{u_{\mathrm{ph}1} v}{c^2} \dfrac{\kappa_1}{\kappa_1 - \kappa_2}}, \tag{11}
$$

where $u_{\mathrm{ph}1} = \omega_1 / \kappa_1$ is the phase velocity of the first wave. The Lorentz contraction is absent in Eq. (11) since $v \ll c$ but the velocity-addition effects are present. The most important difference in the velocity-addition denominator of Eq. (11) as compared to Eq. (2) is that it contains the phase velocity of the *first* wave and is *independent* of the fringe velocity u_{f}. Owing to this, the velocity-addition distortions can be observed for $u_{\mathrm{ph}1} \gg c$ but non-relativistically moving (or even motionless) fringes, $u_{\mathrm{f}} \ll c$. And this is the desired observable relativistic deformation at $v \ll c$, which is described by the denominator of Eq. (11).

Let us illustrate this result by considering the interference of the co-propagating vortex and plane wave, Eq. (10). We set $\omega_1 = \omega_2$ so that the fringes do *not* rotate in the laboratory frame: $u_{\mathrm{f}} = 0$. Figure 4a shows the deformation of the interference fringes upon motion with non-relativistic velocities $v \sim \theta c$, Eq. (5). The characteristic y-distortions of the non-rotating radial fringes appear. One can say that they represent deformations of the superluminal vortex wavefronts in Fig. 2b, revealed by the interference with a plane wave in the moving frame. Thus, we conclude that *non-relativistic motion can produce pronounced relativistic deformations of the intensity pattern when the two waves move with respect to each other in the transverse direction.*

Recall now that in the problem under consideration, the Lorentz transformation of a paraxial plane wave is equivalent to the *tilt* of its wave vector by the angle $\alpha \simeq v / c$, Eq. (9). Therefore, the *same* y-deformation of the radial interference fringes will appear upon a small x-tilt between the interfering vortex and plane wave. This effect is familiar to experimentalists working in singular optics. In Fig. 4b we show experimentally-measured deformations of the radial interference pattern upon a small tilt between the optical vortex beam and a plane wave. Clearly, Figs. 4a and 4b are in perfect agreement with each other.

This is explained by the fact that the Lorentz x-dependent time delay in Eq. (1) is represented (for waves) by the x-dependent *phase* delay, i.e., a tilted wavefront.

Another curious and very close analogy with relativistic deformations occurs in photography, when making pictures of moving objects. Then, the rolling shutter of the camera provides a true x-dependent time-delay effect, and blades of a rotating propeller undergo the y-distortions shown in Fig. 1b[4, 12].

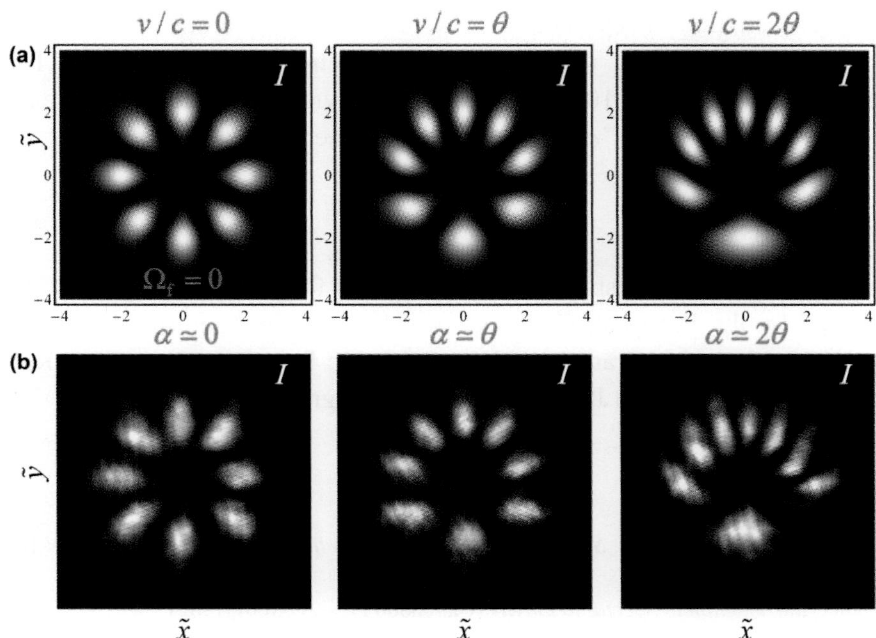

Fig. 4. **(a)** Transverse intensity patterns for interference (10) of a vortex beam ($\ell = 8$) and a plane wave propagating along the z-axis in the *moving* reference frame. Thus, the vortex beam and the wave move with respect to each other with non-relativistic velocity $v / c = 0, \theta, 2\theta \ll 1$ in the x-direction. The pronounced y-distortion of the radial fringes visualize the deformed vortex wavefronts shown in Fig. 2, although the fringes do *not* rotate when the beam and plane wave have the same frequency. **(b)** Experimental pictures of the interference of an optical vortex beam and a plane wave *tilted* by the angle $\alpha = k_x / k \sim \theta \ll 1$. The precise correspondence between **(a)** and **(b)** appears because the Lorentz transformation of time for a transversely moving paraxial wave is equivalent to its tilt (in the approximation $v / c \ll 1$).

4. Conclusion

We have considered relativistic deformations of moving objects observed in a moving reference frame. There are two types of such deformations: the Lorentz contraction and distortions arising from the relativistic velocity addition. Considering transverse Lorentz transformations of paraxial waves, we found that the wavefronts experience significant relativistic velocity-addition deformations at non-relativistic velocities (Fig. 2b). This is because of the superluminal phase velocity in the transverse plane. We have shown that such distortions of the wavefronts reveal themselves in the interference with a plane wave moving with respect to the probed wave (Fig. 4a). It should be noticed that the same deformations appear for a small tilt of the plane wave (Fig. 4b). Therefore, to observe a truly relativistic effect, one has to use paraxial waves with a characteristic propagation angle $\theta \ll 1$, a relative transverse motion of the waves with velocity $v \sim \theta c$, whereas the alignment between the waves should be kept with an accuracy of $|\delta\alpha| < \theta$. Let us estimate the effect described in this paper for electron vortex beams.[7-10] Taking the reasonable paraxial angle $\theta \sim 10^{-6}$, we find that relativistic distortions become noticeable at transverse velocities $v \sim 10^{-6} c \sim 3 \cdot 10^2$ m/s, i.e., at the speed of sound in air.

Acknowledgments

We are grateful to A. Y. Bekshaev and Y. P. Bliokh for fruitful discussions. This work was supported by the European Commission (Marie Curie Action), ARO, JSPS-RFBR contract No. 12-02-92100, Grant-in-Aid for Scientific Research (S), MEXT Kakenhi on Quantum Cybernetics, and the JSPS via its FIRST program.

References

1. E. F. Taylor and J. A. Wheeler, *Spacetime Physics: Introduction to Special Relativity* (W.H. Freeman, New York, 1992).
2. W. Rindler, *Introduction to Special Relativity* (Clarendon Press, Oxford, 1982).
3. R. A. Muller, Thomas precession: Where is the torque?, *Am. J. Phys.* **60**(4), 313–317, (1992).
4. K. Y. Bliokh and F. Nori, Relativistic Hall effect, *Phys. Rev. Lett.* **108**(12), 120403, (2012).
5. L. Allen, S. M. Barnett, and M. J. Padgett, eds., *Optical Angular Momentum* (Taylor & Francis, London, 2003).
6. J. P. Torres and L. Torner, eds., *Twisted Photons* (Wiley, Weinheim, 2011).
7. K. Y. Bliokh, Y. P. Bliokh, S. Savel'ev, and F. Nori, Semiclassical dynamics of electron wave packet states with phase vortices, *Phys. Rev. Lett.* **99**(19), 190404, (2007).
8. M. Uchida and A. Tonomura, Generation of electron beams carrying orbital angular momentum, *Nature* **464**, 737–739, (2010).
9. J. Verbeek, H. Tian, and P. Schattschneider, Production and application of electron vortex beams, *Nature* **467**, 301–304, (2010).
10. B. J. McMorran, A. Agrawal, I. M. Anderson, A. A. Herzing, H. J. Lezec, J. J. McClelland, and J. Unguris, Electron vortex beams with high quanta of orbital angular momentum, *Science* **331**(6014), 192–195, (2011).
11. D. Yang, J. Zhao, T. Zhao, and L. Kong, Generation of rotating intensity blades by superposing optical vortex beams, *Opt. Commun.* **284**(14), 3597–3600, (2011).
12. See, e.g., http://en.wikipedia.org/wiki/Rolling_shutter and http://www.youtube.com/watch?v=17PSgsRlO9

Reprints of Akira Tonomura's Most Important Publications

1. A. Tonomura, A. Fukuhara, H. Watanabe, and T. Komoda, Optical reconstruction of image from Fraunhofer electron-hologram, *Jpn. J. Appl. Phys.* **7**(3), 295, (1968).
   (Demonstrated a possibility of holography with an electron beam)
2. A. Tonomura, T. Matsuda, J. Endo, H. Todokoro, and T. Komoda, Development of a field emission electron microscope, *J. Electron Microsc.* **28**(1), 1-11, (1979).
   (Put electron holography to practical use by developing a "coherent" field-emission electron microscope)
3. A. Tonomura, T. Matsuda, J. Endo, T. Arii, and K. Mihama, Direct observation of fine structure of magnetic domain walls by electron holography, *Phys. Rev. Lett.* **44**(21), 1430-1433, (1980).
   (Invented a method of directly observing microscopic magnetic lines of force as electron phase contours)
4. A. Tonomura, T. Matsuda, R. Suzuki, A. Fukuhara, N. Osakabe, H. Umezaki, J. Endo, K. Shinagawa, Y. Sugita, and H. Fujiwara, Observation of Aharonov-Bohm effect by electron holography, *Phys. Rev. Lett.* **48**(21), 1443-1446, (1982).
   (Produced experimental evidence for the Aharonov-Bohm effect using transparent toroidal magnet)
5. A. Tonomura, N. Osakabe, T. Matsuda, T. Kawasaki, J. Endo, S. Yano, and H. Yamada, Evidence for Aharonov-Bohm effect with magnetic field completely shielded from electron wave, *Phys. Rev. Lett.* **56**(8), 792-795, (1986).
   (Established the physical reality of gauge fields by producing definitive evidence for the Aharonov-Bohm effect using toroidal magnetics covered with superconductors)
6. N. Osakabe, T. Matsuda, T. Kawasaki, J. Endo, A. Tonomura, S. Yano, and H. Yamada, Experimental confirmation of Aharonov-Bohm effect using a toroidal magnetic field confined by a superconductor, *Phys. Rev. A* **34**(2), 815-822, (1986).
   (Full paper of Reference 5)
7. A. Tonomura, J. Endo, T. Matsuda, T. Kawasaki, and H. Ezawa, Demonstration of single-electron buildup of interference pattern, *Amer. J. Phys.* **57**(2), 117-120, (1989).
   (Demonstrated single-electron build-up of an interference pattern)
8. K. Harada, T. Matsuda, J. Bonevich, M. Igarashi, S. Kondo, G. Pozzi, U. Kawabe, and A. Tonomura, Real-time observation of vortex lattices in a superconductor by electron microscopy, *Nature* **360**, 51-53 (5 November 1992).
   (Developed a technique for dynamically observing vortices in superconductors)

9. T. Matsuda, K. Harada, H. Kasai, O. Kamimura, and A. Tonomura, Observation of dynamic interaction of vortices with pinning centers by Lorentz microscopy, *Science* **271**(5254), 1393-1395, (1996).
(Found "intermittent vortex rivers" near pinning centers)

10. T. Kawasaki, T. Yoshida, T. Matsuda, N. Osakabe, A. Tonomura, I. Matsui, and K. Kitazawa, Fine crystal lattice fringes observed using a transmission electron microscope with 1-MeV coherent electron waves, *Appl. Phys. Lett.* **76**(9), 1342-1344, (2000).
(Developed 1-MV field-emission electron microscope having the brightest electron beam and the highest lattice resolution)

11. A. Tonomura, H. Kasai, O. Kamimura, T. Matsuda, K. Harada, Y. Nakayama, J. Shimoyama, K. Kishio, T. Hanaguri, K. Kitazawa, M. Sasase, and S. Okayasu, Observation of individual vortices trapped along columnar defects in high-temperature superconductors, *Nature* **412**(6847), 620-622, (2001).
(Observed the different arrangements of vortex liens trapped and untrapped along tilted columnar defects inside high-Tc superconductors with the 1-MV electron microscope)

12. T. Matsuda, O. Kamimura, H. Kasai, K. Harada, T. Yoshida, T. Akashi, A. Tonomura, Y. Nakayama, J. Shimoyama, K. Kishio, T. Hanaguri, and K. Kitazawa, Oscillating rows of vortices in superconductors, *Science* **294**(5551), 2136-2138, (2001).
(Found an oscillation of a row of vortices reflecting the layered structure of high-Tc superconductors)

13. A. Tonomura, H. Kasai, O. Kamimura, T. Matsuda, K. Harada, T. Yoshida, T. Akashi, J. Shimoyama, K. Kishio, T. Hanaguri, K. Kitazawa, T. Masui, S. Tajima, N. Koshizuka, P.L. Gammel, D. Bishop, M. Sasase, and S. Okayasu, Observation of structures of chain vortices inside anisotropic high-T_c superconductors, *Phys. Rev. Lett.* **88**(23), 237001-1 – 237001-4, (2002).
(Found the formation mechanism of chain vortices in high-Tc superconductors at tilted magnetic fields)

JAPAN. J. APPL. PHYS. **7** (1968) 295

Optical Reconstruction of Image from Fraunhofer Electron-Hologram

Akira TONOMURA, Akira FUKUHARA,
Hiroshi WATANABE and Tsutomu KOMODA

*Central Research Lab., Hitachi Ltd.
Kokubunji, Tokyo*

(Received December 23, 1967)

Electron-beam holography of Fresnel type has been tested by several authors,[1,2] but the reconstructed images are intolerably disturbed by a concurrent virtual image.

This note reports some results in the electron-beam version of Thompson's Fraunhofer holography,[3] which was already mentioned in Gabor's work[4] but has not been tried.

The opaque particles in a plane S are illuminated by a collimated beam of quasi-monochromatic electrons as in Fig. 1(a). Diffraction patterns in a far-field region from the individual particles (A) are magnified and recorded on a photographic film H. The film H thus containing the Fraunhofer hologram is placed in a collimated laser beam as shown in Fig. 1(b) and an image is reconstructed onto a plane B. The effect of the virtual image which has been troublesome[1,2] in the usual Fresnel holography is almost negligible in this scheme. The far-field condition is given by

$$Z \gg \frac{d^2}{\lambda_{el}}$$

where Z is the distance between S and A in Fig. 1(a), d the particle size and λ_{el} the wave length of the electron beam (in the present experiment, $Z =$ 1.9 mm, $d = 100$ A and $\lambda_{el} = 0.037$ A, accordingly $Z = 80\ d^2/\lambda_{el}$). A single-mode He-Ne laser ($\lambda = 6328$ A) was used for the reconstruction and the distance between H and B in Fig. 1(b) was 110 mm.

Figure 2 shows the in-focus E. M. image, the hologram and the reconstructed image of gold particles supported on a carbon film. Although the reconstructed image is inferior to the in-focus E. M. image presumably due to the resolution limit of the photographic film (Fuji FG), the one-to-one correspondence clearly exists between the particles in both images.

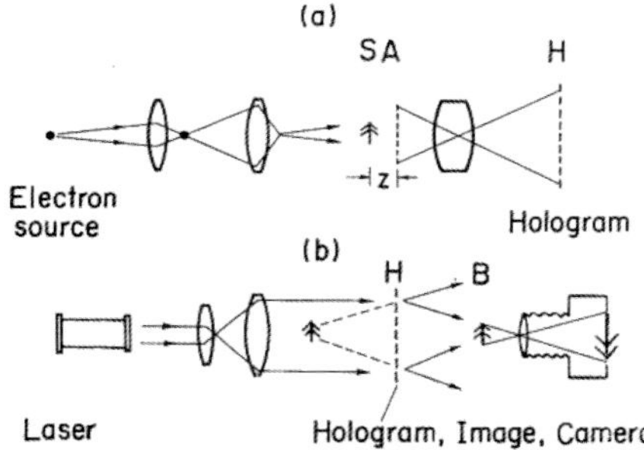

Fig. 1. Arrangement in electron holography.

References

1) M. E. Haine and T. Mulvey: J. Opt. Soc. Amer. **42** (1952) 763.
2) T. Hibi: J. Electronmicroscopy **4** (1956) 10.
3) E. J. Thompson: Japan. J. appl. Phys. **4** Supple. 1 (1965) 302.
4) D. Gabor: Proc. Phys. Soc. B **64** (1951) 449.
   D. Gabor: Proc. Roy. Soc. A **192** (1949) 454.

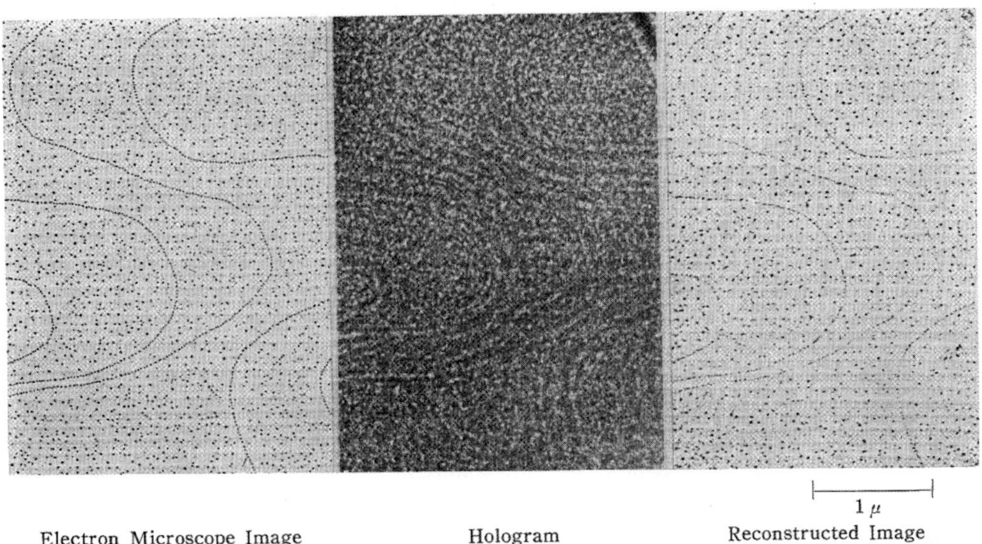

Electron Microscope Image   Hologram   Reconstructed Image

Fig. 2.

J. Electron Microsc., Vol. 28, No. 1, 1–11, 1979

Development of a Field Emission Electron Microscope

Akira TONOMURA, Tsuyoshi MATSUDA, Junji ENDO,
Hideo TODOKORO and Tsutomu KOMODA

Central Research Laboratory, Hitachi Ltd., Kokubunji, Tokyo, 185 Japan

(Received August 1, 1978; accepted December 25, 1978)

A field emission electron microscope with highly stabilized gun and illuminating system was developed.   Several experiments were made to demonstrate the high performance of the microscope.   The brightness of the beam was more than that of the conventional ones, which feature was demonstrated by taking 300 Fresnel fringes and 3,000 bi-prism fringes.   Furthermore, monochromatic feature of the microscope makes the effect of chromatic aberration smaller and the half-spacing fringes of nickel {220} planes ($1/2\ d_{220} = 0.62$ Å) were observed.

Key words=field emission: coherence: lattice resolution: electron bi-prism: electron holography

INTRODUCTION

The ultimate purpose of high resolution electron microscopy lies in observing various kinds of natural phenomena with a measure of atoms and molecules.   For that purpose, improvement of the resolution is still required.   The resolution of an electron microscope was restricted by fundamental limitations, that is, aberrations of the electron lenses.   Meanwhile, subsidiary conditions, such as stabilities of power supplies and mechanics, have been greatly improved for the last ten years, and so the fundamental effect of lens aberrations appear clearly in high resolution images.

In order to remove these image defects proper to an electron microscope, a number of innovative researches and developments have been made.[1]   However, means to overcome the present standstill are not found, up to this time.

We have developed a field emission electron microscope as a tool to break the difficulty.   The field emission electron beam has a monochromatic feature, which can make the chromatic aberration smaller and which can improve the resolution of an electron microscope.   Another feature of the coherent beam makes the phase contrast of an image higher.   In addition, the latter also makes it possible to correct the spherical aberration of an electron lens by realizing electron holography.[1–4]

Field emission electron microscopes so far developed have not made a direct contribution to the performance of an electron microscope.   This may be due to the fact that the current density of illuminating beam onto a specimen was not large enough to observe a high resolution image and also that the illuminating beam is not so highly stabilized as to realize coherent illumination.

Several experiments have been made to confirm that the apparatus has a potential to reach the above-mentioned purpose.

CONSTRUCTION OF FIELD EMISSION ELECTRON MICROSCOPE

Figure 1 shows an outside view of our field emission electron microscope.   The column below the objective lens is that of Hitachi HU-12A electron microscope.   The gun, the illuminating system and the specimen chamber were designed to be magnetically shielded from outer disturbances.   An electron·bi-prism is installed in a field-limiting aperture plane in order to investigate the spatial coherency of

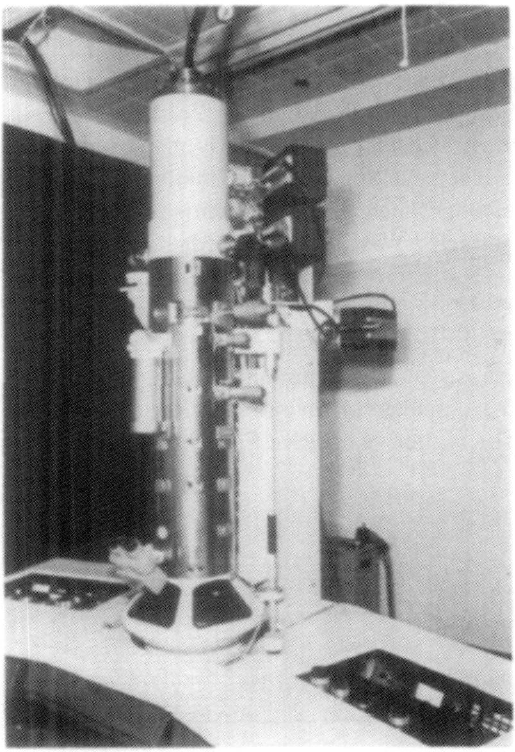

Fig. 1.   Outside view of field emission electron microscope.

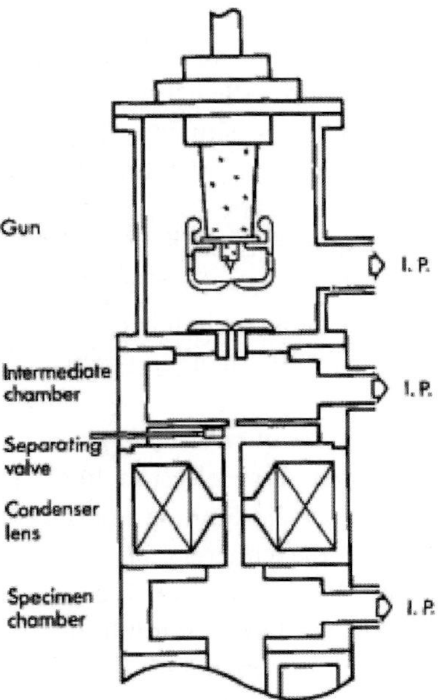

Fig. 2.   Cross-sectional diagram of gun and illuminating system.

the beam. Each part is in detail described below.

1.   Gun and illuminating system

A cross-sectional diagram of the gun and the illuminating system are shown in Fig. 2. Electrons were emitted from $\langle 310 \rangle$-oriented tungsten tip by applying a potential of 3–6 kV between the tip and the first anode. The total emission current was usually 100 μA. About one percent of the electrons is passed through the aperture of the first anode and then is accelerated by Butler type anodes. The shape of the anodes was designed to minimize the spherical aberration of the accelerating electrostatic lens. Far below the specimen plane, the real image of the electron source was formed by the accelerating lens. The diameter of the spot formed on a specimen plane is 300 Å. In general, electrons are not focused on a specimen plane, but are a little defocused

to illuminate a larger area. The accelerating voltage was 70 kV in the present experiment. The gun was able to be operated for more than three hours without thermally flashing the tip.

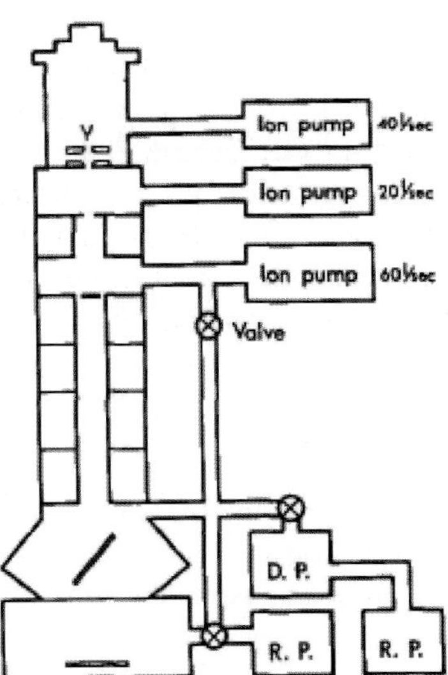

Fig. 3.   Evacuating system of field emission electron microscope.

2. *Evacuating system*

The gun chamber was evacuated by two 20 l/s ion pumps (See Fig. 3). An intermediate chamber is situated for differential pumping between the gun and a condenser lens and is also evacuated by a 20 l/s ion pump. Specimen chamber is evacuated by a 60 l/s ion pump. The final pressures in gun, intermediate and specimen chambers are 5×10^{-10}, 1×10^{-9}, and 1×10^{-8} Torr, respectively.

3. *Interferometer of electron bi-prism*

A Möllenstedt type electron bi-prism[5,6] is installed on the field-limiting aperture plane[7] to investigate the coherent characteristics of the electron beam. Figure 4 shows a perspective drawing of an electron bi-prism. The bi-prism wire was made of 0.3 μm^ϕ quartz fiber, on which gold was evaporated. The fiber was placed between two parallel plates of earth potential. A potential of $0 \sim 200$ V was applied through a spring lead from outside a vacuum. The bi-prism was movable in a plane perpendicular to the direction of electron-beam trajectory, which mechanism was the same as that of the field-limiting aperture.

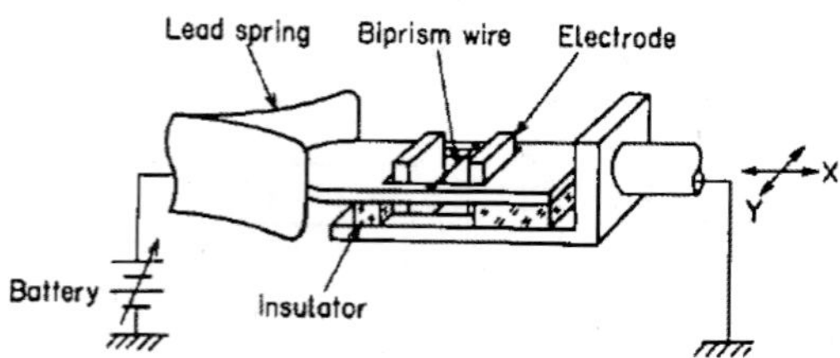

Fig. 4.   Perspective drawing of electron bi-prism.

PERFORMANCE CHARACTERISTICS

The high brightness of a field emission electron beam realizes coherent illumination in an electron microscope. Therefore, the amount of the brightness was estimated backward from the interference experiments. Then, lattice images were taken to investigate the decreased effect of chromatic aberration due to the narrow energy spread of the beam.

1. *Estimation of the beam brightness*

Fresnel fringes were employed to estimate the brightness of the field emission electron beam. The formation of Fresnel fringes is illustrated in Fig. 5. Nearly plane wave with a divergent angle of 2 β illuminates the edge of a specimen. The transmitted wave and the diffracted waves from the edge interfere with each other on a observation plane and then they form Fresnel fringes. The spacing of the fringes becomes smaller with the distance from the edge. The minimum spacing is limited by the diameter of an illumination cone on the observation plane (See Fig. 5).

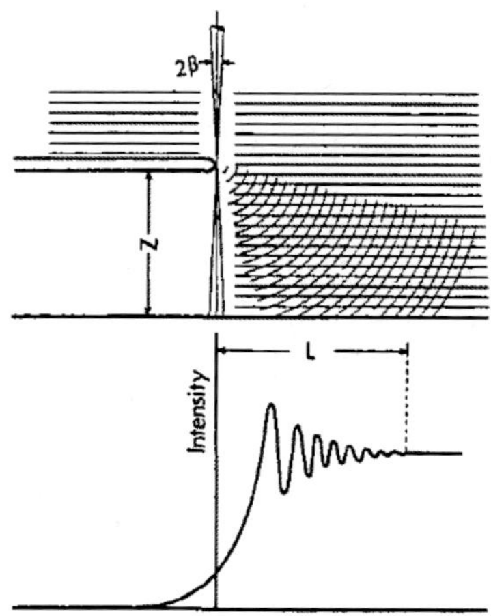

Fig. 5.   Schematic diagram of Fresnel fringes. Nearly plane wave illuminates an edge of a specimen and the interference fringes between the plane wave and the diffracted wave from the edge are observed. The illumination angle 2β limits the minimum fringes observable.

The length of the region where n fringes appear is given by

$$L \sim \sqrt{2n\lambda Z} \qquad (1)$$

where Z is the distance between the specimen and observation planes and λ is the wavelength of electrons.

In order to obtain the coherent region as large as the length L, the illumination angle 2β ought to be as small as $2\beta < \lambda/L$.

Meanwhile, when photographing the fringes, the fringe with the minimum spacing should be

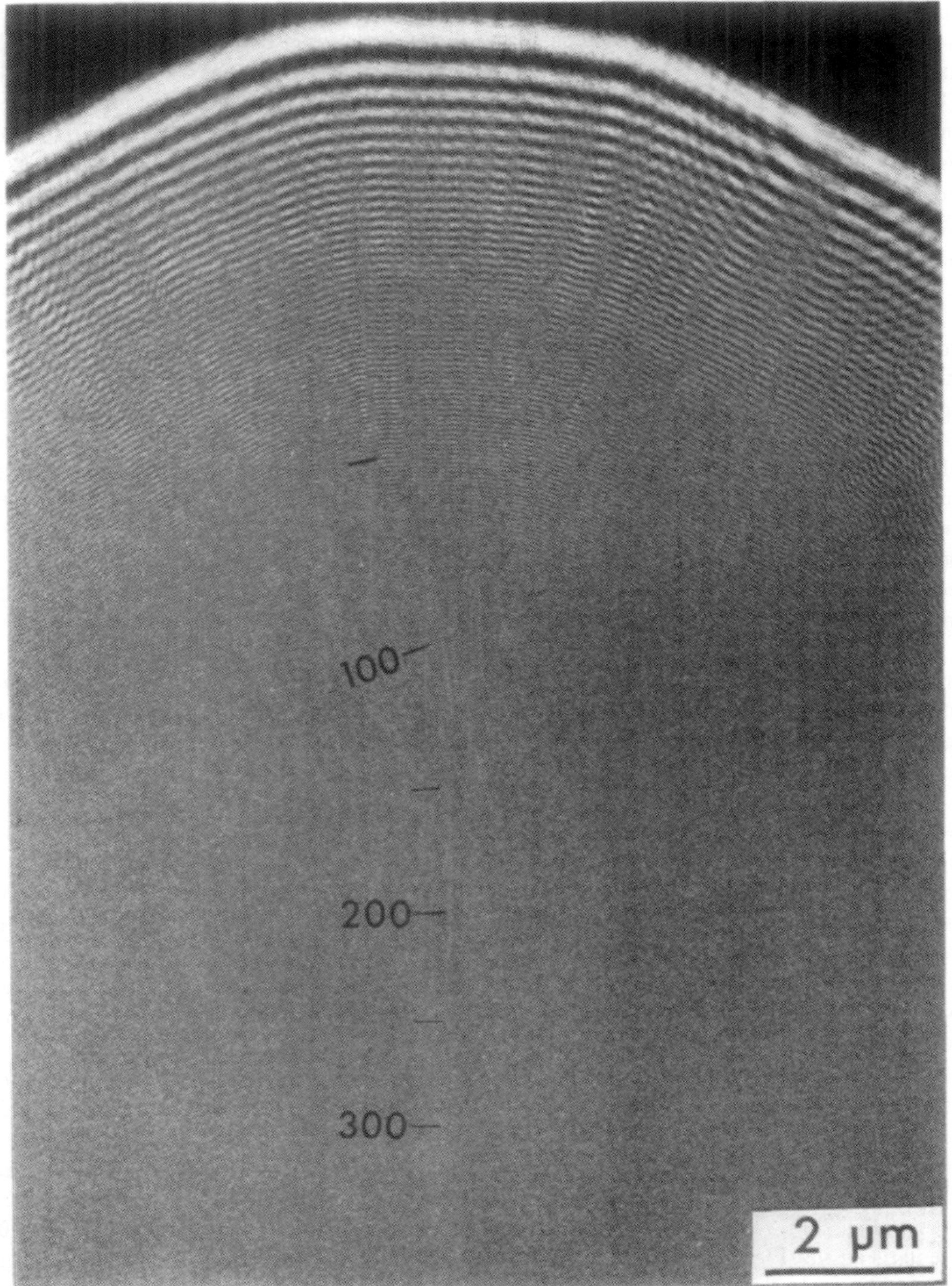

Fig. 6.   Fresnel fringes from a microgrid edge.   The number of the fringes is 300, which is the largest among those reported to date.

exposed with moderate exposure time to record on film. This condition requires large illumination angle 2β.

Therefore, the illumination angle has the optimum value to obtain the largest fringe number. The maximum number for the optimum condition is given by

$$n_{max} = \frac{\lambda}{4d \sqrt{\dfrac{I}{B\pi T}}} \qquad (2)$$

where B is beam brightness, T exposure time, I the amount of charge density necessary to expose a film and d the resolution of the film.

As λ, d and I are constant, n_{max} depends only on B and T. Therefore, beam brightness can be estimated from the maximum number of fringes obtained. Fresnel fringes from a microgrid edge taken with the field emission electron microscope are shown in Fig. 6. The amount of defocus Z is 8 mm, the magnification m is 1,300, and the exposure time is only 4 s. The 300 Fresnel fringes are observed. The number of these fringes is the largest among those reported to date. According to Eq. (2), the resultant brightness is estimated to be 2×10^8 A/cm^2·sr, when $I = 5 \times 10^{-11}$ C/cm^2 and $d = 30$ μm. This result shows that the beam brightness is 100 times larger than that of thermionic electron beam, and that the coherent illumination becomes possible in the microscope, and that the phase contrast will be very high even in highly defocused images, which is the case with biological specimens.

Owing to the high brightness of the electron beam, much more bi-prism fringes are expected to be obtained than hitherto reported. From our experiments, 3,000 bi-prism fringes could be observed with our microscope using a Möllenstedt type bi-prism.[8] The fringes are shown in Fig. 7 (a). Only a part of the pattern is enlarged in Fig. 7 (b). The maximum number of fringes that have been taken is at most 300. Therefore, this result corresponds to the improvement of spatial coherence length by ten times. As an application of this technique, experiments of electron holography were

carried out. An off-axis hologram was formed by the interference between object and reference waves, which was produced by the electron bi-prism. The image hologram of an evaporated gold particle is shown in Fig. 8. The exposure time required was 8 s. The magnification of the hologram was 140,000 times. A potential of 100 V was applied to the bi-prism wire. When an image is exactly focused, Bragg-reflected waves from {111} planes are formed outside the image, due to the spherical aberration of the electron lens ($C_s = 1.7$ mm). Inside the Bragg-reflected images, interference fringes between the reflected and transmitted waves are faintly observed. Furthermore, the oblique interference fringes between the reflected and the reference waves are also observed there. The hologram was illuminated with a He-Ne laser light and then the reconstructed image was obtained under any focusing condition. The exact focused reconstructed image is shown in Fig. 9. Lattice image is reconstructed outside the particle. This shows that the Bragg-reflected waves from {111} planes are certainly reconstructed, and that the lattice resolution of the image reconstructed is better than 2.4 Å.

2. Improvement of lattive resolution

Monochromatic feature of the field emission electron beam is expected to improve the lattice resolution of an electron microscope, because the lattice resolution limit has been restricted by the chromatic aberration. In order to show this, the effect of the field emission electron beam on lattice imaging was investigated. In general, there are three methods of lattice imaging, that is, axial two-wave interference, axial three-wave interference and tilted two-wave interference. The schematic diagram for the former two methods is shown in Fig. 10. In the following, the contrast of lattice image is discussed with respect to chromatic aberration in each case.

1) The method of axial two-wave interference. The contrast of normal lattice fringes formed in this method is decreased by the

6   A. Tonomura, T. Matsuda, J. Endo, H. Todokoro and T. Komoda

(a)

Fig. 7. Interference fringes produced by a Möllenstedt type bi-prism. The number is 3,000, which is 10 times larger than those reported to date. (a) The whole interference pattern. (b) The enlarged interference fringes.

(b)

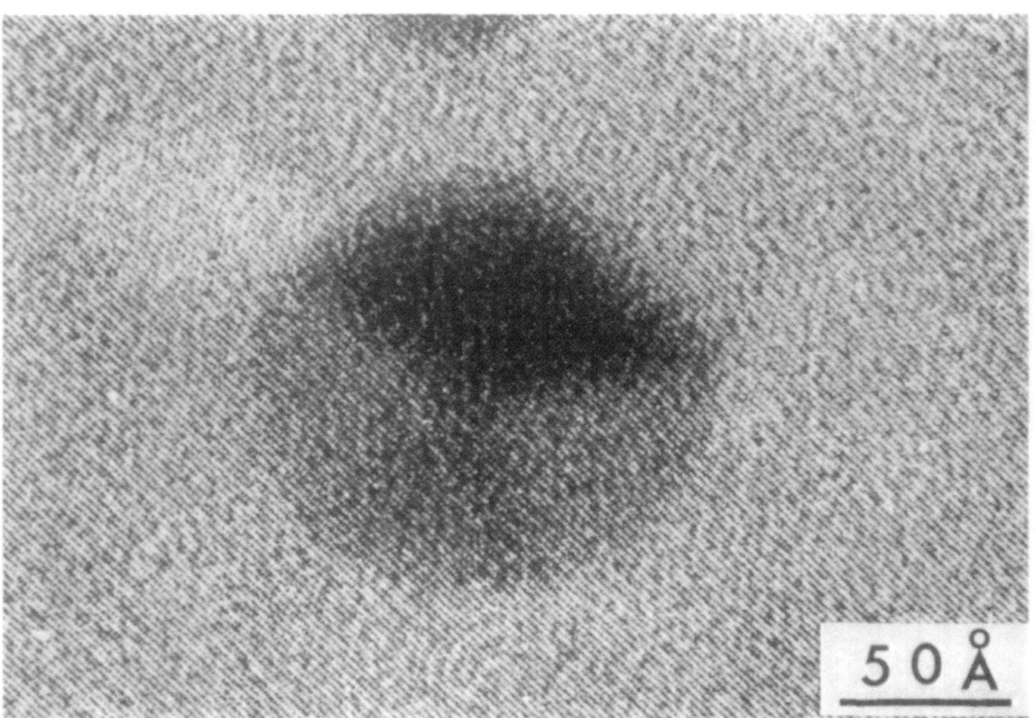

Fig. 8.   Image hologram of an evaporated gold particle.   In addition to bi-prism and lattice fringes, oblique interference fringes between the reference wave and the Bragg-reflected waves are observed where the reflected images are located.

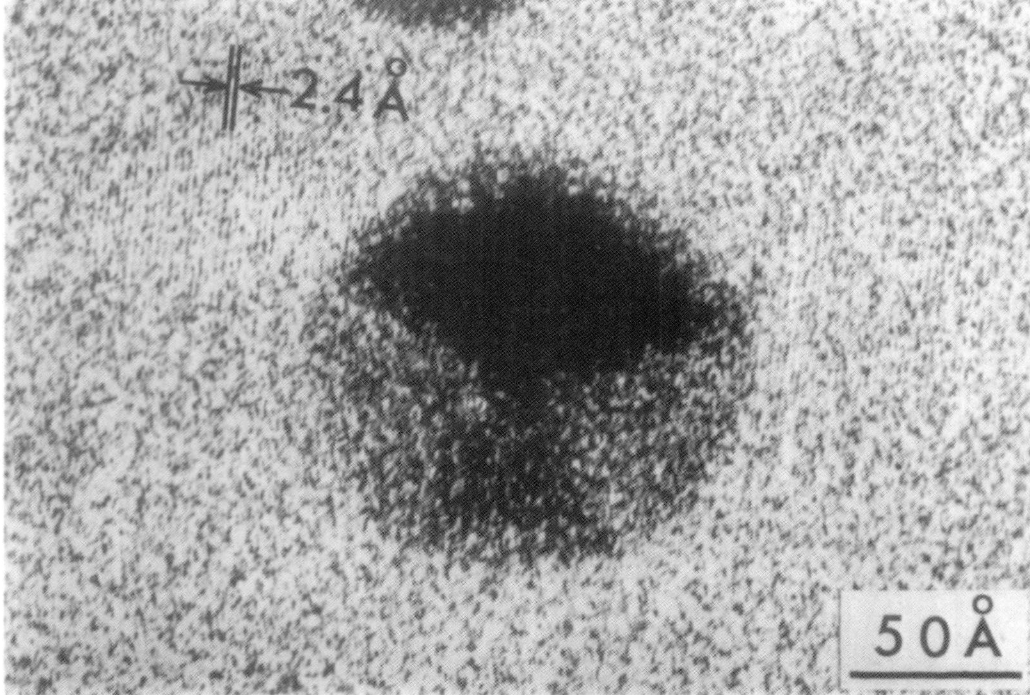

Fig. 9.   Reconstructed image of a gold particle.   The defocused lattice image is observed outside the particle.   This shows that the resolution of the reconstructed image is 2.4 Å.

8 A. TONOMURA, T. MATSUDA, J. ENDO, H. TODOKORO and T. KOMODA

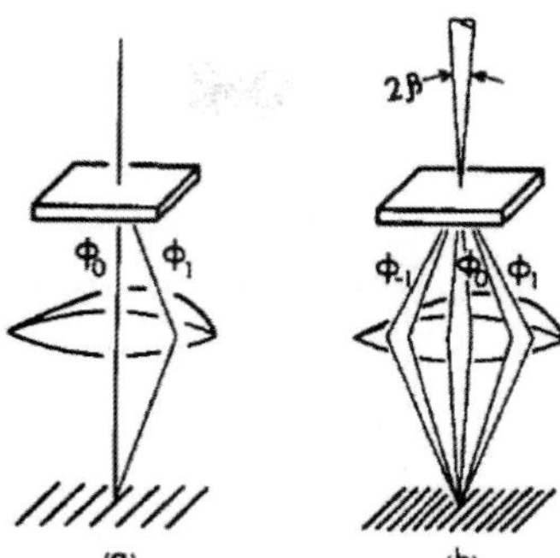

Fig. 10. Schematic diagram of lattice imaging. (a) Axial two-wave interference. (b) Axial three-wave interference.

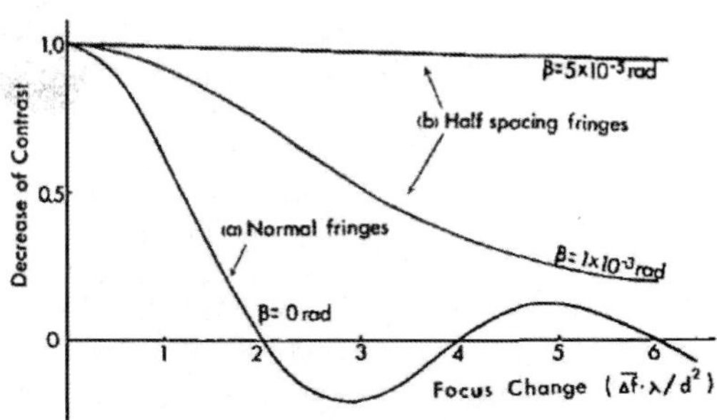

Fig. 11. Decrease of lattice image contrast due to the chromatic aberration. The value in the ordinate is so normalized that it is 1.0 in the case of no chromatic aberration $(\overline{\Delta f}=0)$. The negative value means the inversion of the fringe contrast.

chromatic defect of an electron lens. The energy change of electrons, which is related to the defocusing, makes the lattice fringes move laterally. The contrast of lattice fringes decreases by overlapping the fringes shifted variously. Due to this effect, the contrast decreases sensitively by the instability of high voltage and lens current, and by the energy spread of electron beam. The decrease of

the contrast is estimated under the condition of the uniformed energy distribution corresponding to the defocusing width $\overline{\Delta f}$ to be[9] $\{\sin(\pi\lambda\overline{\Delta f}/2d^2)\}/(\pi\lambda\overline{\Delta f}/2d^2)$. It is plotted against $(\overline{\Delta f}\cdot\lambda/d^2)$ in Fig. 11, where d is the lattice spacing considered. It seems reasonable to assume that lattice fringes are resolved when the contrast of fringes is more than 64%,

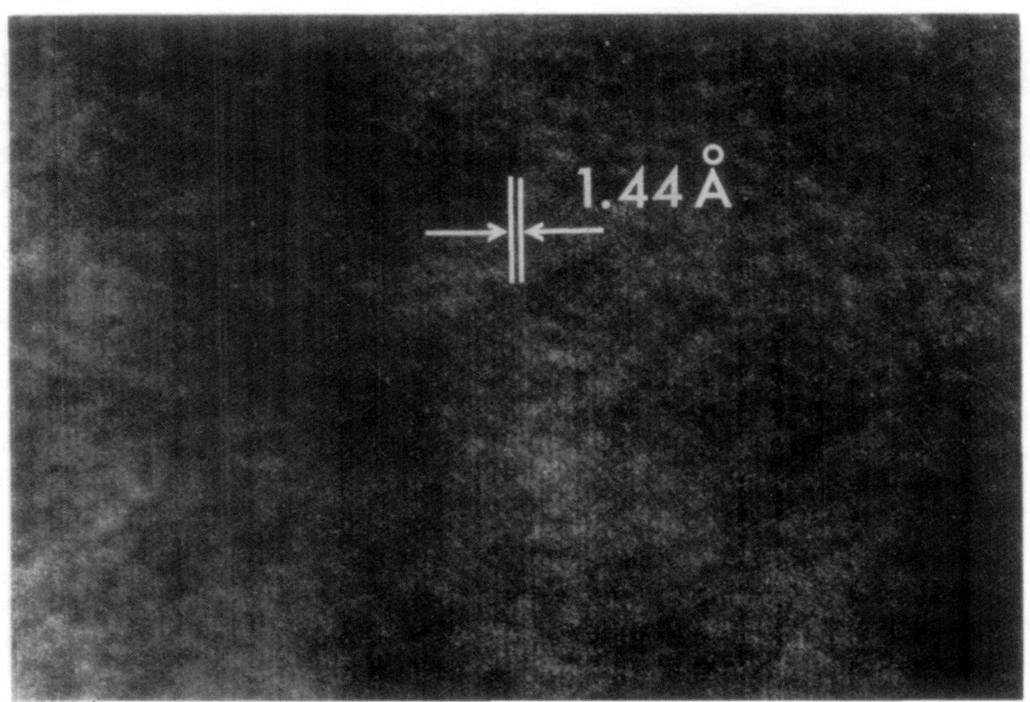

Fig. 12. Lattice fringes of gold {220} planes formed by the axial two-wave interference. The spacing of 1.4 Å is clearly resolved.

which condition corresponds to $(\pi\lambda\overline{\Delta f}/2d^2)=(\pi/2)$. The lattice resolution can be obtained as

$$d=\sqrt{\overline{\Delta f}\cdot\lambda}=\sqrt{C_c\lambda\frac{\Delta E}{E}},$$

where C_c is the coefficient of the chromatic aberration and $\Delta E/E$ is the variation of the electron energy, consisting of the instability of high voltage and lens current, and the energy spread of the electron beam. As the power stability of the apparatus (high voltage and lens current) is better than 2×10^{-6}/min and the energy spreads of electrons from thermionic hairpin and pointed filaments are about 2 and 1 eV, respectively, $\Delta E/E$ is mainly determined by the latter. When $C_c=1.5$ mm and $E=70$ kV, these lattice resolutions become 4 and 3 Å, respectively. Lattice resolution was actually improved using pointed filaments.[10]

Meanwhile, the reported energy spread of field emission electrons is $0.2\sim0.4$ eV and therefore the line resolution taken with the axial illumination is expected to be $1.3\sim1.9$ Å. Actually gold {220} lattice fringes ($d_{220}=1.4$ Å) formed by the axial two-wave interference were clearly resolved with the present microscope as shown in Fig. 12. This result shows that the effect of chromatic aberration decreases with the new microscope and that the lattice resolution becomes higher.

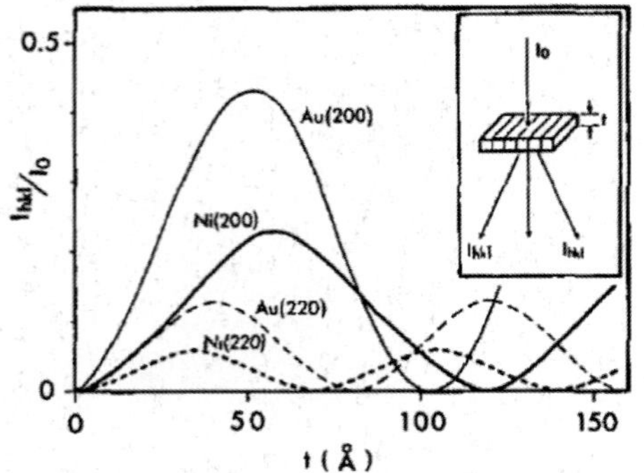

Fig. 13. Calculated intensities of simultaneous Bragg-reflections by three-wave approximation.

2) *The method of axial three-wave inter-ference.* In addition to the normal lattice

fringes, half-spacing fringes appear by the interference of simultaneously reflected waves. The half-spacing fringes do not move laterally by defocusing. Therefore, the energy spread of electrons has no blurring effect on the fringes, in the first order approximation. In the second order approximation, however, illumination angle 2β affects the contrast of the half-spacing fringes, coupling with defocusing width $\overline{\Delta f}$. The decrease of contrast[9] is also calculated to be

$$-\frac{1}{\overline{\Delta f}}\int_{-\overline{\Delta f}/2}^{\overline{\Delta f}/2}\left\{\sin\left(\frac{4\pi\beta\Delta f}{d}\right)\middle/\frac{4\pi\beta\Delta f}{d}\right\}d(\Delta f).$$

It is also shown by curves (b) in Fig. 11, in both cases of $\beta=1\times10^{-3}$ rad and $\beta=5\times10^{-6}$ rad. In actual, the energy spread of the beam makes the contrast of the half-spacing fringes lower if three-wave interference is taken into consideration.

In the conventional electron microscope, β is about 1×10^{-3} rad, while β is as small as 5×10^{-6} rad in the field emission electron microscope. In addition to this effect, the narrow energy spread of electron beam makes the contrast of the fringes further better. With the field emission electron microscope, the contrast of half-spacing fringes is hardly decreased even if 0.5 Å lattice spacing is observed. Based on the above-mentioned consideration for improving the lattice resolution, lattice fringes were observed with the field emission electron microscope. However, there was a difficulty that the intensities of simultaneous Bragg-reflections diminish with the smaller lattice spacing. Intensities of simultaneous Bragg-reflections by three-wave approximation are calculated from the equation (33) in reference 11. The results are shown for gold and nickel in Fig. 13.

To observe half-spacing lattice fringes of {220} planes, the optimum thicknesses for the largest reflected intensities are 40 Å for gold and 35 Å for nickel, these intensities being 13% and 4% of incident electrons, respectively. The value of the optimum thickness for gold is larger than that for nickel, which

10   A. Tonomura, T. Matsuda, J. Endo, H. Todokoro and T. Komoda

tendency is reversed in magnitude, compared with the case of two-wave approximation. This is due to the fact that incident electrons, parallel to lattice plane, are deviated from Bragg-condition and so the extinction distance becomes smaller, especially in the case of the small lattice spacing.

Lattice frings of the $\langle 001 \rangle$-oriented gold film taken with the field emission electron microscope are shown in Fig. 14. Half-spacing fringes of gold $\{220\}$ planes are observed. The half spacing is equal to 0.72 Å, which was also taken by Sieber *et al.*[12] with special imaging technique. Furthermore, lattice fringes of the $\langle 111 \rangle$-oriented nickel film are given in Fig. 15. Half-spacing frings of $\{220\}$ planes ($1/2\, d_{220} = 0.62$ Å) are observed,[13] in addition to normal lattice fringes ($d_{220} = 1.2$ Å). This is the smallest spacing fringes that have been ever taken.

In the same figure, the fringes with 0.72 Å spacing are also observed. They are due to the interference between two $\{220\}$ reflected waves.

CONCLUSION

A field emission electron microscope has newly been developed. Both the gun and illuminating system were designed so as to obtain a highly stable electron beam with high brightness.

Several experiments were made to show the high performance of this microscope. The electron beam was stably obtained, whose brightness was 100 times larger than that of the conventional thermionic electron beam. This feature was successfully exemplified by the experimental results that 300 Fresnel fringes were taken with only 4 s and 3,000 bi-prism fringes were taken with 30 s. The decrease of energy spread of the electron beam made the effect of chromatic aberration considerably smaller in lattice imaging. In

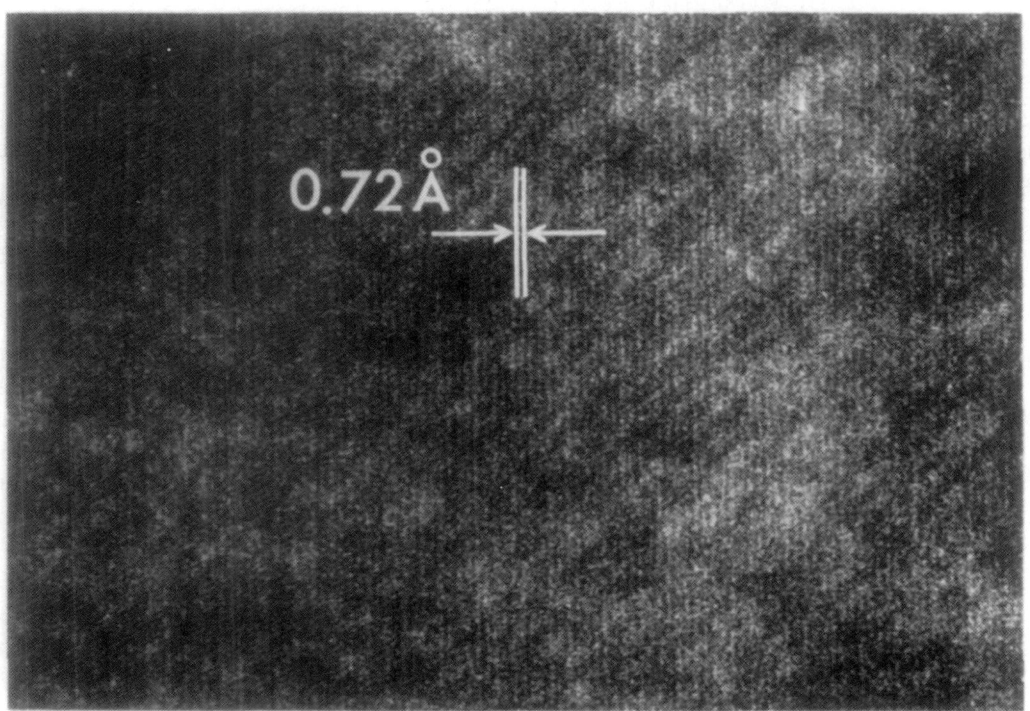

Fig. 14.   Half-spacing fringes of gold $\{220\}$ planes. The spacing of 0.72 Å is resolved, which was also observed by Sieber *et al.* with special imaging technique.[12]

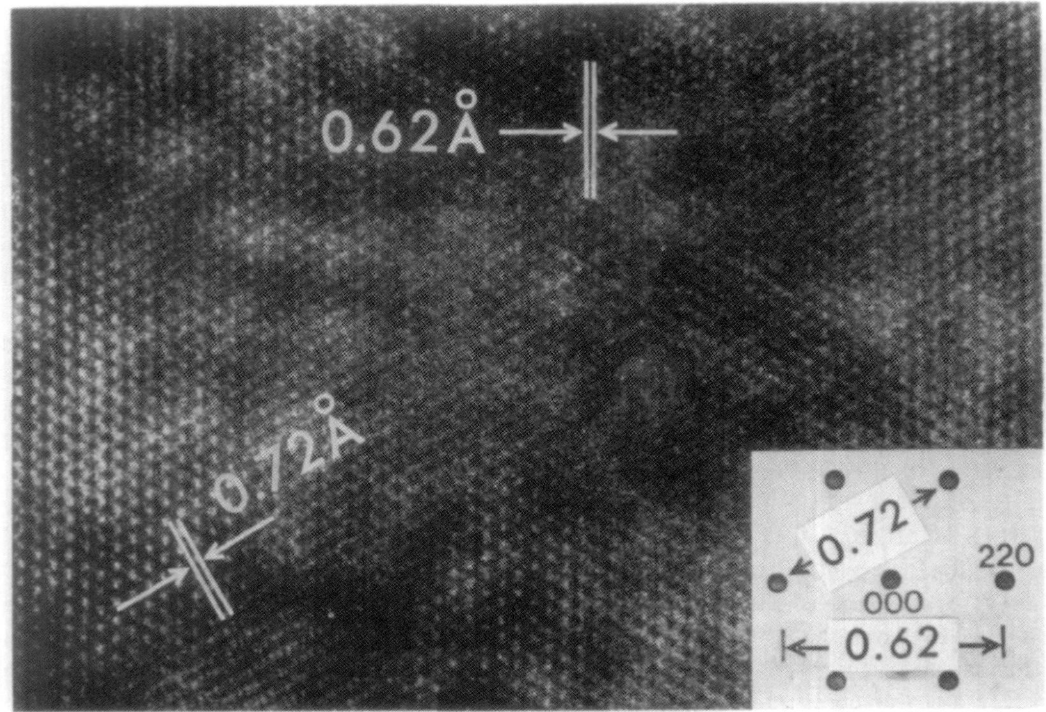

Fig. 15.　Half-spacing fringes of nickel {220} planes.　The spacing of 0.62 Å is resolved, which is the smallest structure that have ever been observed in electron microscopes.

ctual, the half-spacing fringes of nickel {220} lanes (1/2 d_{220}=0.62 Å) were observed for he first time.

REFERENCES

1) Koops, H.: Proc. Int. Congr. Electron Microsc., 1978, Vol. 3, p. 185

2) Tonomura, A., Fukuhara, A., Komoda, T. and Watanabe, H.: *Jap. J. Appl. Phys.*, **7**, 259 (1968)

3) Möllenstedt, G. and Wahl, H.: *Naturwissenschaften*, **55**, 340 (1968)

4) Tomita, H., Matsuda, T. and Komoda, T.: *Jap. J. Appl. Phys.*, **9**, 719 (1970)

5) Möllenstedt, G. and Dücker, H.: *Naturwissen-schaften*, **42**, 41 (1954)

6) Hibi, T. and Takahashi, S.: *J. Electron Microsc.*, **12**, 129 (1963)

7) Yada, K., Shibata, K. and Hibi, T.: *J. Electron Microsc.*, **22**, 223 (1973)

8) Tonomura, A., Matsuda, T. and Komoda, T.: *Jap. J. Appl. Phys.*, **17**, 1137 (1978)

9) Komoda, T.: Doctor thesis (Nagoya Univ., 1974).

10) Yada, K. and Hibi, T.: *J. Electron Microsc.*, **18**, 266 (1969)

11) Fukuhara, A.: J. Phys. Soc. Japan **21**, 2645 (1966)

12) Sieber, P. and Tonar, K.: *Optik*, **42**, 375 (1975)

13) Matsuda, T., Tonomura, A. and Komoda, T.: *Jap. J. Appl. Phys.*, **17**, 2073 (1978)

VOLUME 44, NUMBER 21   PHYSICAL REVIEW LETTERS   26 MAY 1980

Direct Observation of Fine Structure of Magnetic Domain Walls by Electron Holography

Akira Tonomura, Tsuyoshi Matsuda, and Junji Endo
Central Research Laboratory, Hitachi Ltd., Kokubunji, Tokyo 185, Japan

and

Tatsuo Arii and Kazuhiro Mihama
*Central Research Laboratory, Hitachi Ltd., Kokubunji, Tokyo 185, Japan, and Department of Applied Physics,
Faculty of Engineering, Nagoya University, Chikusa-ku, Nagoya 464, Japan*
(Received 28 December 1979)

Holographic interference electron microscopy is presented for investigating the structure of domain walls in plate-shaped cobalt particles. Circular magnetic lines of force are directly observed as contour fringes which overlap individual particle micrographs. These fringes show at a glance how the spin rotates across domain walls. It is also suggested from holographic electron diffraction that the spin stands up in the center of the particle.

PACS numbers: 75.70.Kw, 61.16.Di

The fine structure of a domain wall in a ferromagnetic thin film plays an important role in determining the fundamental characteristic of the film. However, only a limited number of experimental investigations have been reported, in contrast to many theoretical calculations of wall structures. This is because magnetization cannot be quantitatively measured even by Lorentz microscopy, which has been the only effective method for observing the fine structures of magnetic domain walls to date.

A new method has been developed for observing magnetization in thin films. In this method, the phase distribution of electrons transmitted through a specimen is observed as a contour map by means of holographic interference electron microscopy.[1] Although the electron phase itself is not uniquely determined, the phase difference $\Delta\Phi$ between two points P_1 and P_2 in the specimen plane is given by the following equation[2]:

$$\Delta\Phi(P_1, P_2) = (e/\hbar) \int \vec{B} \cdot d\vec{S}. \qquad (1)$$

Here the integral is performed over the surface enclosed by the two electron trajectories passing through points P_1 and P_2. From this equation, an important result is deduced. The phase difference is equal to zero if P_1 and P_2 lie along a magnetic line of force in the film. Therefore, magnetic lines of force can be directly observed as a contour map of the electron wave front by means of electron holography.[3,4] The holographic method for magnetization measurement was first proposed by Cohen.[5] However, no practical results have been reported except some preliminary experiments.[6-8]

Electron holograms were formed in a newly

developed field-emission electron microscope.[9] The schematic diagram is shown in Fig. 1. An object is illuminated with a collimated electron beam and its image is formed through the objective lens. A reference beam is projected on the image plane by a Möllenstedt-type electron biprism,[10] forming the off-axis image hologram. The total magnification in the electron microscope was 40 000 times. The electron accelerating voltage was 80 kV.

Reconstruction was carried out in the optical system shown in Fig. 2, where phase-amplification technique was employed for detailed observation.[11] Laser beams A and B illuminate the hologram and each beam produces a reconstructed image and its conjugate, whose phase distributions are opposite in sign. Only the reconstructed image of beam A and the conjugate

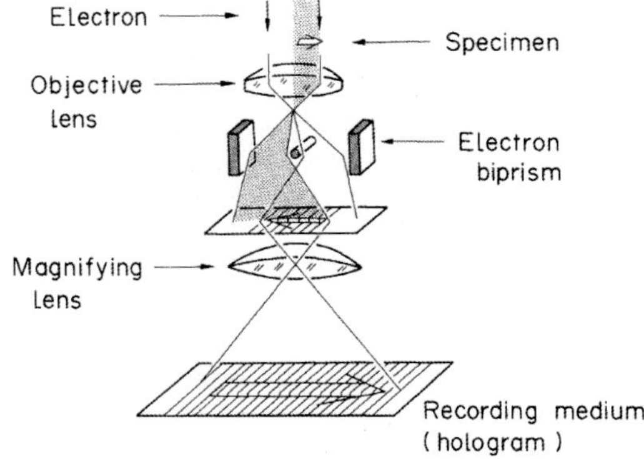

FIG. 1. Schematic diagram of hologram formation.

Volume 44, Number 21　　　**PHYSICAL REVIEW LETTERS**　　　26 May 1980

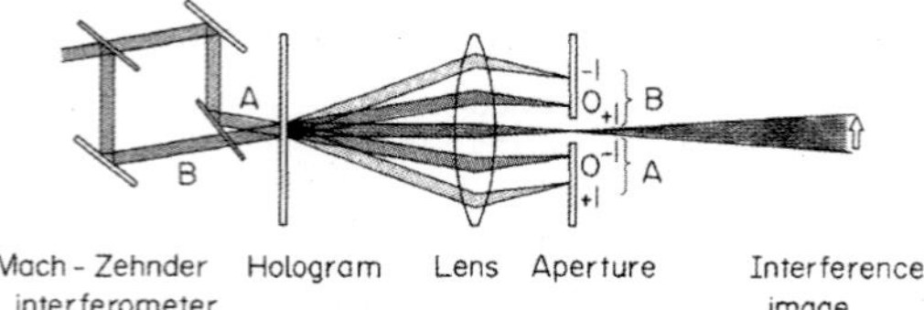

FIG. 2. Schematic diagram of optical reconstruction.

image of beam B pass through an aperture and an interference image with a doubled phase distribution is formed. The background of the interferogram is even when the interfering beams travel in the same direction, but a slight tilt between them produces a reference system of regular fringes over the background.

Cobalt specimens[12] were prepared by gas evaporation in a 10-Torr inert gas atmosphere. The particles studied are plate shaped and have $\{111\}$ surfaces and $\langle 110 \rangle$ edges.

Interference micrographs of a triangular cobalt particle are shown in Fig. 3. No contrast is observed inside reconstructed image (a), which is equivalent to the electron microscopic image. On the other hand, many contour lines appear in contour map (b). Contour lines parallel to the three edges show that thickness increases linearly to 55 nm. The inner contour lines are magnetic lines of force, since the inner region is uniform in thickness due to the typical crystal habit of fcc particles.[13] This was also ascertained by observing the extinction contour lines in the electron microscopic image of the particle in which Bragg reflection was excited. The effect of stray field was estimated to be small from the contour map, showing that the magnetic field was closed inside the particle image.

The contour lines in Fig. 3(b) clearly show how the spin direction rotates across the three domain walls. It cannot be determined from this contour map whether the magnetization rotates clockwise or counterclockwise. The direction can be decided from interferogram (c) obtained by changing the angle between beams A and B in the reconstruction stage. Interference fringes are displaced downward at particle edges and they go further downward inside the particle. This can be interpreted as follows. As an electron beam travels faster in the crystal than in vacuum and consequently has a shorter wavelength, the wave front of the transmitted electron beam through the particle is retarded. Furthermore, the wave front is either advanced or retarded depending on whether magnetization is clockwise or counterclockwise, as known from Eq. (1). Therefore, the magnetization direction proves to be counterclockwise. Particles with clockwise magnetization were also observed. These two kinds of particles have already been observed by Lorentz microscopy.[14]

The magnetization rotation distribution across a domain wall was measured from the contour map. The measured value corresponds to the average magnetization over a film thickness. Wall width obtained was 40 nm.

From the contour map [Fig. 3(b)], magnetization was found to be circular in the central part of the particle. However, it is not reasonable to consider that the spin still remains circular in the extreme center. This is because exchange energy increases infinitely with this magnetization configuration. It is speculated that the spin stands up, but this has not been supported by experimental evidence.

In order to clarify spin behavior in the center of triangular particles, low-angle electron dif-

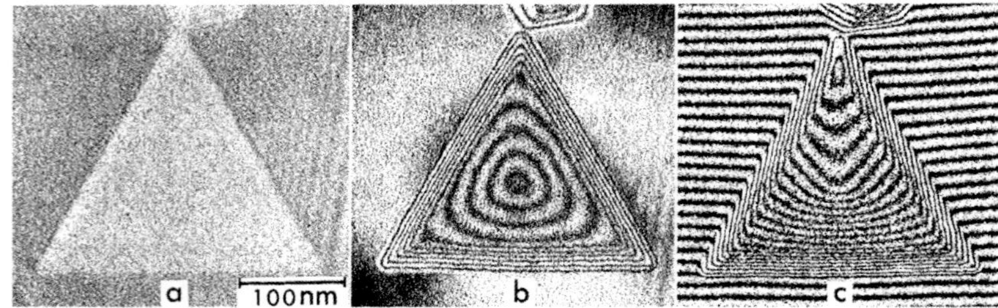

FIG. 3. Triangular cobalt particle: (a) reconstructed image, (b) contour map, and (c) interferogram. Contour lines in contour map (b) are magnetic lines of force. Magnetization direction is counterclockwise from interferogram (c).

VOLUME 44, NUMBER 21   PHYSICAL REVIEW LETTERS   26 MAY 1980

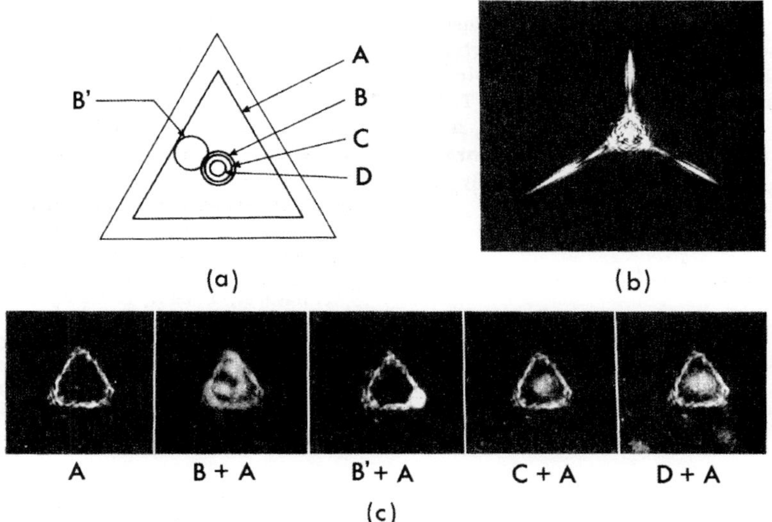

FIG. 4. Low-angle electron diffraction patterns from areas in a triangular cobalt particle: (a) selected areas,
(b) diffraction pattern from the particle, and (c) diffraction patterns from selected areas. Diffraction patterns
from areas A and B consist of three deflection spots corresponding to the three magnetic domains. However, a
single spot appears in diffraction patterns from central areas less than 15 nm in radius (C and D). These results
suggest that the spin stands up in the center.

fraction patterns were obtained from the central
regions as shown in Fig. 4. These diffraction
patterns were optically formed from holograms,
since electron holography can reproduce all the
information provided by electron beam scattering. The three streaks in diffraction pattern (b)
correspond to the three peripheral regions (wedges) of the particle. Actually the streaks are absent in the electron diffraction patterns from the
inner region excluding the peripheral wedges as
shown in Fig. 4(c). The selected areas are designated A, B, B', C, and D in Fig. 4(a). The
radii of selected areas B, C, D are 20, 15, and
10 nm, respectively, and are superimposed over
the diffraction pattern from region A to make precise location measurement possible. A diffraction pattern from a large area such as region A
consists of three diffraction spots and streaks
connecting them. The three spots are due to the
three magnetic domains. Actually the measured
angle between the center and one vertex of the
triangle structure in Fig. 4(c) is 1.4×10^{-4} rad.
This agrees fairly well with the calculated deflection angle of 1.1×10^{-4} rad assuming a spontaneous magnetization of 1450 Oe and particle thickness of 55 nm. In the case of selected area B,
three spots are still observed. When the areas

become smaller, as C and D, only one spot is
observed in the center of the triangle. A diffraction effect becomes appreciable with decreasing
areas. However, if the area of B is moved to
another location, B', in one of the magnetic domains, the diffraction pattern becomes a single
spot situated at a vertex of the triangle. This
was also true for both areas C and D. In addition, the diffraction spots for areas C and D are
not so widely spread as to cover the triangle
structure. Therefore, it can be concluded that
few electrons are deflected in regions C and D,
and that the measured magnetization component
in the specimen plane is smaller than the spontaneous magnetization.

These results support the assumption that the
spin stands up in the center and furthermore give
the information that the area where the spin
stands up is less than 15 nm in radius. This is
also consistent with the interferogram in Fig.
3(c), which indicates that the electron wave front
is not pointed in the center, but curved in a central region about 10 nm in radius.

We are very grateful to Professor Dr. R. Uyeda
in Meijō University for first suggesting the holographic observation of magnetic fine particles.
The authors thank Dr. S. Tsukahara at the Elec-

VOLUME 44, NUMBER 21   PHYSICAL REVIEW LETTERS   26 MAY 1980

trotechnical Laboratory Headquarters and Professor Dr. E. Zeitler at the Fritz-Haber-Institut der Max-Planck-Gesellschaft for their participation in valuable discussions. The invaluable advice and stimulation of Dr. H. Okano, Dr. T. Komoda, Dr. A. Fukuhara, Dr. S. Taniguchi, and Dr. Y. Sugita at the Central Research Laboratory, Hitachi Ltd., are also most gratefully acknowledged.

[1] A. Tonomura, J. Endo, and T. Matsuda, Optik **53**, 143 (1979).

[2] Y. Aharanov and D. Bohm, Phys. Rev. **115**, 485 (1959).

[3] D. Gabor, Proc. Roy. Soc. London, Ser. A **197**, 454 (1949).

[4] D. Gabor, Proc. Phys. Soc. London, Ser. B **64**, 449 (1951).

[5] M. S. Cohen, J. Appl. Phys. **38**, 4966 (1967).

[6] A. Tonomura, Jpn. J. Appl. Phys. **11**, 493 (1972).

[7] G. Pozzi and G. F. Missiroli, J. Microsc. (Paris) **18**, 103 (1973).

[8] B. Lau and G. Pozzi, Optik **51**, 287 (1978).

[9] A. Tonomura, T. Matsuda, J. Endo, H. Todokoro, and T. Komoda, J. Electron Microsc. **28**, 1 (1979).

[10] G. Möllenstedt and H. Dücker, Z. Phys. **145**, 377 (1956).

[11] K. Matsumoto and M. Takashima, J. Opt. Soc. Am. **60**, 30 (1970).

[12] T. Arii, S. Yatsuya, N. Wada, and K. Mihama, Jpn. J. Appl. Phys. **17**, 259 (1978).

[13] T. Hayashi, T. Ohno, S. Yatsuya, and R. Uyeda, Jpn. J. Appl. Phys. **16**, 705 (1977).

[14] T. Arii, S. Yatsuya, N. Wada, and K. Mihama, in *Proceedings of the Fifth International Conference on Electron Microscopy, Kyoto, Japan, 1977*, edited by T. Imura and H. Hashimoto (Japanese Society of Electron Microscopy, Tokyo, 1977), p. 203.

PHYSICAL REVIEW
LETTERS

VOLUME 48   24 MAY 1982   NUMBER 21

Observation of Aharonov-Bohm Effect by Electron Holography

Akira Tonomura, Tsuyoshi Matsuda, Ryo Suzuki, Akira Fukuhara, Nobuyuki Osakabe,
Hiroshi Umezaki, Junji Endo, Kohsei Shinagawa, Yutaka Sugita, and Hideo Fujiwara
Central Research Laboratory, Hitachi Ltd., Kokubunji, Tokyo 185, Japan
(Received 16 February 1982)

In this experiment, an electron- and optical-holographic technique is employed with
small toroidal ferromagnets each forming a magnetic-flux closure. The holographic interferometry proves that a phase difference between two electron beams having passed
through the field-free regions agrees well with the fundamental relation known as the
Aharonov-Bohm effect. It is also confirmed from the same hologram that flux leakage
from the toroids does not affect the conclusion.

PACS numbers: 03.65.Bz, 41.80.Dd, 42.40.Mg

The existence of the Aharonov-Bohm effect[1]
(AB effect) has recently been questioned by
Bocchieri et al.[2] and Roy.[3] The AB effect states
that a phase difference between two electron
beams is produced proportional to the enclosed
magnetic flux, even if they never touch the magnetic field. Bocchieri et al. asserted that the
AB effect is purely of mathematical origin. Experiments in the past[4] were also questioned from
the standpoint that electrons were affected by inevitable leakage magnetic fields from finite
whiskers or solenoids used in these experiments.[4]
Although these assertions have since then been
disputed theoretically by many authors,[5] the controversy has still not fully abated.[6]

Our experiment employs electron holography.[7]
In order to avoid the questioned leakage effects,[8]
tiny toroidal magnets[9] were used instead of
whiskers or solenoids to make complete flux circuits. Furthermore, a new method of holographic interference microscopy[10] was employed, both
to obtain contour maps of the electron phase and
to detect quantitatively the amount of leakage that
might have, by some chance, come from the mag-

nets.

The toroidal magnets were prepared in the following way. Permalloy thin films (80% Ni and
20% Fe) were prepared by vacuum evaporation.
The substrate was a glass plate covered with an
evaporated thin film of NaCl. Permalloy toroids
of various sizes were formed by means of electron-beam lithography. These toroids were
floated off on a water surface, and applied to
thin carbon films approximately 100 Å thick.

An electron-microscopic image and an under-
focused Lorentz micrograph of such a toroidal
magnet are shown in Figs. 1(a) and 1(b). The
Lorentz micrograph shows that magnetization is
closed within the magnet. This is the case with
most toroids presumably as a result of the shape
effect.

Off-axis electron homograms of the toroidal
magnets were formed in a 100-kV field-emission
electron microscope.[11] The schematic diagram
for hologram formation is shown in Fig. 2. A
toroidal magnet was illuminated with a collimated electron beam. Its demagnified image
(magnification $\sim \frac{1}{2}$) was formed through both ob-

                1443

VOLUME 48, NUMBER 21   PHYSICAL REVIEW LETTERS   24 MAY 1982

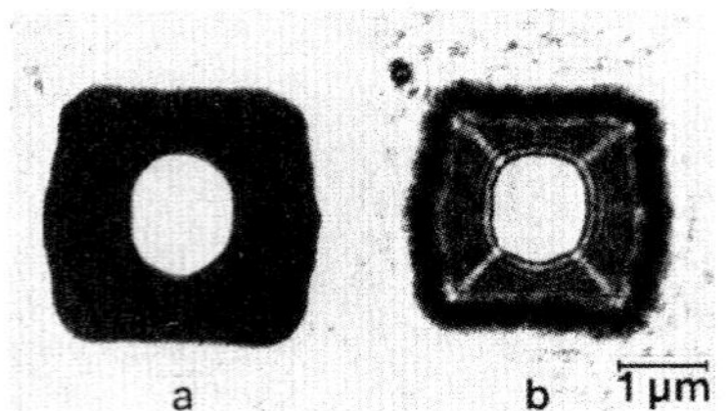

FIG. 1. Toroidal magnet. (a) Electron-microscopic image. (b) Lorentz micrograph.

jective and intermediate lenses. A reference beam was projected on the image plane by an electron biprism[12] to form the image hologram. Final magnification at the recorded hologram was 2 000.

Reconstruction was performed in the optical system shown in Fig. 3. A collimated laser beam from a He-Ne laser was split into two beams by beam splitter A. One beam illuminated the electron hologram to reconstruct the image, which was focused again by lenses E and F on the observation plane. The other beam (comparison beam) from the splitter was superposed on the observation plane to form the interference image. The advantage of the holography technique is that it makes it possible to obtain phase-amplified interference images.[13]

Interference micrographs for the toroidal magnet (Fig. 1) are shown in Fig. 4. The phase contour map of an electron beam transmitted through a magnet, shown in Fig. 4(a), was obtained with the comparison beam parallel to the object beam. It cannot be determined from the contour map whether the wave front of the object beam is advanced or retarded. Therefore, interferogram (b) was taken with a tilted comparison beam to determine this. The wave front obtained is schematically shown in Fig. 4(c).

The photographs reveal that a phase difference really exists between two electron beams that have passed through the inner and outer spaces of a toroidal magnet, where there were no magnetic fields in those spaces. In addition, the phase difference of 5.5λ, measured from the interference micrograph, agrees with the theoretical value of 6.0λ to 20%. This is estimated from data where 4π times magnetization was 9500 ± 500 Oe, film thickness was 400 ± 30 Å, and toroid width was 6400 ± 500 Å. Phase shifts at the magnet edges are partly due to the refraction effect.[13] However, this effect can be ignored in the present estimation, because the effect of

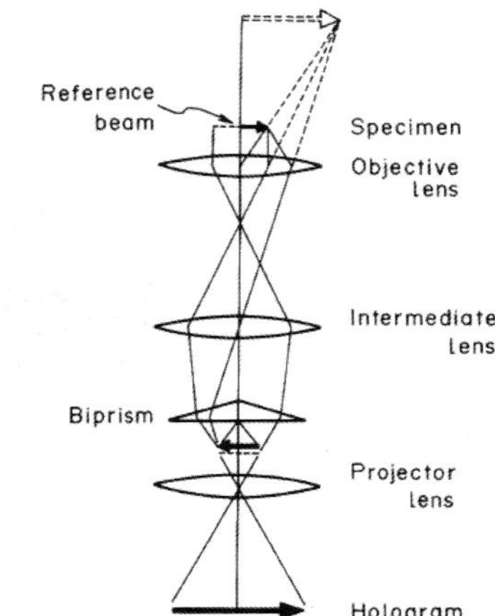

FIG. 2. Schematic arrangement for electron hologram formation.

the phase shift made at the outer edge is cancelled at the inner edge. In addition the shift value itself is smaller than one λ, as is also explained from the thickness and inner potential (~20 V) of the sample.

Another example with a slightly larger magnet on the same carbon film is shown in Fig. 5. In contrast to the previous example, the phase here is retarded in the inner space of the magnet. Correspondingly, the magnetization direction is counterclockwise. The number of contour lines, i.e., the phase difference, is increased compared with that in Fig. 4. This is in proportion to the magnet width since film thicknesses are the same. The deviation from the proportional relation was measured to be less than 10% for

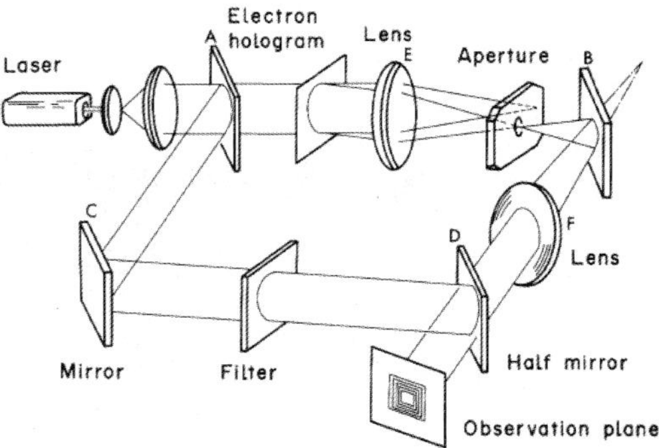

FIG. 3. Optical reconstruction system for interference microscopy.

VOLUME 48, NUMBER 21   PHYSICAL REVIEW LETTERS   24 MAY 1982

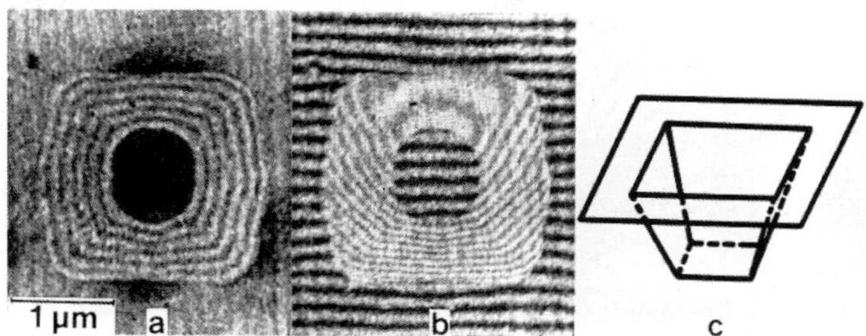

FIG. 4. Interference micrographs of the toroidal magnet shown in Fig. 1. (a) Contour map of electron phase. (b) Interferogram of electron phase. (c) Schematic form of the wave front.

various toroidal sizes.

These experimental results verify the existence of the AB effect. Quantitative agreement[14] is achieved with the fundamental AB effect relation.

Leakage-field effects were confined to be sufficiently small in the cases of Figs. 4 and 5. Contour lines in interference micrographs were verified to follow magnetic lines of force as viewed along the direction of the electron beam.[15] Therefore, contour lines must exit from the toroid if magnetic fields are leaking from the magnet. An example of field leakage is shown in Fig. 6. Leakage fields do not show up in the Lorentz micrograph, Fig. 6(a), but can be clearly observed in the interference micrograph, Fig. 6(b). The magnetic flux between two adjacent contour lines is equal to a constant, h/e, irrespective of electron energy. It can be concluded from the contour maps shown in Figs. 4(a) and 5(a) that the leakage flux was less than h/e and that the resultant phase change is too small to conceal the AB effect.

In this experimental arrangement, the electron beam partly touched and even penetrated the magnet. This point is open to criticisms, but our argument for this is as follows. In the present experiment, the shape of a magnetic sample is reproduced as a clear image on the interferogram. Consequently, the part of the beam transmitted through the magnetic flux in the sample does not contribute to points outside the sample image. The beams reaching these points must have felt only the magnetic vector potential, if any.

It was for the measurement of the phase difference by tracing the interference fringes that the penetrable toroidal magnets were adopted in our experiment. This is an advantage of our experiment over former experiments.[4] If the fringes on the images of the toroids are not observed, the phase difference is determined by only a fraction of a wavelength unit.[16]

The different electron energy causes an appreciable change in electron penetrability, but no change in phase difference. This fact was confirmed at 80, 100, and 125 kV. If there were an essential difference between an absolutely inaccessible field and a negligibly accessible field,[6] then the AB effect could be neither confirmed nor denied experimentally.

Regardless of the strength of penetrability, our experimental results of the interference

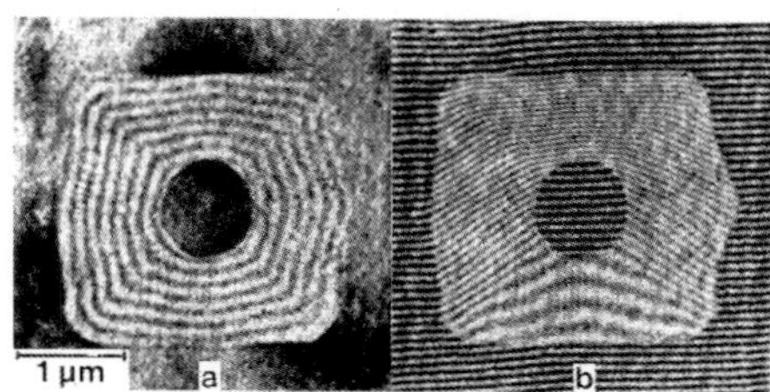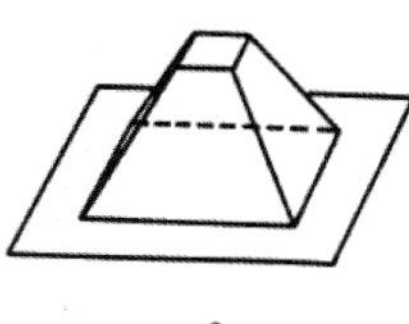

FIG. 5. Interference micrographs of magnet having a magnetization direction opposite to that in Fig. 4. (a) Contour map. (b) Interferogram. (c) Schematic form of the wave front.

Volume 48, Number 21   PHYSICAL REVIEW LETTERS   24 May 1982

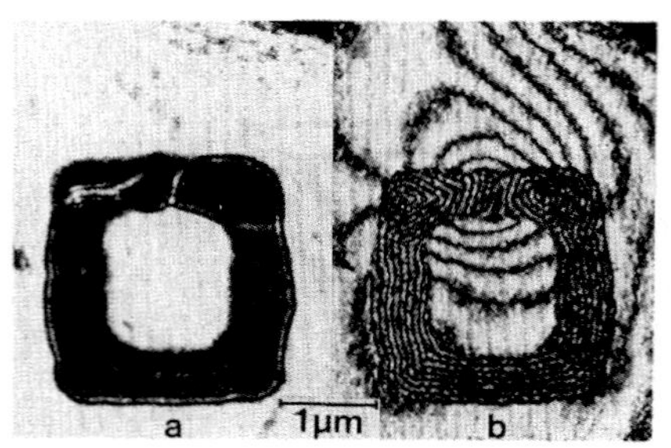

FIG. 6. Example of leakage magnetic fields. (a) Lorentz micrograph. (b) Interference micrograph. (Phase amplification by factor two.)

pattern, e.g., Fig. 4(b), can be fully explained with the Stokes vector potential. This cannot be expected if vector potential is zero everywhere outside the toroid, as Bocchieri *et al.* proposed in case of a solenoid.[2]

We are very grateful to Professor C. N. Yang of State University of New York, Stony Brook, for useful discussions at the planning stage. We would like to express our sincere thanks to Dr. T. Doi for his support and encouragement of such a fundamental research work. We also gratefully acknowledge the valuable advice and stimulation given by Professor R. Uyeda of Meijo University and by Dr. H. Okano, Dr. M. Kudo, Dr. T. Komoda, and Dr. F. Nagata of our laboratory.

---

[1]Y. Aharonov and D. Bohm, Phys. Rev. 115, 485 (1959).

[2]P. Bocchieri and A. Loinger, Nuovo Cimento A 47, 475 (1978); P. Bocchieri, A. Loinger, and G. Siragusa, Nuovo Cimento A 51, 1 (1979); P. Bocchieri and A. Loinger, Lett. Nuovo Cimento 25, 476 (1979); P. Bocchieri, A. Loinger, and G. Siragusa, Nuovo Cimento A 56, 55 (1980).

[3]S. M. Roy, Phys. Rev. Lett. 44, 111 (1980).

[4]R. G. Chambers, Phys. Rev. Lett. 5, 3 (1960); H. A. Fowler, L. Marton, J. A. Simpson, and J. A. Suddeth, J. Appl. Phys. 32, 1153 (1961); H. Boersch, H. Hamisch, K. Grohmann, and D. Wohlleben, Z. Phys. 165, 79 (1961); G. Möllenstedt and W. Bayh, Phys. Bl. 18, 299 (1962).

[5]For example, D. Bohm and B. J. Hiley, Nuovo Cimento A 52, 295 (1979); H. J. Lipkin, Phys. Rev. D 23, 1466 (1981).

[6]P. Bocchieri and A. Loinger, Nuovo Cimento A 66, 164 (1981).

[7]D. Gabor, Proc. Roy. Soc. London, Ser. A 197, 454 (1949), and Ser. B 64, 449 (1951).

[8]The fringing field from a solenoid was reduced by completing the flux circuit with a high-permeability return strip. However, the amount of the fringing field remained quantitatively unknown: W. Bayh, Z. Phys. 169, 492 (1962).

[9]D. M. Greenberger, Phys. Rev. D 23, 1460 (1981); D. M. Greenberger, D. K. Atwood, J. Arthur, and C. G. Shull, Phys. Rev. Lett. 47, 751 (1981).

[10]A. Tonomura, J. Endo, and T. Matsuda, Optik (Stuttgart) 53, 143 (1979).

[11]A. Tonomura, T. Matsuda, J. Endo, H. Todokoro, and T. Komoda, J. Electron Microsc. 28, 1 (1979).

[12]G. Möllenstedt and H. Dücker, Naturwissenschaften 42, 41 (1954).

[13]J. Endo, T. Matsuda, and A. Tonomura, Jpn. J. Appl. Phys. 18, 2291 (1979).

[14]See Ref. 8. Better accuracy was obtained in the experiment by Bayh.

[15]A. Tonomura, T. Matsuda, J. Endo, T. Arii, and K. Mihama, Phys. Rev. Lett. 44, 1430 (1980); A. Tonomura, T. Matsuda, H. Tanabe, N. Osakabe, J. Endo, A. Fukuhara, K. Shinagawa, and H. Fujiwara, to be published; T. Matsuda, A. Tonomura, R. Suzuki, J. Endo, N. Osakabe, H. Umezaki, H. Tanabe, Y. Sugita, and H. Fujiwara, to be published.

[16]T. T. Wu and C. N. Yang, Phys. Rev. D 12, 3845 (1975).

VOLUME 56, NUMBER 8　　PHYSICAL REVIEW LETTERS　　24 FEBRUARY 1986

Evidence for Aharonov-Bohm Effect with Magnetic Field Completely Shielded from Electron Wave

Akira Tonomura, Nobuyuki Osakabe, Tsuyoshi Matsuda, Takeshi Kawasaki, and Junji Endo
Advanced Research Laboratory, Hitachi Ltd., Kokubunji, Tokyo 185, Japan

and

Shinichiro Yano and Hiroji Yamada
Central Research Laboratory, Hitachi, Ltd., Kokubunji, Tokyo 185, Japan
(Received 4 December 1985)

Evidence for the Aharonov-Bohm effect was obtained with magnetic fields shielded from the electron wave. A toroidal ferromagnet was covered with a superconductor layer to confine the field, and further with a copper layer for complete shielding from the electron wave. The expected relative phase shift was detected with electron holography between two electron beams, one passing through the hole of the toroid, and the other passing outside. The experiment gave direct evidence for flux quantization also.

PACS numbers: 03.65.Bz, 41.80.Dd

The Aharonov-Bohm (AB) effect[1] has recently received much attention as an unusual but important quantum effect.[2] The predicted effect is the production of a relative phase shift between two electron beams enclosing a magnetic flux even if they do not touch the magnetic flux. Such an effect is inconceivable in classical physics and directly demonstrates the gauge principle of electromagnetism.[3]

Although the affirmative experimental test was offered[4] soon after its prediction, Bocchieri et al.[5] and Roy[6] questioned the validity of the test, attributing the phase shift to leakage fields. The authors' recent experiment[7] using a toroidal magnet established the existence of the AB effect, under the condition of complete confinement of the magnetic field in the magnet; electron holography confirmed quantitatively the expected relative phase shift between the two beams. Bocchieri, Loinger, and Siragusa[8] still argued that the phase shift could be due to the Lorentz-force effect on the portion of the electron beam going through the magnet.[9]

The present experiment[10] is designed to provide a crucial test of the AB effect. A tiny toroidal magnet covered entirely with a superconductor layer and further with a copper layer is fabricated. The two layers prevent the incident electron wave from penetrating the magnet. In addition, the magnetic field is confined to the toroidal magnet by the Meissner effect of the covering superconductor. Then the relative phase shift between two electron beams, one passing through a region enclosed by the toroid and the other passing outside the toroid, is measured by means of electron holography. The experimental results detected the predicted relative phase shift, giving conclusive evidence for the AB effect. This experiment also demonstrated the flux quantization.[11]

Tiny toroidal samples were fabricated by use of photolithography. A Permalloy (80% Ni and 20% Fe) thin film, 200 Å thick, was prepared by vacuum evaporation on a silicon wafer covered with Al (3000 Å thick), Nb (2500 Å thick), and SiO (500 Å thick); the SiO layer serves to reduce the coercive force of the Permalloy. After evaporation of a 2000-Å-thick layer of SiO on the Permalloy, the toroidal shape was cut out to the depth of the Nb surface. The NbO produced by the lithography processes at the Nb surface had to be removed to ensure a perfect contact with the Nb layer (2500 Å thick) that was subsequently sputtered on the whole structure (see Fig. 1). The superconducting contact was confirmed by another experiment. We note that the thickness of the upper SiO layer decreased to 500 Å after the ion sputtering.

A toroidal sample with a tiny support bridge (see the scanning electron micrograph in Fig. 2) was then cut so that the Permalloy toroid was completely covered

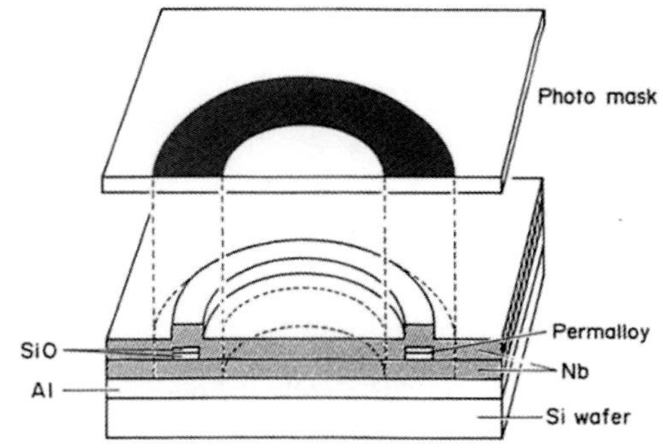

FIG. 1. Schematic diagram for fabrication of the toroidal magnet.

VOLUME 56, NUMBER 8   PHYSICAL REVIEW LETTERS   24 FEBRUARY 1986

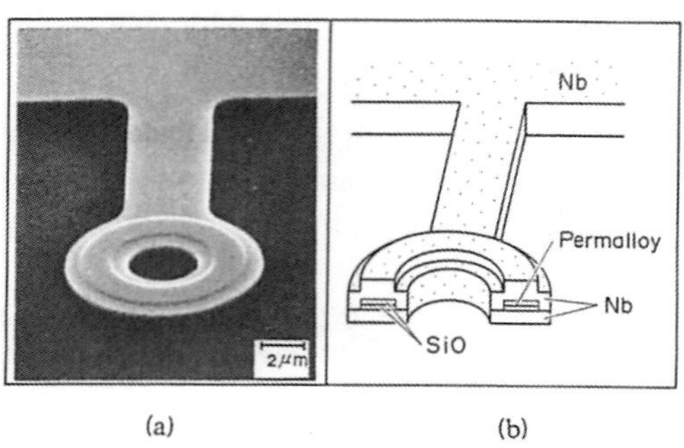

(a)   (b)

FIG. 2. Toroidal magnet. (a) Scanning electron micrograph; (b) diagram. The toroid is connected to a Nb plate by a tiny bridge for high thermal conductivity.

by the superconducting bulk Nb. The toroidal sample was peeled off the wafer by dissolving the Al in NaOH solution, and was placed on a Cu mesh. Finally, a copper film 500–2000 Å thick was evaporated on all of its surfaces; the film serves to prevent penetration of the electron wave, and to keep the sample from experiencing charge-up and contact-potential effects.

Electron holograms were formed in a 150-kV field-emission electron microscope (wavelength, 0.030 Å) that had a liquid-He–cooled specimen stage attached. The object wave, phase shifted by the sample, and the reference wave were brought together by the electron biprism to form an interference pattern, as shown in Fig. 3. The pattern was enlarged 1000 times by electron lenses and recorded on film as a hologram.

The phase shift due to the sample was reconstructed by means of He-Ne laser light (wavelength, 6328 Å) in the optical system shown in Fig. 4. Two waves, A and B, illuminated the hologram. Each wave produces two diffracted waves, one which reconstructs the phase shift due to the sample, and the other, its conjugate.

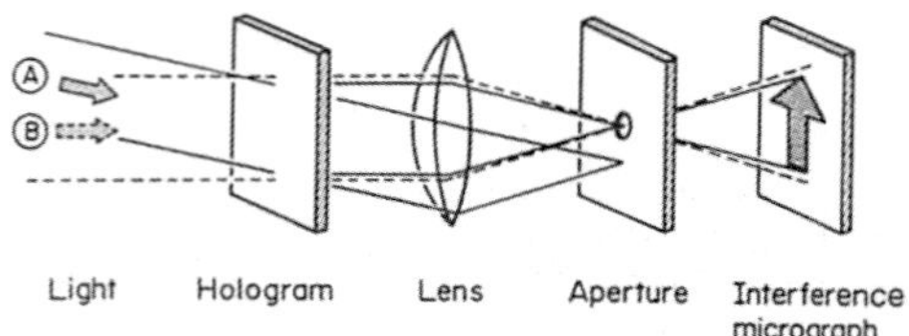

FIG. 4. Optical reconstruction system for interference microscopy.

An interference micrograph is obtained when the reconstructed image of beam A is superposed with beam B after passage through an aperture. Moreover, a twice–phase-amplified interference micrograph[12] is formed when the reconstructed image of beam A and the conjugate image of beam B are superposed by the tilting of beam B.

The leakage fluxes of fabricated samples at room temperature were quantitatively measured[13] by interference electron microscopy, and only samples with flux less than $h/20e$ [14] were selected for this experiment. Figure 5 shows an example of a twice–phase-amplified interference micrograph, which indicates a very large leakage flux of $\sim 2h/e$.

Now, the AB effect is the production of a relative phase shift of $\pi\Phi/(h/2e)$ between two electron beams enclosing magnetic flux Φ. The interference micrograph in Fig. 6(a) is clear evidence for the AB effect. Each interference fringe inside the ring, i.e., the image of the toroidal sample, lies just in the middle of two fringes outside the ring. This shows that there is a relative phase shift $n\pi$ (n odd), as expected from the quantized magnetic flux $nh/2e$ enclosed within the superconducting Nb. That the relative phase shift here is an integral multiple of π can be seen precisely from the twice–phase-amplified micrograph obtained from the same hologram [Fig. 6(b)], in which there are no relative displacements between the fringes inside and outside the ring. We emphasize that the magnetic flux is confined within the superconductor and that the

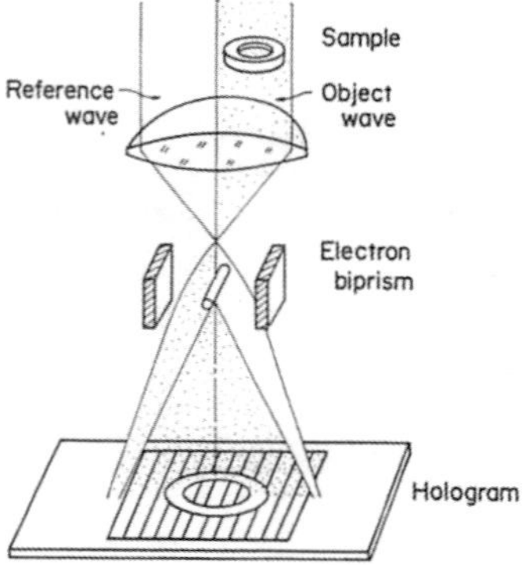

FIG. 3. Electron-optical system for hologram formation.

FIG. 5. Leakage fields from a toroidal magnet (phase amplification, 2×). Leakage flux can be quantitatively measured since a constant flux of $h/2e$ flows between two adjacent interference fringes.

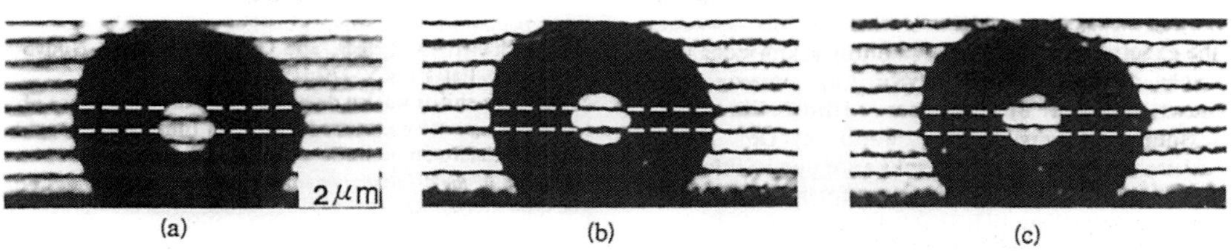

FIG. 6. Interference micrographs of a toroidal magnet at low temperatures. (a) $T = 4.5$ K (phase amplification, 1×); (b) $T = 4.5$ K (phase amplification, 2×); (c) $T = 15$ K (phase amplification, 2×). The enclosed flux is quantized in units of $h/2e$ when $T < T_c$ ($= 9.2$ K). The number of fluxons is odd.

field is shielded from the electron wave by the Cu and Nb covering. It is estimated that the leakage flux is far less than $h/20e$, since the leakage flux at room temperature is less than $h/20e$ and the minimum thickness and penetration depth of Nb are 2500 and 1100 Å, respectively. Only a slight portion, approximately 10^{-6}, of the incoming electron wave is estimated to reach the magnetic field coherently, since a 150-kV electron beam has to penetrate through the Cu (~ 1000 Å) and Nb (2500 Å) layers for it. The sufficient shielding of electron penetration is also supported by the experimental result that the change in the Cu-layer thickness from 500 to 2000 Å had no effect on the interference fringes around the quantized magnetic flux.

If the temperature T of the sample is raised, the interference pattern changes abruptly when T crosses the superconducting critical temperature T_c; the relative phase shift is no longer an integral multiple of π. In the case of Figs. 6(a) and 6(b), it in fact becomes $(0.32 + n)\pi$ as can be seen from Fig. 6(c). The transition was confirmed to be reversible. This behavior is evidence for the effect of the superconductor that confines the magnet flux quanta below T_c.

Of course, there are cases of even n, in which no relative displacements are observed, as shown in Figs. 7(a) and 7(b). With this sample, the relative displacement can be seen only when its temperature is raised above T_c; the displacement in Fig. 7(c) represents a relative phase shift of $(0.25 + n)\pi$ (n even).

When the temperature T of the sample was further raised to room temperature, the relative displacement changed by half the fringe spacing in a twice–phase-amplified interference micrograph; this corresponds to the estimated decrease ($\sim 5\%$) in the magnetization of the Permalloy. This temperature dependence supports our view that the relative phase shift is controlled by the magnetic flux of the Permalloy.

The experimental results described above provide crucial evidence for the existence of the AB effect. Furthermore, the quantization of the flux trapped by a superconductor was directly observed with use of the AB effect of an electron beam.

The most controversial point in the dispute over experimental evidence for the AB effect has been whether or not the phase shift would be observed when both electron intensity and magnetic field were extremely small in the region of overlap. Since experimental realization of absolutely zero field is impossible, the continuity of physical phenomena in the transition from negligibly small field to zero field should be accepted instead of perpetual demands for the ideal; if a discontinuity there is asserted, only a futile agnosticism results.

The authors are grateful for the idea for this experiment, which was proposed by Professor Chen Ning Yang of the State University of New York.[15] Also deserving of thanks are Dr. Ushio Kawabe of Hitachi, Ltd., for his advice and stimulation, Mr. Mikio Hirano of Hitachi, Ltd., for his help in preparing samples, and

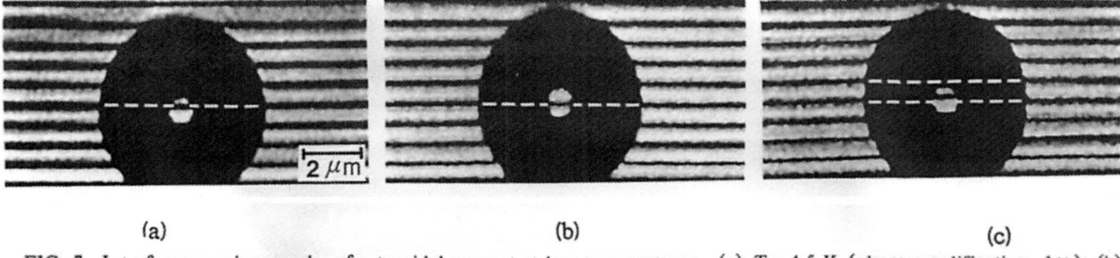

FIG. 7. Interference micrographs of a toroidal magnet at low temperatures. (a) $T = 4.5$ K (phase amplification, 1×); (b) $T = 4.5$ K (phase amplification, 2×); (c) $T = 15$ K (phase amplification, 2×). The number of fluxons is even.

VOLUME 56, NUMBER 8   PHYSICAL REVIEW LETTERS   24 FEBRUARY 1986

Mr. Shuji Hasegawa of Hitachi, Ltd., for his assistance in the experiment. We also gratefully acknowledge the valuable discussions and advice in preparing this manuscript given by Professor Hiroshi Ezawa of Gakushuin University, and also by Dr. Akira Fukuhara of Hitachi, Ltd. Thanks are due to Professor Ryozo Aoki of Kyushu University for his cooperation in developing a liquid-He–cooled specimen stage.

[1]Y. Aharonov and D. Bohm, Phys. Rev. **115**, 485 (1959).

[2]For example, S. Olariu and I. I. Popescu, Rev. Mod. Phys. **57**, 339 (1985).

[3]T. T. Wu and C. N. Yang, Phys. Rev. D **12**, 3845 (1975).

[4]R. G. Chambers, Phys. Rev. Lett. **5**, 3 (1960); H. A. Fowler, L. Marton, J. A. Simpson, and J. A. Suddeth, J. Appl. Phys. **32**, 1153 (1961); H. Boersch, H. Hamisch, K. Grohmann, and D. Wohlleben, Z. Phys. **165**, 79 (1961); G. Möllenstedt and W. Bayh, Phys. Bl. **18**, 299 (1962).

[5]P. Bocchieri and A. Loinger, Nuovo Cimento Soc. Ital. Fis. **47A**, 475 (1978); P. Bocchieri, A. Loinger, and G. Siragusa, Nuovo Cimento Soc. Ital. Fis. **51A**, 1 (1979); P. Bocchieri and A. Loinger, Lett. Nuovo Cimento Soc. Ital. Fis. **30**, 449 (1981).

[6]S. M. Roy, Phys. Rev. Lett. **44**, 111 (1980).

[7]A. Tonomura *et al.* Phys. Rev. Lett. **48**, 1443 (1982).

[8]P. Bocchieri, A. Loinger, and G. Siragusa, Lett. Nuovo Cimento Soc. Ital. Fis. **35**, 370 (1982).

[9]The phase shift was also detected when the top surface of a toroidal magnet was covered with gold film thick enough to prevent electron penetration. See A. Tonomura *et al.*, in *Proceedings of the International Symposium on Foundations of Quantum Mechanics, Tokyo, 1983*, edited by S. Kamefuchi *et al.* (Physical Society of Japan, Tokyo, 1984), p. 20.

[10]A similar experiment using a hollow toroidal superconductor was proposed by C. G. Kuper, Phys. Lett. **79A**, 413 (1980).

[11]The quantization of the trapped flux in a hollow superconducting cylinder has been detected by electron interferometry. See H. Wahl, Optik **28**, 417 (1968/1969); B. Lischke, Phys. Rev. Lett. **22**, 1366 (1969).

[12]J. Endo, T. Matsuda, and A. Tonomura, Jpn. J. Appl. Phys. **18**, 2291 (1979).

[13]A. Tonomura *et al.*, Phys. Rev. Lett. **44**, 1430 (1980); T. Matsuda *et al.*, J. Appl. Phys. **53**, 5444 (1982); N. Osakabe *et al.*, Appl. Phys. Lett. **42**, 746 (1983).

[14]Holographic interference microscopy is estimated to be as precise as $\frac{1}{50}$ of a wavelength in an ideal case, which corresponds to a magnetic flux of $h/50e$. See A. Tonomura *et al.*, Phys. Rev. Lett. **54**, 60 (1985).

[15]C. N. Yang, in *Proceedings of the International Symposium on Foundations of Quantum Mechanics, Tokyo, 1983*, edited by S. Kamefuchi *et al.* (Physical Society of Japan, Tokyo, 1984), p. 27.

PHYSICAL REVIEW A   VOLUME 34, NUMBER 2   AUGUST 1986

Experimental confirmation of Aharonov-Bohm effect using a toroidal magnetic field confined by a superconductor

Nobuyuki Osakabe, Tsuyoshi Matsuda, Takeshi Kawasaki, Junji Endo, and Akira Tonomura
Advanced Research Laboratory, Hitachi Ltd., Kokubunji, Tokyo 185, Japan

Shinichiro Yano and Hiroji Yamada
Central Research Laboratory, Hitachi Ltd., Kokubunji, Tokyo 185, Japan
(Received 21 January 1986)

The electron holography technique was employed to make a crucial test of the existence of the Aharonov-Bohm (AB) effect. The relative phase shift was measured between two electron waves passing through spaces inside and outside a tiny toroidal ferromagnet, covered completely with a superconductor layer and a Cu layer. Below the transition temperature the relative phase shift was measured to be 0 or π due to the magnetic-flux quantization in units of $h/2e$. The results directly demonstrated the existence of the AB effect even when the magnetic field was confined by the surrounding superconductor due to the Meissner effect and an electron beam was prevented from penetrating the magnet.

I. INTRODUCTION

The Aharonov-Bohm (AB) effect[1] has become significant as an important quantum effect.[2] Aharonov and Bohm predicted that an electron phase shift $\Delta\phi$ due to an electromagnetic vector potential $\mathbf{A}$ should be observed:

$$\Delta\phi = \frac{e}{\hbar} \oint \mathbf{A} \cdot d\mathbf{s} , \tag{1.1}$$

even when the electron wave travels through a magnetic-field-free region. Since the potential $\mathbf{A}$ is the simplest example of a gauge field, the AB effect directly shows that electromagnetism conforms to the theory of gauge fields.[3-5] The AB effect has also received considerable attention from experimentalists. In the year following the prediction, the first experiment[6] was done, followed by several other tests.[7]

Although the experimental tests supported the AB effect, the existence of the AB effect was strongly denied by Bocchieri and co-workers.[8] They asserted that the AB effect was of purely mathematical origin. Their argument was based upon the choice of the gauge representation. A gauge transformation is generated by a scalar function f in the following manner:

$$\mathbf{A} \rightarrow \mathbf{A} + \nabla f , \tag{1.2}$$

$$\Psi \rightarrow \Psi \exp(ief/\hbar) , \tag{1.3}$$

where Ψ is a wave function. They claimed that the vector potential can be eliminated, by means of a function f which generates a multivalued phase factor $\exp[(ie/\hbar)f]$, in a multiply connected field-free region. Their assertions were criticized on the basis of the single-valuedness of a wave function or on the basis of interpretation of boundary conditions in a multiply connected region.[9] Bocchieri *et al.* also appealed to the hydrodynamical

representation of the Schrödinger equation without potentials.[10] However, it was pointed out that their analysis did not represent the topological conditions. In spite of these objections the discussion of the effect continued.

They also pointed out[11] possible questions about the leakage of magnetic flux in previous experimental tests. Several ideas for refined experiments have been proposed. Lyuboshits *et al.*[12] suggested the use of a toroidal solenoid. Kuper[13] proposed an experiment using a hollow superconducting torus to confine the magnetic flux. A toroidal ferromagnet was suggested by Greenberger.[14]

The present authors established the existence of the AB effect using a toroidal magnet without leakage flux;[15] the relative phase shift observed by electron holography agreed with the theoretical estimate. However, Bocchieri *et al.*[16] still argued that phase shift could be due to the Lorentz force on the portion of the electron beam passing through the magnet. In our recent experiment,[17] the AB effect was confirmed using an impenetrable toroidal magnet: Au 350 nm thick was deposited on the toroid surface facing the incident electron beam. The experiment to be described here is designed to provide a crucial test of the AB effect.[18] The conceptual diagram is shown in Fig. 1. An electron is strictly excluded from the magnetic field: A superconductor completely covering a toroidal magnet confines the magnetic flux by the Meissner effect. In addition, the superconductor layer and a Cu layer are thick enough to prevent electron penetration.

II. EXPERIMENTAL

A. Fabrication of toroidal magnets (Ref. 19)

A toroidal magnet covered by a superconductor is shown in Fig. 2. Three kinds of toroidal samples were fabricated for this experiment. Their sizes are listed in Table I. A small supporting branch attached to each toroid ensures thermal conductivity for cooling even

34   815   ©1986 The American Physical Society

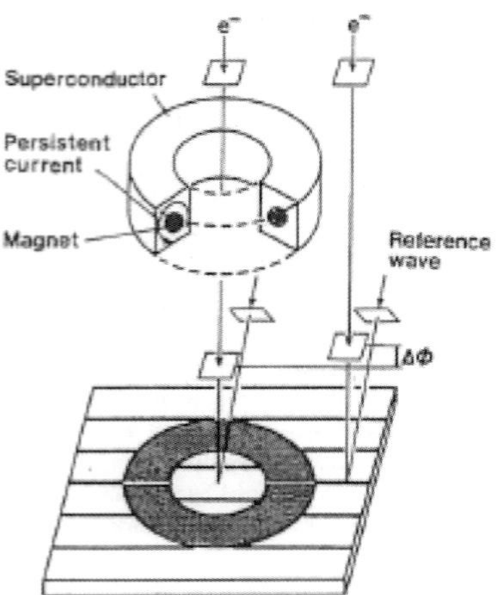

FIG. 1. Conceptual diagram of the experiment. A Cu layer for shielding from an electron wave is not shown.

under electron-beam radiation. Nb was chosen as a superconductor for the following reasons. The superconducting transition temperature is relatively high, $T_c = 9.2$ K, and the penetration depth is as small as 100 nm. Further, the processing methods for Nb have been well developed in conjunction with Josephson junction device.

The fabrication process is shown in Fig. 3. Permalloy (83 wt. % Ni–17 wt. % Fe) thin film 20 nm thick was evaporated onto a Si wafer covered with Al (300 nm thick), Nb (200 nm thick), and SiO (50 nm thick). The Al film is dissolved to peel off the completed toroid. The

TABLE I. Sizes of the toroidal magnets covered with a superconductor.

	Magnet		Superconductor	
Symbol	Inner diam (μm)	Outer diam (μm)	Inner diam (μm)	Outer diam (μm)
R1	3.0	5.0	1.5	6.5
R2	3.0	5.0	2.0	6.0
R3	3.5	5.5	2.0	7.0

SiO film reduces the coercive force of Permalloy to as low as 10 Oe, which is favorable for making a closed magnetic circuit. After deposition of the SiO layer (200 nm thick), a toroidal shape of the Permalloy sandwiched between the SiO layers was cut by ion etching. The Nb surface was cleaned by Ar^+ sputtering in order to remove a possible oxide layer [Fig. 3(a)]. The upper SiO layer protects the Permalloy toroid from Ar^+ sputtering. The film was, therefore, thinned to 50 nm in the cleaning process. After deposition of the 300-nm-thick Nb layer [Fig. 3(b)], an outer toroidal shape with a small supporting branch was cut by plasma etching [Fig. 3(c)].

Superconducting contact between the two Nb layers was checked as follows. A test device was prepared, having a contact fabricated under the same conditions as that

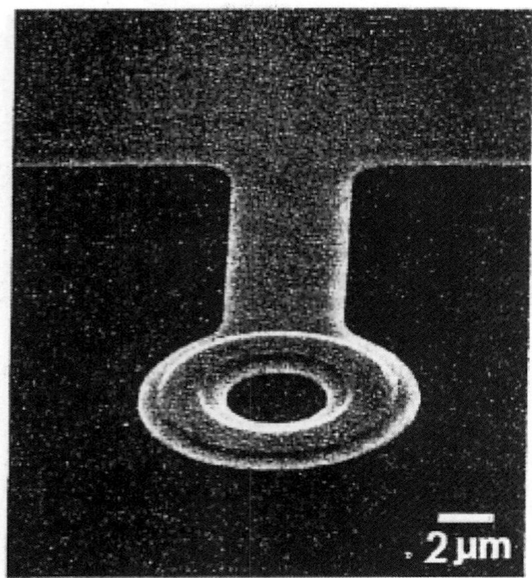

FIG. 2. Scanning electron micrograph of toroidal magnet covered by a superconductor.

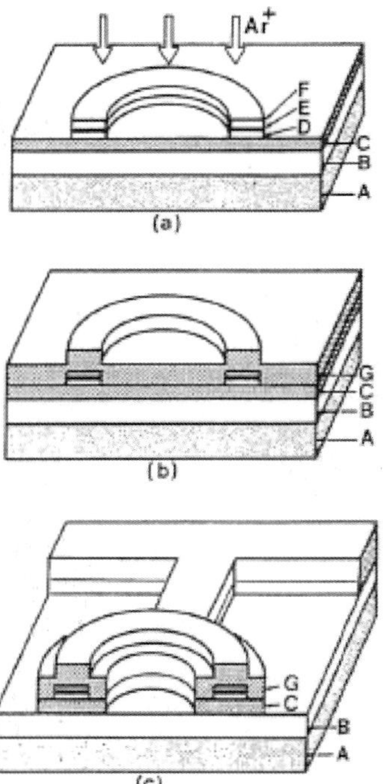

FIG. 3. Fabrication process of toroidal magnet. (a) Ar^+ sputtering for surface cleaning. (b) Sputter deposition of upper Nb layer. (c) Completed toroidal magnet. *A*, Si wafer; *B*, Al (300 nm); *C*, Nb (200 nm); *D*, SiO (50 nm); *E*, Permalloy (20 nm); *F*, SiO (200 nm); *G*, Nb (300 nm).

in which the toroidal sample was made. The measured critical current density through the contact at 4.2 K was 40 mA/μm^2. On the other hand, the persistent current density to quantize the magnetic flux was numerically calculated to be less than 10 mA/μm^2. The completed toroid was peeled off from the Si wafer by dissolving the Al layer in NaOH solution and was placed on a Cu mesh.

Finally, Cu was evaporated onto all the surfaces to avoid both charging-up and contact-potential effects: (a) An oxide layer of the Nb surface formed in the air would destroy a charge balance under electron beam radiation possibly because of the charge storage due to its low conductivity or overemission of secondary electrons. It produces electric fields around the toroidal sample. (b) The potential difference appears when the two different metals, Nb of the toroid and Cu of the mesh, are connected. The value is equal to the difference between the two work functions. The potential difference of 1 V between the toroid and the mesh produces the phase shift of π between two electron beams, one near the edge of the toroid and the other 3 μm apart from it.

B. Cooling apparatus

A cross-sectional diagram of an electron microscope with cooling apparatus is shown in Fig. 4. A liquid-He reservoir B is attached to the column. The toroidal sample is placed at the cooling stage C in the column, which is shown in Fig. 5. The stage consists of three cooling parts for both radiation and condensation protection. Part A is cooled by liquid N$_2$ and the two inner parts, B and C, are cooled by liquid He. Each part is connected to a conduction rod D from the reservoir through a cylindrical Ag foil with many slits for flexibility E. A specimen

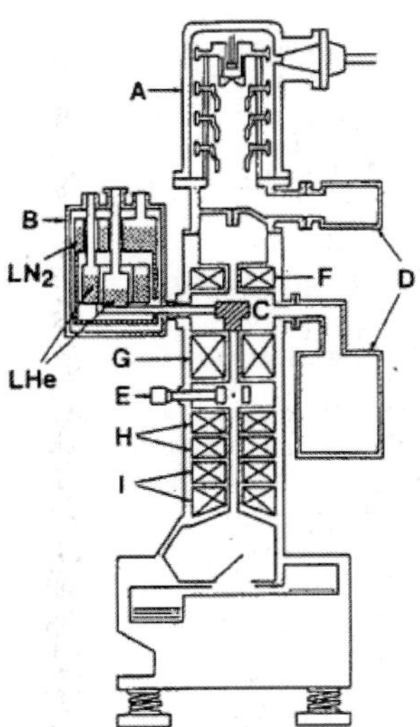

FIG. 4. Cross section of electron microscope with cooling apparatus. A, field emission gun; B, liquid-He reservoir; C, cooling stage; D, ion pumps; E, biprism; F, condenser lens; G, objective lens; H, intermediate lenses; I, projector lenses.

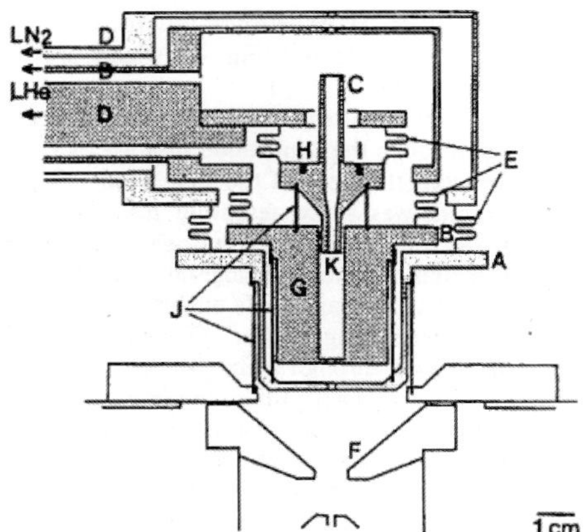

FIG. 5. Cross section of cooling stage. A, second shield; B, first shield; C, specimen holder; D, conducting rods; E, heater; F, objective pole-piece; G, superconducting coil; H, heater; I, Ge resistor; J, insulating supports; K, specimen.

is placed in part C (specimen holder). Thermal contact is given careful attention. A tiny In ring is inserted between the specimen and the holder. They are clamped by a screw cap. The temperature is controlled by a heater H, and measured by a calibrated Ge resistor I (Scientific Instruments, Inc.). The temperature can be controlled with the precision $\Delta T/T < 10^{-3}$ in the temperature range below 50 K. However, the apparatus is incapable of keeping a specimen at a constant temperature higher than 50 K because of rapid evaporation of liquid He. A magnetic field can be applied using a superconducting coil G.

The temperature difference between the toroid and the holder is estimated to be less than 0.1 K under the following assumptions: The current density of a 150-kV electron beam is 1×10^{-5} A/cm^2, the energy of all the electrons hit the toroid changes into the thermal energy, and the thermal conductivity of Nb is 0.2 W/cm K. It was also confirmed experimentally that the difference was less than 0.5 K. The superconducting transition temperature $T_c = 9.2$ K of Nb was compared to that of a holder when a trapped magnetic flux in a hole of Nb film leaked.

Careful attention was paid to the cooling procedure to prevent gas condensation onto the sample, which would create potential differences through charging up. The specimen holder was kept at room temperature until part B cooled to 20 K, where all the residual gases except Ne, H, and He were condensed onto part B. Further, part B was cooled to 5 K while the specimen holder was kept at 20 K for Ne condensation. Finally the specimen was cooled to liquid-He temperature. The gas from an electron beam line was reduced by inserting two apertures cooled by liquid N$_2$ into objective and projector lenses.

C. Electron holography

Relative phase-shift detection was carried out by measuring that of an optically reconstructed wave from an electron hologram. The off-axis image hologram was formed in a Hitachi H-800 electron microscope installed with a field emission gun[20] (Fig. 4), which provided a coherent electron beam.

Electrons were emitted from a cold (310)-oriented tungsten tip, through the application of a 3—5-kV electric potential between the tip and the first anode. The total emission current was a few tenths of μA, while only a several nA electron beam along the optical axis was fed through apertures to form a hologram in order to eliminate electrons having trajectories distorted by the aberrations of the accelerating lenses. The electron was accelerated to 150 kV through a six-stage accelerating tube to form a collimated beam. Illuminating condition (beam divergence angle α and current density j) was controlled by a condensor lens and an aperture.

Electron optical system for the hologram formation is shown in Fig. 6. Only a half plane of a specimen position was used. The other half was used for the reference beam. The image of the specimen was formed on a photographic film with the magnification of 1000 times. A Möllenstedt-type electron biprism, located between an objective and intermediate lenses, was employed to overlap an object wave and a reference wave to form a hologram.

The beam current density and exposure time were restricted by coherence conditions of the electron beam and stability of the apparatus. The beam current density at the specimen was set to range from 1×10^{-6} to 5×10^{-6} A/cm^2 and the exposure time was 10—30 sec. With these parameters, the density of the electron beam is such that on the average there is only one electron in the microscope column. The specimen temperature is not influenced by the beam radiation within the beam current range as

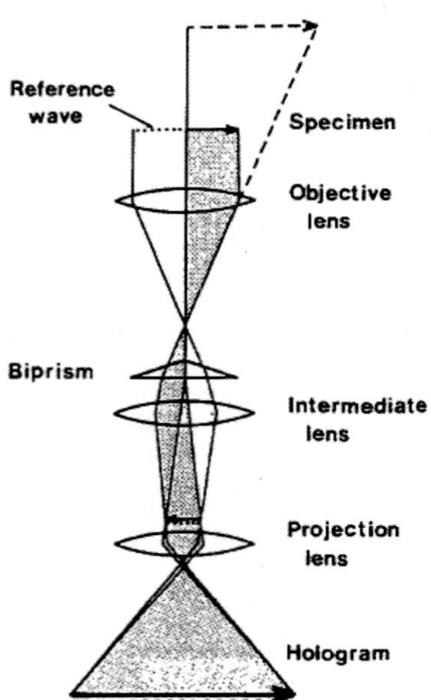

FIG. 6. Schematic diagram of hologram formation.

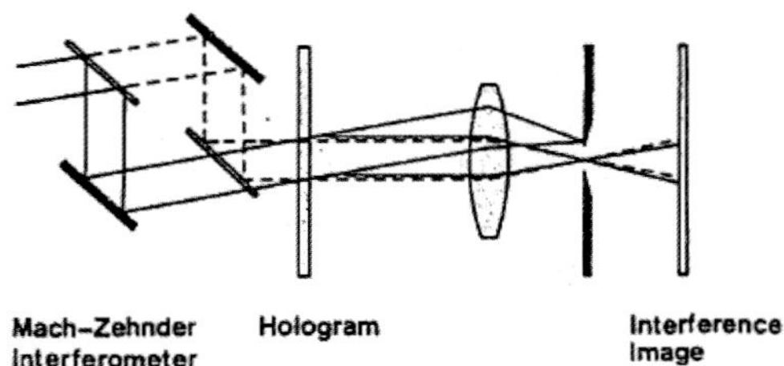

FIG. 7. Optical reconstruction system for interference microscopy.

described in Sec. II B. Actually the result was independent of the value of the current density.

Optical reconstruction and relative phase-shift measurement were carried out through use of the setup shown in Fig. 7. A pair of collimated He-Ne laser beams (wave length, 633 nm) made by Mach-Zehnder interferometer illuminated the hologram. Each beam produced the reconstructed image and its conjugate. The reconstructed image of one beam was superposed onto the transmitted wave of the other wave to produce a standard interference micrograph. The reconstructed and the conjugate images of each beam were used for a doubly phase-amplified interference micrograph.[21]

Here, two types of interference images are used: a contour map and an interferogram. The former is obtained by setting the reconstructed and the transmitted (conjugate, in case of double phase amplification) waves parallel as illustrated in Fig. 7, which represents a contour map of relative phase-shift distribution. If the object is a magnetic field, the contour lines indicate projected magnetic lines of force.[22] There, constant fluxes of h/e and $h/2e$ flow between two adjacent contour lines in standard and doubly phase-amplified images, respectively. The latter, i.e., the interferogram, is obtained by tilting the two waves to each other. The interference fringe spacing depends on the angle between the beams ($\sim 10^{-3}$ rad), however it does not affect the result because the relative phase shift was measured as a relative displacement to a regular fringe spacing.

III. RESULTS

A. Studies above T_c

Fabricated toroidal magnets were examined at room temperature to select ones with closed magnetic circuits. Contour maps in Fig. 8 show examples of toroids with and without leakage magnetic flux. In Fig. 8(a) the presence of the magnetic poles are clearly seen at the upper right of the toroid and its opposite position, where the magnetic lines of force emerge. The toroidal magnet consists of two magnetic domains having opposite magnetization directions. On the other hand, no fields can be seen around the toroid 8(b), substantiating a closed magnetic circuit.

Whether a closed magnetic circuit is formed or not depends on the toroid's thickness and diameter. It was confirmed from the experiment that optimum thickness was

FIG. 8. Contour maps of toroidal magnets at room temperature. (a) Toroid with leakage field (phase amplification, ×2). (b) Toroid without leakage field (phase amplification, ×2).

20 nm, and smaller diameter was favorable for producing a closed circuit. Approximately 60% of all the toroids fabricated formed closed circuit for the dimension employed in this experiment.

When the toroid was cooled from room temperature, the phase-shift value was detected to change. This is explained using the Weiss theory of ferromagnetism, predicting that spin orientation in the Permalloy is aligned due to the reduction of thermal vibration. Therefore, spontaneous magnetization increases to change the phase-shift value. The change of measured phase shift for 20 toroidal magnets is shown in Fig. 9. The cross sections of the toroids were so designed as to have the same dimensions, 1 μm wide and 20 nm thick. The width fluctuates ± 0.1 μm due to the precision of a photo mask. This is the reason why phase shift varies from sample to sample. The measured difference between the phase shift at 15 K, $\Delta\phi_{15\,K}$, and the phase shift at 300 K, $\Delta\phi_{300\,K}$, is given by the following equation:

$$\Delta\phi_{15\,K} - \Delta\phi_{300\,K} = \pm 0.45\pi + 2n\pi ,$$

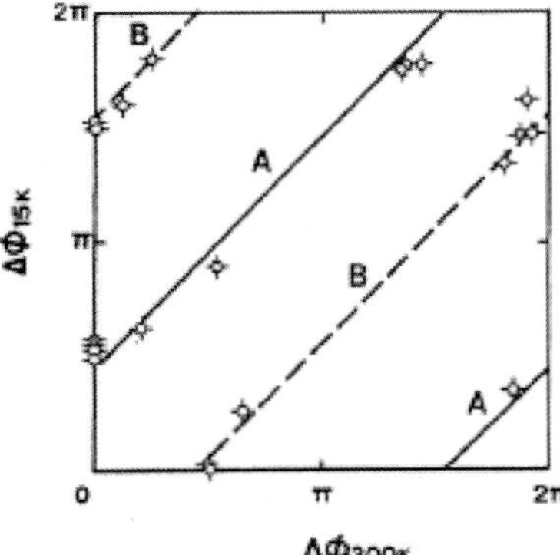

FIG. 9. Relation between relative phase shift at 300 K ($\Delta\phi_{300\,K}$) and at 15 K ($\Delta\phi_{15\,K}$).

where the term $2n\pi$ is trivial, and can be eliminated by gauge transformation. The positive and negative signs in the first term of the right side corresponding to the lines A and B in Fig. 9 are due to the different directions of the rotation of the magnetization. The value $0.45\pi\pm0.05\pi$ can be well explained as follows: According to the experimental data by Crangle and Hallam,[23] the magnetization of the Permalloy of the composition employed here increases by 5% when cooled from 300 to 15 K. The amount of the total flux confined in a toroid at 300 K is approximately $5(h/e)$ ($M_s = 0.95$ T; cross section is 1 μm$\times$20 nm.) Then the relative phase-shift change is 0.5π. Thus it is clearly demonstrated that the relative phase shift is controlled by the magnetic flux confined in the toroid.

B. Studies below T_c

The toroids were further cooled below T_c for magnetic shielding by the Meissner effect. The superconducting state of Nb was confirmed through observation of the flux quantization.[24]

Figures 10(b) and 10(e) are doubly phase-amplified interferograms of two toroids cooled at 5 K. There is no relative displacement between the fringes inside and outside the rings, i.e., the images of the toroids, while displacements are seen in the interferograms of the toroids cooled at 15 K [Figs. 10(a) and 10(d)]. Therefore, the phase shifts at 5 K are integral multiples of π, consistent with the quantization of the confined flux in units of $h/2e$.[25] The transition was observed to take place reversibly at 9 K $\pm$1 K, which is good agreement with the superconducting transition of Nb ($T_c = 9.2$ K).

In standard interferograms, i.e., interferograms without phase amplification, it can be counted whether the n, the multiplication number of the unit flux $h/2e$, is even or odd. In case of odd n, the relative phase shift is π, and the shift is 0 for even n. Thus, they are distinguished as just half-spacing displacement [Fig. 10(f)] and no displacement [Fig. 10(c)]. Further evidence for the supercon-

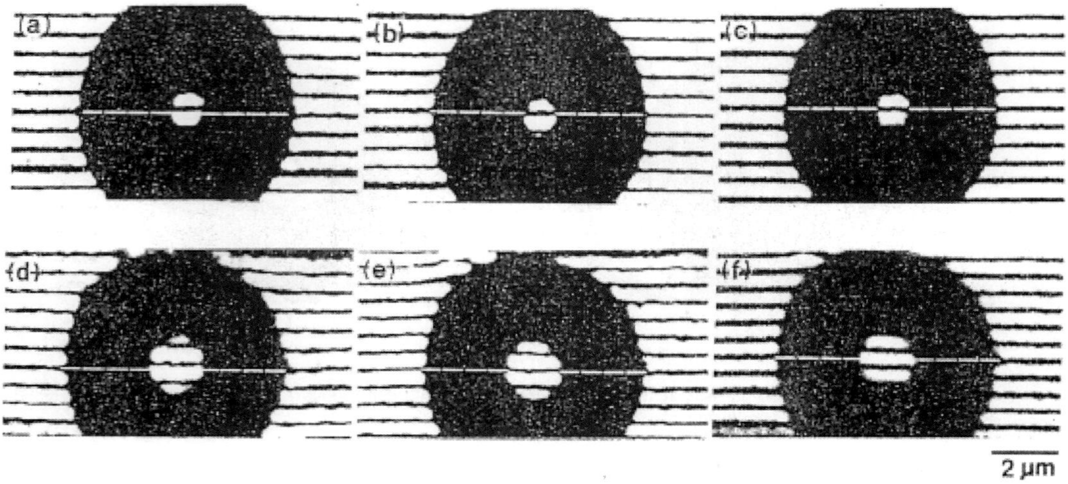

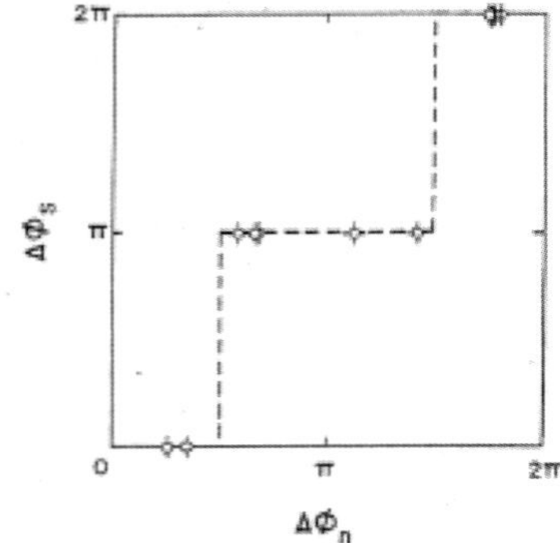

FIG. 10. Interferograms of toroids at 15 and 5 K. (a), (b), and (c): magnetic flux is quantized in $n(h/2e)$ (n is even) below T_c. The toroid is $R1$ (see Table I); (d), (e), and (f): magnetic flux is quantized in $n(h/2e)$ (n is odd) below T_c. The toroid is $R2$. (a) and (d), $T=15$ K (phase amplification, $\times 2$). (b) and (e), $T=5$ K (phase amplification, $\times 2$). (c) and (f), $T=5$ K (phase amplification, $\times 1$).

ducting transition are given by comparing the phase shift below T_c (0 or π) and above T_c ($0-2\pi$). In Fig. 11, the phase-shift change is consistent with the theoretically predicted relationship shown by the dotted line. The flux quantization occurs so as to minimize the Gibbs free energy. Energy of the magnetic field to add for quantizing the flux is minimized.

The observations described above substantiate the superconducting state of the surrounding Nb layer, which confines the magnetic flux. Then the observed relative phase shift π in the case of odd n clearly demonstrates the existence of the A.B effect.

IV. DISCUSSION AND CONCLUSION

A magnetic field produced by the objective pole piece would be the primary magnetic source except the toroidal magnet. To reduce the field, the specimen was separated from the gap center of the pole piece by 6 cm, and the objective lens was used at weak excitation as small as 1000 ampere turn. Estimated magnetic flux flowing through the toroidal hole is less than $h/100e$. The magnetic field originated outside the microscope was shielded by a 5-cm-thick μ-metal column, and only a field of an order of μG remained.

Further, we confirmed experimentally that interference fringes were not shifted in our setting if nonmagnetic specimen were introduced at 5 K and at room temperature. A test specimen was fabricated, which was identical to the specimen except that it had no Permalloy toroidal magnet. It was prepared in the same way as the specimen for holographic measurement. Resultant interferograms of both double and standard phase amplifications are shown in Fig. 12. No relative phase shift was detected between inside and outside the toroid in the measurement precision of $h/50e$.

Electron beam touching the magnetic field was noted[16] as possible cause of the fringe shift in our previous experiment.[15] The portion of electrons penetrating through the Cu and Nb layers is estimated. Only coherently transmitting electrons are evaluated here, as possible contributors to the formation of the interference fringes. Further, both plasmon and core-electron excitations are considered to be incoherent scattering. Under these assumptions, the optical potential method[26] is employed with the potential value numerically calculated by Radi.[27] The relative intensity of electrons penetrating to meet the magnetic flux is thus estimated to be 5×10^{-6} from thicknesses 300 and 100 nm of Nb and Cu layers and from the penetration depth 100 nm of the magnetic flux into Nb; the shielding by the Nb and Cu layers is considered to be satisfactory.

FIG. 11. Relation between relative phase shift below T_c ($\Delta\phi_r$) and above T_c ($\Delta\phi_n$).

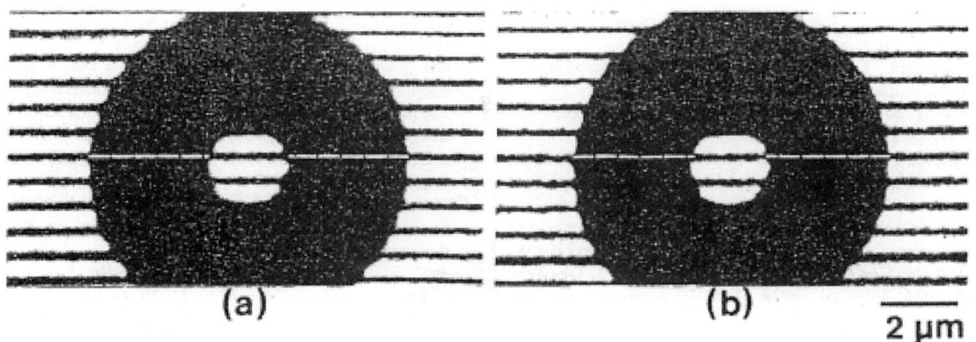

FIG. 12. Interferograms of toroid without magnet. (a) Phase amplification, ×1. (b) Phase amplification, ×2.

Sufficient shielding from electron penetration was also supported by the experimental results in which the change in Cu layer thickness from 50 to 200 nm had no effect in inteference fringes around the quantized magnetic flux.

The leakage flux from the magnet, if any, would cause interaction between electron waves and the magnetic field. However, it could be shielded by a superconducting Nb layer by the Meissner effect. It is estimated that the leakage flux is far less than $h/200e$, since the leakage flux just above the transition temperature is less than $h/20e$ and the minimum thickness and London penetration depth of Nb are 250 and 100 nm, respectively.

In conclusion, the existence of the AB effect was confirmed under the conditions of no leakage field and of no electron penetration. It was also supported by observation of the expected dependence of the relative phase shift to the magnetic flux confirmed through temperature dependence of spontaneous magnetization.

ACKNOWLEDGMENTS

The authors are grateful to Professor Chen Ning Yang of the State University of New York for giving them the idea for this experiment, which was proposed at the International Symposium on Foundations on Quantum Mechanics, 1983. They wish to acknowledge Professor Tatsuo Arii of The National Institute for Physiological Sciences for his cooperation and advice on the magnetism of Permalloy ring. They are indebted to Dr. Ushio Kawabe of Hitachi, Ltd. for his valuable discussion on experiment design and results. They also wish to acknowledge Mr. Mikio Hirano of Hitachi, Ltd. for his valuable advice on the toroid fabrication and preparation of electron holography measurements. They gratefully acknowledge the valuable discussions and advice in preparing this manuscript given by Dr. Akira Fukuhara of Hitachi, Ltd. and Professor Hiroshi Ezawa of Gakushuin University.

[1] Y. Aharonov and D. Bohm, Phys. Rev. 115, 485 (1959).

[2] S. Olariu and I. I. Popescu, Rev. Mod. Phys. 57, 339 (1985). This paper reviewed the recent discussions on the AB effect.

[3] T. T. Wu and C. N. Yang, Phys. Rev. D 12, 3845 (1975).

[4] J. Anandan, *Proceedings of the Conference on Differential Geometric Method in Theoretical Physics, Trieste, 1981* (World Scientific, Singapore, 1983).

[5] M. Peshkin, I. Talmi, and L. J. Tassie, Ann. Phys. 16, 426 (1960).

[6] R. G. Chambers, Phys. Rev. Lett. 5, 3 (1960).

[7] H. A. Fowler, L. Marton, J. A. Simpson, and J. A. Suddeth, J. Appl. Phys. 32, 1153 (1961); H. Boersch, H. Hamisch, K. Grohmann, and D. Wohlleben, Z. Phys. 165, 79 (1961); G. Möllenstedt and W. Bayh, Phys. Bl. 18, 299 (1962).

[8] P. Bocchieri and A. Loinger, Nuovo Cimento 47A, 475 (1978); P. Bocchieri, A. Loinger, and G. Siragusa, *ibid.* 56A, 55 (1980).

[9] U. Klein, Lett. Nuovo Cimento 25, 33 (1979); D. Bohm and B. J. Hiley, Nuovo Cimento 52A, 295 (1979); D. M. Greenberger, Phys. Rev. D 23, 1460 (1981).

[10] F. Strocchi and A. S. Wightman, J. Math. Phys. 15, 2189 (1974).

[11] S. M. Roy, Phys. Rev. Lett. 44, 111 (1980).

[12] V. H. Lyuboshits and Y. A. A. Smorodinskii, Zh. Eksp. Teor. Fiz. 75, 40 (1978) [Sov. Phys.—JETP 48, 19 (1978)].

[13] C. G. Kuper, Phys. Lett. 79A, 413 (1980).

[14] D. M. Greenberger, Phys. Rev. D 23, 1460 (1981).

[15] A. Tonomura, T. Matsuda, R. Suzuki, A. Fukuhara, N. Osakabe, H. Umezaki, J. Endo, K. Shinagawa, Y. Sugita, and H. Fujiwara, Phys. Rev. Lett. 48, 1143 (1982).

[16] P. Bocchieri, A. Loinger, and G. Siragusa, Lett. Nuovo Cimento 35, 370 (1982).

[17] A. Tonomura, H. Umezaki, T. Matsuda, N. Osakabe, J. Endo, and Y. Sugita, *Proceedings of the International Symposium on Foundations on Quantum Mechanics, Tokyo, 1983* (Physical Society of Japan, Tokyo, 1984), p. 20. Mollenstedt *et al.* tested the AB effect using a solenoid in a glass tube: G. Möllenstedt, H. Schmid, and H. Lichte, *Proceedings of the International Congress on Electron Microscopy, Humburg, 1982* (Deutsche Gesellschaft fur Elektronen mikroskopie eV., Frankfurt, 1982), Vol. 1, p. 733.

[18] A. Tonomura, N. Osakabe, T. Matsuda, T. Kawasaki, J. Endo, S. Yano, and H. Yamada, Phys. Rev. Lett. 56, 792 (1986).

[19] The detailed description of the toroid will be submitted by S. Yano *et al.* (unpublished).

[20] A. Tonomura, T. Matsuda, J. Endo, H. Todokoro, and T. Komoda, J. Electron Microsc. 28, 1 (1979).

[21] J. Endo, T. Matsuda, and A. Tonomura, Jpn. J. Appl. Phys. 18, 2291 (1979).

[22] A. Tonomura, T. Matsuda, H. Tanabe, N. Osakabe, J. Endo, A. Fukuhara, K. Shinagawa, and H. Fujiwara, Phys. Rev. B 25, 6799 (1982); T. Matsuda, A. Tonomura, R. Suzuki, J. Endo, N. Osakabe, H. Umezaki, H. Tanabe, Y. Sugita, and H. Fujiwara, J. Appl. Phys. 53, 5444 (1982); N. Osakabe, K. Yoshida, Y. Horiuchi, T. Matsuda, H. Tanabe, T. Okuwaki, J. Endo, H. Fujiwara, and A. Tonomura, Appl. Phys. Lett. 42, 746 (1983).

[23] J. Crangle and G. C. Hallam, Proc. R. Soc. London 272, 119 (1963).

[24] The quantization of the trapped flux in a hollow superconducting cylinder was detected by electron interferometry: B. Lischke, Phys. Rev. Lett. 22, 1366 (1969); H. Boersch and B. Lischke, Z. Phys. 237, 449 (1970); B. Lischke, *ibid.* 237, 469 (1970); 239, 360 (1970); H. Wahl, Optik (Stuttgart) 30, 508; H. Wahl, *ibid.* 30, 577 (1970).

[25] The Nb layer was confirmed from the numerical calculation to be thick enough that the fluxoid quantization results in the flux quantization.

[26] H. Yoshioka, J. Phys. Soc. Jpn. 12, 618 (1957).

[27] G. Radi, Acta Crystallogr. A 26, 41 (1970).

Demonstration of single-electron buildup of an interference pattern

A. Tonomura, J. Endo, T. Matsuda, and T. Kawasaki
Advanced Research Laboratory, Hitachi, Ltd., Kokubunji, Tokyo 185, Japan

H. Ezawa
Department of Physics, Gakushuin University, Mejiro, Tokyo 171, Japan

(Received 17 December 1987; accepted for publication 22 March 1988)

The wave–particle duality of electrons was demonstrated in a kind of two-slit interference experiment using an electron microscope equipped with an electron biprism and a position-sensitive electron-counting system. Such an experiment has been regarded as a pure thought experiment that can never be realized. This article reports an experiment that successfully recorded the actual buildup process of the interference pattern with a series of incoming single electrons in the form of a movie.

I. INTRODUCTION

The two-slit interference experiment with electrons is frequently discussed in textbooks on quantum mechanics, and is referred to as "impossible, absolutely impossible to explain in any classical way, and has in it the heart of quantum mechanics." [1] In this experiment (see Fig. 1), electrons incident on a wall with two slits pass through the slits and are detected one by one on a screen behind them. Accumulation of successive single electrons detected at the screen builds up an interference pattern. According to the interpretation in quantum mechanics, a single electron can pass through both of the slits in a wave form called "probability amplitude" when the uncertainty of the electron position in the wall plane covers the two slits, and when no observation is made of the electron at either one of the slits. The electron is then detected as a particle at a point somewhere on the screen according to the probability distribution of the interference pattern. However, if the electron is caught when passing through the slits, it takes place at either one of the two slits, never both, and the probability distribution on the screen will be completely different.

Although in textbooks this experiment is talked about as a matter of fact, "this experiment has never been done in just this way, since the apparatus would have to be made on an impossibly small scale," as Feynman points out. [1] However, this is not necessarily true. In fact, several attempts have been made up to now; Zeilinger *et al.* [2] confirmed the

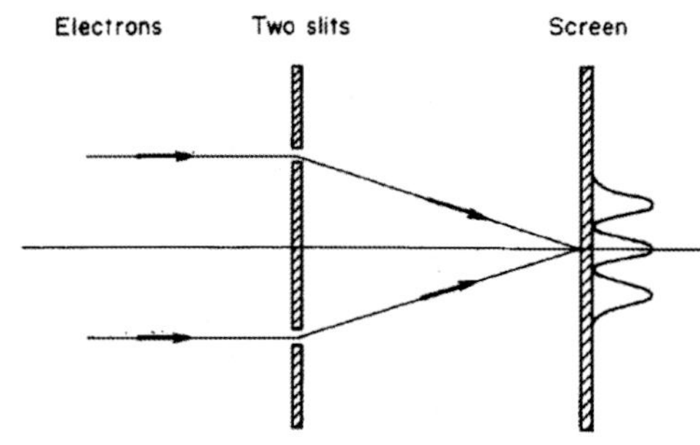

Fig. 1. Two-slit electron interference experiment.

© 1989 American Association of Physics Teachers

formation of the neutron interference pattern, just as quantum mechanics predicts, by counting arriving neutrons with a scanning counter. In the case of electrons, two groups, one at Tübingen University[3] and the other at Bologna University,[4] demonstrated, in the form of a movie using a highly sensitive TV camera, the observability of the electron interference pattern as it appears when the frequency of incident electrons increases; they showed the electron arrival in each frame without recording the cumulative arrivals. In the case of photons, the buildup process of the interference pattern was recorded on a movie film by Tsuchiya *et al.*[5] with a position-sensitive counter to accumulate the arrival of single photons on the screen. We note that the typical wavelengths of photons are much larger than those of electrons. Therefore, the difficulty Feynman attributes to the two-slit experiments for electrons does not exist for photons.

The present experiment aims at realizing the two-slit thought experiment for electrons in the form of biprism interference.

II. THEORY OF THE BIPRISM INTERFERENCE PATTERN[6]

The principle of the electron biprism invented by Möllenstedt and Dücker[7] in 1956 has been investigated from both geometric- and wave-optical aspects.[7-10] Here, a brief account of the biprism interferometer is given for the reader's convenience.

The biprism consists of two parallel grounded plates with a fine filament between them, the latter having a positive potential relative to the former. If, in the coordinate system shown in Fig. 2, the electrostatic potential is given by $V(x,z)$ and the incoming electron wave by $e^{ik_z z}$, the deflected wave is given by

$$\psi(x,z) = \exp i\left(k_z z - \frac{me}{\hbar^2 k_z}\int_{-\infty}^{z} V(x,z')dz'\right),\qquad (1)$$

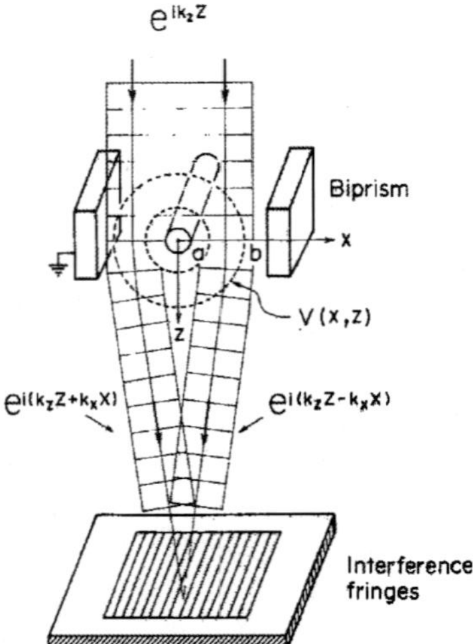

Fig. 2. Deflection of electron waves by biprism—the case of plane-wave incidence.

when $\hbar^2 k_z^2/2m \gg e|V(x,z)|$, as is the case in the present experiment.

The two waves having passed on each side of the filament can be approximated by

$\exp i(k_z z \pm k_x x)$ up to a constant factor, where

$$k_x = -\frac{me}{\hbar^2 k_z}\int_{-\infty}^{\infty}\left(\frac{\partial V(x,z')}{\partial x}\right)_{x=a} dz',\qquad (2)$$

and the symmetry $V(x,z) = V(-x,z)$ has been taken into account. Therefore, the wave fronts of the two waves are deflected as shown in Fig. 2 and, consequently, the waves propagate toward the center, since $k_x > 0$.

This can be interpreted classically also: $-e[\partial V(x,z')/\partial x]_{x=a}$ is the x component of the force exerted on the electron. Its integral with respect to $dz/v_z = dt$, $v_z = \hbar k_z/m$) gives the impulse imparted to it, which is the same in absolute value but reversed in sign, depending on which side of the filament the electron passes.

If the two waves overlap in the observation plane to give

$$\psi(x,z) = e^{ik_z z}(e^{-ik_x x} + e^{ik_x x}),\qquad (3)$$

then this leads to the interference fringes

$$|\psi(x,z)|^2 = 4\cos^2 k_x x.\qquad (4)$$

If the potential in the neighborhood of the filament is approximated by

$$V(x,z) = V_a\left[\ln(\sqrt{x^2+z^2}/b)/\ln(a/b)\right],\qquad (5)$$

then

$$k_x = \pi e V_a/\hbar v_z \ln(b/a).\qquad (6)$$

For $v_z = c/2 = 1.5\times10^8$ m/s, $V_a = 10$ V, $a = 0.5$ μm, $b = 5$ mm, $k_x = (\pi/900)$ Å^{-1}, and fringe spacing $d = 900$ Å. In the actual experiment, a spherical wave instead of a plane wave is incident on the biprism and, consequently, the fringe spacing becomes larger, as described in Sec. III.

III. EXPERIMENTS

Experiments were carried out using an electron microscope equipped with an electron biprism and a position-sensitive electron-counting system.

Coherent electron waves from a sharp field-emission tip were, after collimation, sent to an electron biprism. The biprism interference pattern was enlarged by the electron lenses and the single-electron buildup of the interference pattern was observed in time sequence on the TV monitor of a two-dimensional position-sensitive electron-counting system, which was connected to a storage memory. Electrons could be detected one by one, since the detection efficiency was approximately 100% and the detection error was less than 1%.

The detailed experimental arrangement is shown in Fig. 3. Electrons are emitted from a field-emission tip by an applied electrostatic potential $V_1 = 3 - 5$ kV, and then accelerated to the anode of potential $V_0 = 50$ kV. The electron beam accelerated to V_0 is associated with a wave of wavelength

$$\lambda = h/\sqrt{2meV_0(1 + eV_0/2mc^2)},\qquad (7)$$

which, in the present case, is 0.054 Å. The total emission current is intentionally limited to ~ 1 μA, only 10^{-4} of which passes through the anodes. The electrons are focused through the condenser lens into fine probe P_1, and

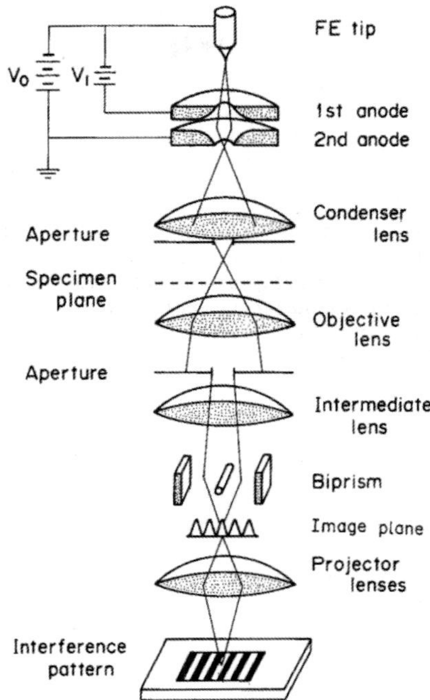

Fig. 3. Electron-optical diagram of the interference experiment.

then illuminate the specimen plane. At this stage, $\frac{1}{10}$ of the electrons pass through the condenser aperture. The electron amplitude in the specimen plane is reproduced through the objective and intermediate lenses onto the image plane of the intermediate lens. Only the central portion of the beam passes through the field-limiting aperture located just above the intermediate lens and is focused by the lens into fine probe P_2. The total current is controlled by changing the focal length of the intermediate lens and, in a typical case, is 1.6×10^{-16} A or 10^3 electrons/s.

The beam issuing from the probe is then incident on the biprism, and the two partial beams are deflected by the angles $\pm k_x/k_z$ [see Eq. (2)] on each side of the biprism. The divergence angle 2α of the incident beam is approximately 4×10^{-8} rad. Consequently, the transverse coherence length given by $\lambda/2\alpha$ is 140 μm, which is larger than the diameter of the biprism filament ($< 1\ \mu$m) but smaller than the distance between the two grounded electrodes (~ 10 mm). The two beams interfere on the image plane to form interference fringes, one-half of the angle β between the interfering beams being given by

$$\beta = l/(l + l')(k_x/k_z)\,, \tag{8}$$

and the fringe spacing by

$$d = \lambda/2\beta\,, \tag{9}$$

which is larger by a factor $(l + l')/l$ than the value given at the end of Sec. II.

In this experiment, $\lambda = 0.054$ Å, $V_a = 10$ V, $l'/l = 6$, and $2\beta = 8 \times 10^{-6}$ rad, so that $d = 7000$ Å. The interference pattern is finally magnified 2000 times through two projector lenses onto the detector plane. The detector is

approximately 12 mm in diameter and the enlarged fringe spacing is 1.4 mm.

Electrons are detected by a two-dimensional position-sensitive electron-counting system, which is schematically illustrated in Fig. 4. This system is a combination of a fluorescent film and the photon-counting image acquisition system (PIAS) produced by Hamamatsu Photonics K.K. We paid special attention to suppressing both counting loss and detection noise to less than 1%. When a 50-kV electron hits the fluorescent film, approximately 500 photons are produced from the spot. The photons excite the photo cathode through the fiber plate and photo electrons are produced. They are accelerated to 3 kV through the electrostatic lens and the point image of electrons is formed at the upper surface of the multichannel plate (MCP). The number of electrons is multiplied there and the position is then measured by the position sensor. The signal of the electron arrival at each channel is transferred to the storage memory and the accumulated electron image is displayed on the TV monitor.

The experiment was performed at the electron arrival rate of approximately 10^3 electrons/s in the whole field of view so that the interference fringes could be formed in a reasonable time, say, 20 min. The distance from the source to the screen is 1.5 m, while the average interval of successive electrons is 150 km. In addition, the length of the electron wave packet is as short as $\sim 1\ \mu$m. Therefore, there is very little chance for two electrons to be present simultaneously between the source and the detector, and much less chance for two wave packets to overlap.

An example of the buildup process of the interference pattern is shown in Fig. 5 in the form of a time series of photographs. The photographs were taken from single frames in a TV display. Electrons were detected one by one, and the total number of accumulated electrons increases with time. At first, electrons appear to be distributed quite at random. A dim figure of the biprism fringes begins to emerge in Fig. 5(c). The fringes can finally be clearly observed in Fig. 5(e), where the total number of electrons is approximately 70 000, i.e., 14 000 electrons per fringe.

These results unambiguously demonstrate the wave-particle duality of electrons. On the one hand, a single electron passes through the two slits as a wave and forms a probability interference pattern; electron–electron interaction plays no role in this process since the subsequent elec-

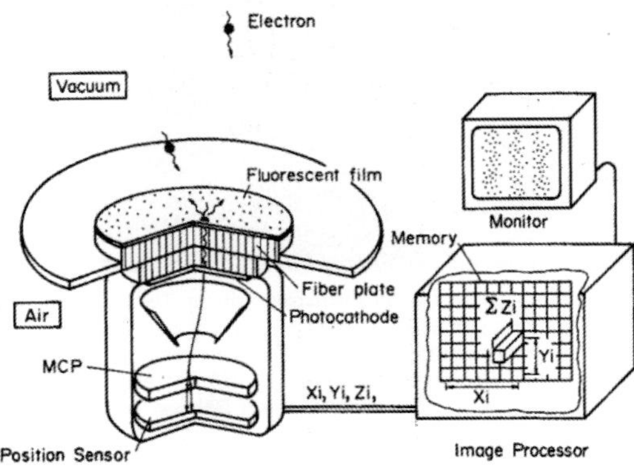

Fig. 4. Schematic diagram of position-sensitive electron-counting system.

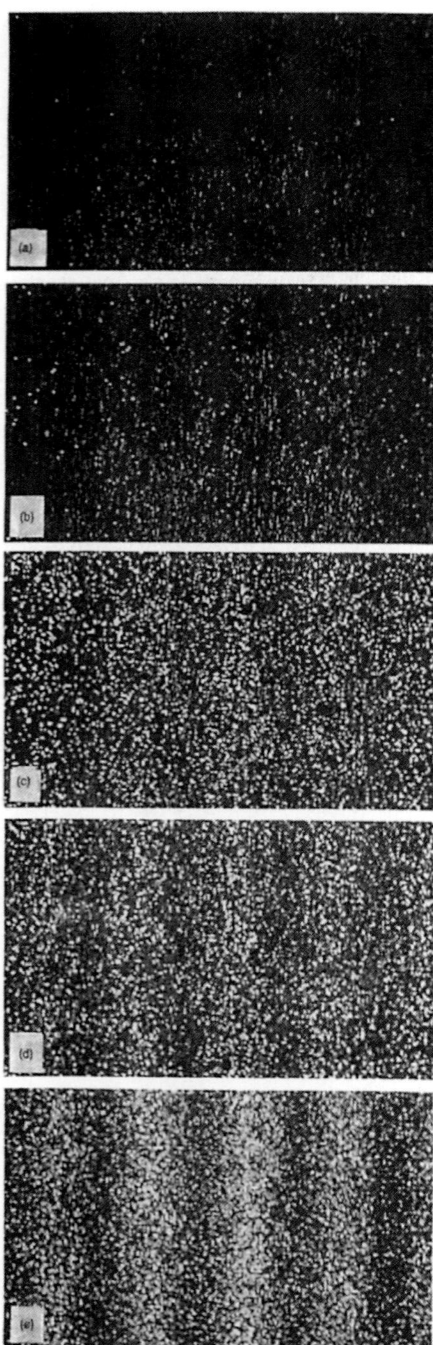

Fig. 5. Buildup of the electron interference pattern. The central field of view, ½ width and ½ length, of the whole field of the detector plane is shown here. The picture extends similarly to the whole field: (a) Number of electrons = 10; (b) Number of electrons = 100; (c) Number of electrons = 3000; (d) Number of electrons = 20 000; and (e) Number of electrons = 70 000.

tron is not even produced from the cathode till long after the preceding electron is detected. At the detector, on the other hand, an electron is observed as a localized particle. We must conclude that a certain position on the screen is selected, onto which the electron wavefunction collapses. The position cannot be predicted, but occurs in the probabilistic way dictated by the probability amplitude.

A series of similar experiments was carried out for different electron intensities ranging from 5000 to 200 electrons/ s. The contrast of the fringes obtained remains the same within experimental error of 10%. At the smaller intensity, the error often became large due to the long exposure time, since the error originates mainly from the drift of the biprism filament.

IV. CONCLUSION

We realized a two-slit interference experiment, once regarded as a pure thought experiment with no hope of precise execution, with a combination of both electron-counting and magnifying techniques. The resultant buildup of the interference pattern is exactly as predicted by quantum mechanics.

ACKNOWLEDGMENTS

We should like to thank Dr. Yoshiji Suzuki, Dr. Yutaka Tsuchiya, and Nobuyuki Hirai of Hamamatsu Photonics K.K. for their cooperation in developing the electron detector. Thanks are also due to Dr. Hideo Todokoro of Central Research Laboratory and Shuji Hasegawa of Advanced Research Laboratory, Hitachi, Ltd., for their helpful discussions and assistance in the experiment.

[1] R. P. Feynman, R. B. Leighton, and M. Sands, *The Feynman Lectures on Physics* (Addison-Wesley, Menlo Park, CA, 1965), Vol. III, pp. 1-1–1-5.
[2] A. Zeilinger, R. Gaehler, C. G. Shull, and W. Treimer, in *Proceedings of the Conference on Neutron Scattering, Argonne, 1981*, edited by J. Faber, Jr. (AIP, New York, 1982), p. 93.
[3] The movie was shown in the International Symposium on Foundations of Quantum Mechanics held at Tokyo in 1983, by H. Lichte, Institute of Applied Physics, University of Tübingen, 74 Tübingen, West Germany; see also H. Lichte, in *New Techniques and Ideas in Quantum Measurement Theory*, edited by D. M. Greenberger (New York Academy of Sciences, New York, 1988), p. 175.
[4] The movie was produced by G. Pozzi and G. F. Missiroli, Department of Physics, University of Bologna, 40126 Bologna, Italy.
[5] Y. Tsuchiya, E. Inuzuka, T. Kurono, and M. Hosoda, in *Advances in Electronics and Electron Physics*, edited by P. Hawkes (Academic, New York, 1982), Vol. 64A, p. 21.
[6] H. Ezawa (in preparation).
[7] G. Möllenstedt and H. Dücker, Z. Phys. **145**, 375 (1956).
[8] J. Faget and C. Fert, Cah. Phys. **83**, 285 (1957).
[9] T. Hibi and S. Takahashi, J. Electron Microsc. **12**, 129 (1963).
[10] J. Komrska, V. Drahos, and A. Delong, Opt. Acta **14**, 147 (1967).

Real-time observation of vortex lattices in a superconductor by electron microscopy

K. Harada, T. Matsuda, J. Bonevich, M. Igarashi, S. Kondo, G. Pozzi*, U. Kawabe† & A. Tonomura

Advanced Research Laboratory, Hitachi, Ltd, Hatoyama, Saitama 350-03, Japan
* Department of Material Science, University of Lecce, via Arnesano, 73100 Lecce, Italy
† Central Research Laboratory, Hitachi, Ltd, Kokubunji, Tokyo 185, Japan

THE dynamic behaviour of the quantized vortices of magnetic flux that penetrate type II superconductors, and specifically their interaction with pinning sites, is a topic of both scientific and technological interest, particularly since the discovery of high-transition-temperature (high-T_c) superconductors[1]. Until now, however, it has not been possible to study this behaviour in 'real time'. The Bitter technique[2], scanning tunnelling microscopy[3] and scanning electron microscopy[4] have so far provided only static images, whereas a magneto-optical technique that provides time-resolved information[5] does not resolve individual vortices. Here we report the real-time observation of vortices in a thin film of niobium, using the technique of Lorentz microscopy[6]. The coherent and penetrating beam of a recently developed 300-kV field-emission electron microscope[6] allows us to observe, at 30 frames per second, the motion of thermally activated vortices, and their response to an applied magnetic field.

Lorentz microscopy[6] has been used to observe magnetic domain structures in ferromagnetic thin films. The domain walls, unseen in an in-focus electron micrograph, become visible as black or white lines upon defocusing. Two electron beams transmitted through neighbouring domains are deflected in different directions so that the defocusing makes the two beams overlap or separate, thus increasing or decreasing the intensity along the domain wall. If the magnetic flux to be observed is less than h/e (where h is Planck's constant) as in the present experiment, then this description in terms of geometric optics must be replaced by a wave-optical one[8] taking into account the quantum-mechanical phase shifts or the rotation of the wavefront of the electron waves caused by the magnetic vector potential (compare with the Aharonov-Bohm effect[9]).

Previous attempts[10–15] to apply this method to vortex observation have had no success, because the wavefront of an electron passing through a single vortex is rotated through a very small angle ($\sim10^{-6}$ rad). Our electron microscope[7] produces a coherent electron beam with a divergence angle much less than 10^{-6} rad and has high energy, 300 keV, sufficient to penetrate the sample without losing the beam collimation.

Our experimental arrangement is shown in Fig. 1. A thin film of niobium ($T_c = 9.2$ K, resistance ratio $R(300\ \text{K})/R(10\ \text{K}) \approx 20$) for transmission observation was prepared by chemically etching a piece 2×2 mm^2 by 30 μm thick cut from a rolled Nb film, which had been annealed at 2,200 °C in a vacuum of 1×10^{-8} torr to increase its grain size up to 200–300 μm. The film had a [110] surface and one or two holes in the middle, its thickness tapering off towards the hole; we observed the area near the hole edge which was 700 ± 200 Å thick.

The sample was positioned on a low-temperature stage, and tilted at 45° to the vertically incident electron beam, so that the electrons could sense the vortex magnetic fields penetrating the sample perpendicularly to its surface; the penetrating field was due to an external magnetic field applied horizontally. The objective lens was turned off to remove its vertical magnetic field, and the intermediate lens was used instead for focusing.

Wave-optical calculation[15] shows that the vortices should be visible in an appropriately defocused image (the Lorentz micro-graph) as a tiny globule, one side bright and the other side dark with the dividing line parallel to the projection of the vortex field on the image plane. In the present case, the optimum defocusing distance was 10 mm.

Increasing the applied magnetic field B. The sample was first cooled to 4.5 K, and B was then gradually increased. A few vortices appeared at $B = 32$ G, which gives the lower critical field H_{c1} of our film sample. When B was increased further, the number of vortices increased almost instantly, as expected from electron microscopy indicating that the sample was free from strong pinning sites such as crystal dislocations, voids and grain boundaries.

Some vortices (10–20% of those seen) continued to move for a few minutes before calming down, the length of relaxation time depending on the region of the film observed. They fluctuated around their own sites with amplitudes of ~0.2 μm and frequencies of ~1 s^{-1}, occasionally hopping a distance of ~0.5 μm. These weak pinning sites are thought to be interstitial or substitutional atoms. The vortices increased in number with B, to become hexagonally arranged at $B \approx 100$ G.

An example of the equilibrium Lorentz micrographs at $B = 100$ G is shown in Fig. 2. The sample film has a fairly uniform thickness in the region shown, but is bent along the black curves ('bend contours') which are due to Bragg reflections at atomic planes brought to a favourable angle by bending. Each globule with black and white contrast is the image of a vortex (see Fig. 1) as we explained before. We also confirmed by measuring the electron phase shift due to a vortex that the image corresponded to a single vortex carrying one flux quantum $h/2e$. This contrast was reversed as it should when B was reversed. The density of the vortices is about 5 μm^{-2} as expected for $B = 100$ G; in fact $(100\ \text{G})/(h/2e) = 4.8$ μm^{-2}. As the black (or white) sides of all

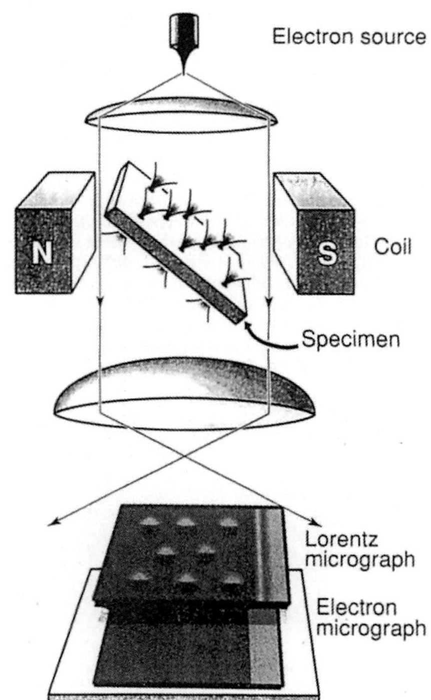

FIG. 1 Schematic diagram of vortex-lattice observation.

Reprinted with permission. doi: 10.1038/360051a0

LETTERS TO NATURE

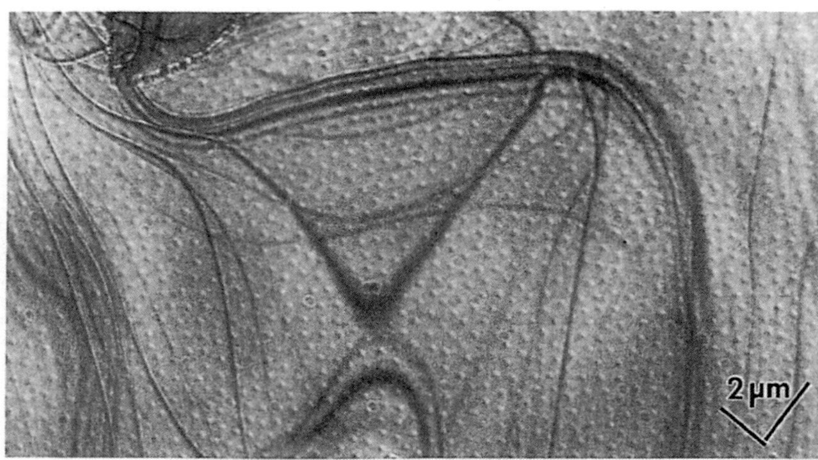

FIG. 2 Lorentz micrograph of a Nb thin film. A magnetic field of 100 G is applied to the film in the superconducting state at 4.5 K. Vortices are observed as small globules, each with black and white contrast. The magnification differs in the two directions by $\sqrt{2}$, because the film is tilted by 45°. The dark lines are bend contours.

the globules have the same orientation, the polarities of all the vortices in the region are the same.

When B was increased to 150 G, the vortex lattice was compressed to a more regular hexagonal lattice with a lattice spacing of 0.35 μm. Apparently the vortices were influenced in configuration as well as in movements by weak pinning sites in the Nb sample.

Raising the sample temperature T. As T was raised gradually from 4.5 K to 8 K at fixed magnetic field $B = 150$ G, vortices were released from pinning sites and hopped to form a more and more regular hexagonal lattice, as shown in Fig. 3. The vortex diameters increased with T because of the increase in the penetration depth, and eventually the vortices disappeared at $T_c = 9.2$ K.

When the temperature of the same sample was gradually lowered from 10 K to the initial 4.5 K, the vortex lattice, although disturbed by hopping, was more regular than it had been initially, showing that an irreversible transformation had occurred.

Removal of the applied field B. It took 6 s for the applied field to decay to zero from $B = 150$ G when the coil current was switched off. During this period the electron beam was deflected out of the field of view. When it returned, 90% of the vortices were gone. Those which remained, being trapped at pinning sites, slowly moved towards the hole in the sample, where they disappeared. They moved by hopping in the direction of decreasing vortex density, seemingly along preset routes according to local variations in thickness. Besides these, there were some that

oscillated repeatedly around fixed centres, with amplitudes ~ 0.3 μm and frequencies ~ 10 s^{-1}. Examples of the vortex movements as reproduced from our videotape are shown in Fig. 4. Figure 4a and b shows the instants when a vortex happened to start and stop hopping, respectively, as indicated by the arrows; the interval of time is 0.1 s. In the succeeding 1.3 s between b and c, three vortices hopped, as indicated by the arrows. That each arrow in c stands for a single hop was checked by examining all the frames in between.

We conclude that Lorentz microscopy with a 300-kV field-emission electron microscope opens up new opportunities for real-time observation of the microscopic dynamics of individual vortices. We can observe not only flux creep, pinning and so on in motion, but also tell polarities and flux values of individual vortices. The spatial resolution of the technique is a few hundred ångströms in this experiment, but can be improved to a few ångströms by adjusting the image defocusing. Time resolution is 1/30 s at present, and may be improved to 1/300 s. Magnetic flux can be resolved down to $h/6e$, one-third of the elementary flux. The applied magnetic field could be varied up to 150 G in our experimental arrangement with its horizontal field, but the upper limit can be 10 kG if the (vertical) field of the objective lens is used.

We hope to apply our method to observe flux melting and other novel features of vortices in high-T_c superconductors. There are some technical problems to solve. First, the sample must be a self-supporting thin film with thickness of $\sim 1,000$ Å for transmission observation. Some high-T_c materials can be

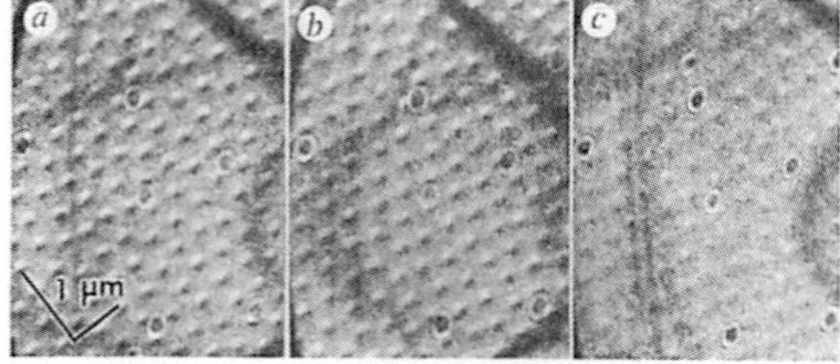

FIG. 3 Dependence of the vortex images on temperature. a, $T = 4.5$ K; b, $T = 6.0$ K; c, $T = 8.0$ K. The vortices become larger as the penetration depth increases with temperature. The vortex configuration, random at 4.5 K, turns into a more and more regular hexagonal lattice at higher T, which can be recognized more easily when the photographs are seen obliquely.

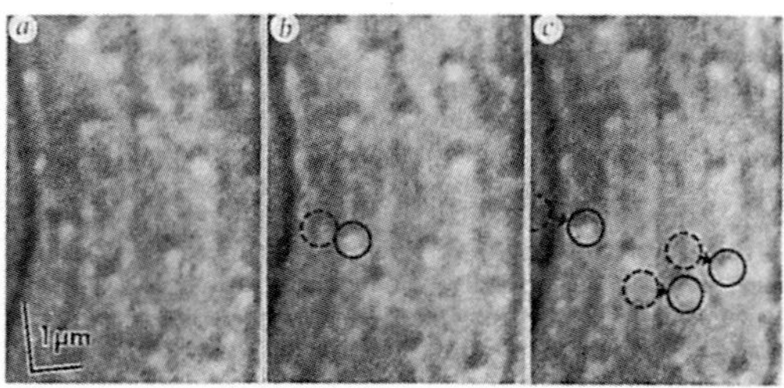

FIG. 4 Dynamics of the vortices as seen at times t after magnetic field has been switched off from 150 G. a, $t = t_0 + 0$ s; b, $t = t_0 + 0.1$ s; c, $t = t_0 + 1.4$ s ($t_0 = 170$ s). Dotted and solid circles show vortex positions before and after hopping, and the hopping directions are indicated by the arrows.

LETTERS TO NATURE

grown only on a single-crystalline substrate, and its removal could destroy them. Second, the sample must be placed in a high vacuum, where it may suffer oxygen desorption, to which the high-T_c materials are very sensitive. Third, the larger penetration depths in such materials may require substantial changes in the observation conditions. As Lorentz microscopy relies on defocusing, its spatial resolution is necessarily less than that of electron holography[16,17], which, however, is not suitable for time-resolved, real-time observations. These two techniques are thus complementary. Experiments using electron holography are now in progress to study the interaction of the vortices with small defects. □

Received 14 August; accepted 30 September 1992.

1. Bishop, D. J., Gammel, P. L., Huse, D. A. & Murray, C. A. *Science* **255**, 165–172 (1992).
2. Essman, V. & Träuble, H. *Phys. Lett.* **A24**, 526–527 (1967).
3. Hess, H. F., Robinson, R. B., Dynes, R. C., Valles, J. M. Jr & Waszczak, J. V. *Phys. Rev. Lett.* **69**, 214–216 (1989).
4. Mannhart, J., Bosch, J., Gross, R. & Huebener, R. P. *Phys. Rev.* **B35**, 5267–5269 (1987).
5. Durán, C. A. *et al. Nature* **357**, 474–477 (1992).
6. Hirsch, P. B., Howie, A., Nicholson, R. B., Pashley, D. W. & Whelan, M. J. *Electron Microscopy of Thin Crystals*, (Butterworths, London, 1965).
7. Kawasaki, T., Matsuda, T., Endo, J. & Tonomura, A. *Jap. J. appl. Phys.* **29**, L508–L510 (1990).
8. Wohlleben, D. J. *Phys. Lett.* **22**, 564–566 (1966).
9. Peshkin, M. & Tonomura, A. *The Aharonov-Bohm Effect, Lecture Notes in Physics*, Vol. 340 (Springer, Heidelberg, 1989).
10. Hirsch, P. B. Howie, & Jakubovics, J. P. *Proc. int. Conf. Electron Diffr. Crystal Defects*, Vol. 1, 8-5 (Pergamon, Oxford, 1965).
11. Yoshioka, H. *J. phys. Soc. Jpn* **21**, 948–954 (1966).
12. Goringe, M. J. & Jakubovics, J. P. *Phil. Mag.* **15**, 393–403 (1967).
13. Wohlleben, D. *J. appl. Phys.* **38**, 3341–3352 (1987).
14. Colliex, C., Jouffrey, B. & Kleman, M. *Acta crystallogr.* **A24**, 692–696 (1968).
15. Capiluppi, C., Pozzi, G. & Valdrè, U. *Phil. mag.* **26**, 865–883 (1972).
16. Matsuda, T. *et al. Phys. Rev. Lett.* **62**, 2519–2522 (1989).
17. Tonomura, A. *Adv. Phys.* **41**, 59–103 (1992).

ACKNOWLEDGEMENTS. We thank H. Ezawa, T. Fujita, A. Fukuhara, C. N. Yang, Y. Murayama, F. Nagata, M. Tachiki and T. Onogi for discussions, and N. Osakabe, S. Kubota, H. Kasai, T. Yoshida, J. Endo, N. Miyamoto, H. Kajiyama and S. Matsunami for technical advice and assistance.

Observation of Dynamic Interaction of Vortices with Pinning Centers by Lorentz Microscopy

Tsuyoshi Matsuda, Ken Harada, Hiroto Kasai, Osamu Kamimura, Akira Tonomura*

When a magnetic field penetrates a superconductor, it forms lattices of thin filaments called magnetic vortices. If a current is applied, these vortices move if not pinned down, destroying the superconductivity. The spatiotemporal behavior of the vortices was observed in a niobium film with a square lattice of defects made by ion irradiation. The vortices formed a domain of lattices. When the intensity of the applied magnetic field was decreased, the vortices were driven out of the film across its edges; when the intensity was increased, the vortices were driven into the film. The lattice exhibited brief and intermittent flow in "rivers" along domain boundaries. The rivers did not always flow along the same paths because new lattice domain configurations emerged whenever the flow stopped.

$\mathbf{A}$n applied magnetic field penetrates a type II superconductor in the form of thin filaments called vortices. An electric current can flow without resistance only when the vortices are fixed by pinning centers against the Lorentz forces exerted by the current; otherwise, the motion of the vortices generates voltage, and the resultant heat is likely to destroy the superconducting state. For instance, no net pinning would arise if pinning centers were distributed randomly and were so weak as to leave the vortex lattice rigid; in this case, the forces on the whole lattice at different centers would add up to zero. Thus, the whole lattice would translate freely if it were not deformed, either elastically (without tearing) or plastically (split into domains).

What happens when strong pinning centers exist? Does the vortex lattice flow only with elastic distortions that fluctuate with time, large at some instants and small at others? Or does the lattice break down plastically into blocks, some moving and others standing still? Obtaining a clear picture of the pinning is not only important for practical applications of superconductors but also relevant to other physics problems, such as charge- and spin-density waves and Wigner crystals, in which a driven elastic medium exhibits plastic flows with tears and rips as it interacts with a rigid substrate.

Several investigations into this problem of plastic flow have been reported. Anderson (1) introduced the concept of the flux bundle, a collection of vortices thermally depinned from pinning centers and moving in a bundle, in flux creep; this concept was used to interpret the voltage noises observed during flux flow (2). Studies of such collective depinning and creep of vortices

Advanced Research Laboratory, Hitachi Ltd., Hatoyama, Saitama 350-03, Japan.

*To whom correspondence should be addressed.

are summarized in (3). The vortex dynamics that should arise when a superconductor is slowly driven to the threshold of instability were recently characterized as vortex avalanches, in analogy to the avalanches observed in sandpiles (4, 5). More recently, the spatial disorder in the vortex lattice that takes place when a transport current is applied was observed through small-angle neutron scattering (5). However, the microscopic behaviors of vortices have been inaccessible to direct spatiotemporal observations. Here, we observed the dynamics of vortices interacting with pinning centers in real time by Lorentz microscopy (6) as they were driven by a gradient in the vortex density that arose when a decrease in the applied magnetic field intensity drove the vortices out of a superconducting film across its edges, or when an increase in the field intensity had the opposite effect. The features of the vortex motion mentioned above were seen.

Niobium samples were prepared by

chemically etching a rolled film, 15 μm thick, which was annealed at 2200°C for 10 min to increase the grain size to ~300 μm in diameter. A region of the film that was ~1000 Å thick was chosen for observation. To clearly see the effects of pinning centers on the vortex dynamics, we produced a 4 × 4 rectangular lattice (spacing, 3.3 μm) of strong centers by irradiating the chosen region with a focused 30-keV Ga ion beam (diameter, 200 Å) from a Hitachi Focused Ion Beam Machine (FB-2000). Transmission electron microscopy revealed that at each of the 4 × 4 sites, the ion beam produced a pit 400 Å in diameter and a few hundred angstroms in depth, surrounded by a region of entangled dislocations 3000 Å in diameter. The whole region acts as a pinning center and is termed a defect.

The film sample was placed in a low-temperature specimen stage in a 300-kV field-emission transmission electron microscope so that its surface formed a 45° angle with both the incident electron beam and the applied magnetic field B. The objective lens of the microscope was replaced by an intermediate lens to make an out-of-focus image—that is, a Lorentz micrograph—in which the vortices appear as spots of dark-bright contrasts (7) because the incident parallel beam of electrons is slightly deflected when passing through a vortex and is displaced on the observation plane (Fig. 1). The out-of-focus distance appropriate for the Lorentz microscopy was 10 to 30 mm, and the image magnification was ×1000 to ×2000.

Examples of such Lorentz micrographs are shown in Fig. 2. In Fig. 2A, the sample was in the normal state, and only the black image of a defect can be seen. The black or white fringes running vertically are called bend contours; they arise from the Bragg reflections around a line along which the film is slightly bent. The shift of the contours between A and B in Fig. 2 is

Fig. 1. Experimental arrangement for observation of vortices and defects by Lorentz microscopy. Electrons transmitted through a vortex are deflected by its magnetic field and are displaced in the observation plane. The orientation of B can be discerned from the orientation of the line dividing the dark-bright contrast.

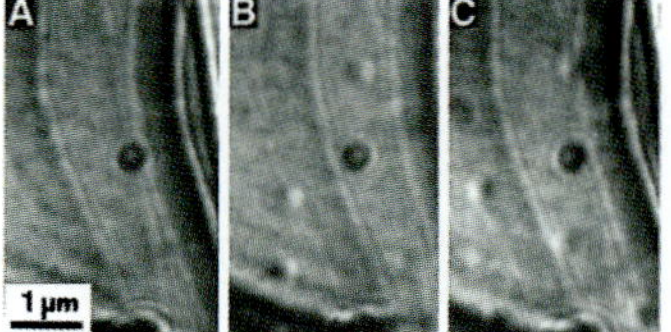

Fig. 2. Lorentz micrographs of a Nb thin film (T_c = 9.2 K) with a defect. (**A**) B = 5 G, T = 10 K; (**B**) B = 5 G, T = 4.5 K; and (**C**) B = −5 G, T = 4.5 K. Only the out-of-focus image of the defect can be seen in (A), whereas vortex images of dark-bright contrasts can also be seen in (B) and (C). Although the intensity distribution of the defect image is symmetric in (A), it is asymmetric in (B) and (C), thus showing that a vortex is trapped in the defect. The orientation of the dark-bright contrast of the image in (C) is reversed from that in (B), as expected when the direction of B is reversed.

Reprinted with permission. DOI: 10.1126/science.271.5254.1393 attributable to a change in the incident angle of the electron beam to the crystal plane, which is caused by changes in experimental conditions such as B and the sample temperature T.

At $T = 4.5$ K and $B = 5$ G, three vortex images appeared, all with the bright side on the right (Fig. 2B). Because the image of the defect is also bright on its right side (though only slightly), the implication is that the defect has trapped a vortex. This interpretation was confirmed by reversing the orientation of B; in Fig. 2C, the bright sides of both the defect and vortex images are now on the left.

Using the technique of simultaneous observation of vortices and defects, we first studied how defects affected the static configuration of vortices, and then how vortices flowed through defects. With increasing B, the vortices formed lattices that became more regular; that is, the vortices repelling each other became more tightly packed (see Fig. 3). However, the vortices tended to be trapped by the defects, so that they did not form a single regular lattice but rather several domains of lattices because of the existence of the defects (Fig. 3, B and C).

The lattice became more regular as T increased. The pinning centers that existed in regions without irradiation became ineffective

above $T = 7$ K, which implied that the pinning forces weakened at higher temperatures. Although the pinning by the defects also became weaker at $T = 9$ K, its effect was still recognizable. This is clearly seen in Fig. 4, where T was increased from 4.5 to 9 K with B fixed at 75 G. The random vortex configuration at 4.5 K was attributable to both kinds of pinning centers (Fig. 4A). At 7.5 K, the vortex lattice was disturbed only in the neighborhoods of the defects, which showed that the

pinning centers other than those attributable to the defects became ineffective (Fig. 4B). At $T = 9$ K, the vortices formed a nearly perfect lattice with only slight distortions around the defects (Fig. 4C), which showed that the pinning by the defects became weak as well (the vortex images here are appreciably blurred because of the high temperature).

We next observed the dynamics of vortices in two experiments (8). The first experiment

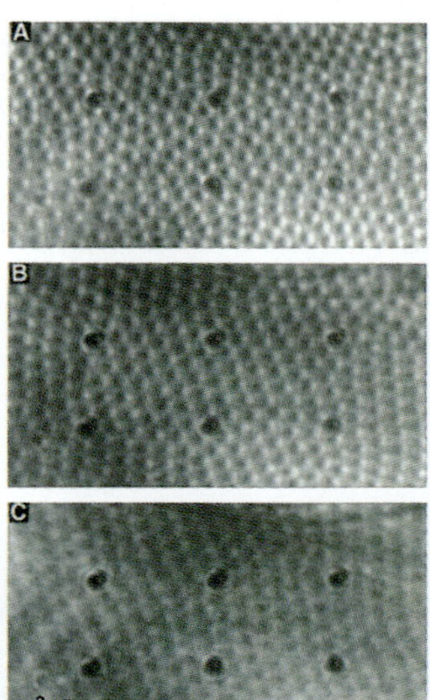

Fig. 4. Temperature dependence of vortex configuration near defects at $B = 75$ G. (**A**) $T = 4.5$ K; (**B**) $T = 7.5$ K; and (**C**) $T = 9$ K. When T is increased, the pinning effect decreases. In (A), vortices are scattered randomly. In (B), the pinning effect of the irradiation defects becomes dominant, and the vortex lattice domains have boundaries near the defects. In (C), vortices form an almost perfect lattice, but the lattice lines are still tilted at the defects.

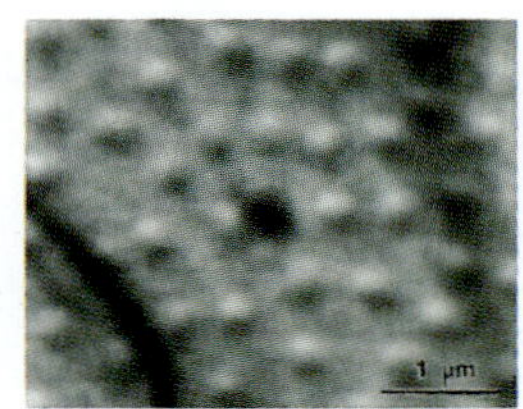

Fig. 5. Video frame of sparse vortices detouring a defect. Vortex images of dark-bright contrasts can be seen surrounding the dark image of the defect in the center of the picture. The white spot on the left side of the defect image indicates that a vortex is trapped there. The vortices hopping from the upper left to the lower right must detour this defect [see the video clip (8)].

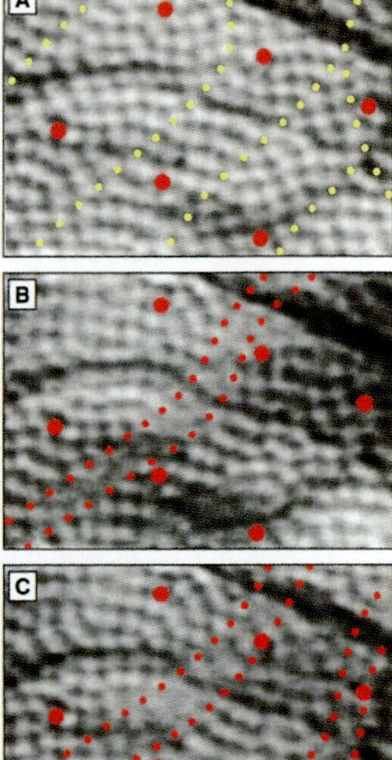

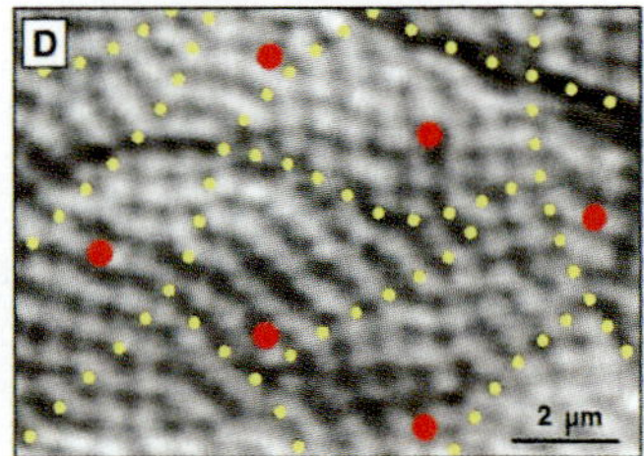

Fig. 6. Video frames of vortex lattices. (**A**) $t = 0.0$ s. Vortex lattices have domain boundaries near the defects (red); some of these boundaries are shown by yellow dotted lines. (**B**) $t = 0.27$ s. Vortices suddenly start to flow like avalanches, forming a river (between two red dotted lines) along one of the domain boundaries shown in (A). (**C**) $t = 0.43$ s. Another river starts flowing. (**D**) $t = 0.80$ s. The vortex flow stops, and a new configuration of vortex lattices is formed.

Fig. 3. Magnetic field dependence of vortex configuration near defects at $T = 7$ K. (**A**) $B = 41$ G; (**B**) $B = 78$ G; and (**C**) $B = 106$ G. When B is increased, vortices tend to form closely packed lattice domains, but the lattices have domain boundaries and dislocations near the defects.

concerned the motion of sparse vortices. In the initial state, $T > T_c = 9.2$ K and $B = 0$. After cooling the sample film below T_c, we applied $B = 80$ G and observed the vortices entering the film from its edges. These sparse vortices flowed on average in the direction perpendicular to the sample edges; hence, they were driven by the gradient in the vortex density, although individual vortices often hopped in scattered directions. The effect of the gradient in film thickness was appreciable only in the case of gentle vortex flows such as flux creep. The vortex hoppings were not completely independent from each other because a vortex hopping was often triggered by an approaching vortex. In the cases where a vortex was trapped in a defect (Fig. 5), the ensuing vortices would typically make detours to avoid the defect. Only rarely did collisions occur; sometimes an oncoming vortex recoiled after collision, and sometimes an oncoming vortex was trapped for a few seconds, followed by a release of a single vortex. These observations showed that the state of two vortices trapped in one defect should be unstable, as expected. No such correlations were found in the direction orthogonal to the direction of motion.

The second experiment concerned the case in which vortices are closely packed to form lattice domains. We applied $B = 180$ G to the sample at $T > T_c$, cooled the sample to $T = 4.5$ K to find the lattice-domain formation, and then reduced B to 85 G. After the lattice domains relaxed, the sample temperature was gradually raised to let the lattice domains move freely; they then exhibited more complicated and seemingly irreproducible motion in some cases.

A typical example is shown in Fig. 6. Just before the vortex flow began, the vortices formed lattice domains at 6 K (Fig. 6A). The domain size depended on experimental conditions; it was typically a few micrometers in diameter and contained 5×5 vortices. Each domain appeared to be pinned by the defects (Fig. 6A). The vortices remained stationary for a while, then suddenly began moving in "rivers" like avalanches near some domain boundaries (Fig. 6B). When vortices flowed in rivers, they became disordered. With an exposure time of 1/30 s (Fig. 6B), those vortices that moved were blurred and could only be seen as a river. The appearance of the river changed a little after 0.16 s (Fig. 6C). When the flow stopped, a new configuration of vortex lattices emerged (Fig. 6D), and domain boundaries were formed at different locations, which in turn triggered new avalanches. The duration of the flow depended on conditions but was generally <1 s. The velocity of a vortex hopping was on the order of 1 μm s^{-1}.

The rivers flowed along the domain boundaries that were located near the defects; the vortex river in Fig. 6B emerged along the domain boundary located near the upper middle defect in Fig. 6A. At higher temperatures, the rivers became wider until the whole vortex lattice started to flow, keeping its form unchanged. These observations indicate that we visually monitored the transition from plastic to elastic flow.

REFERENCES AND NOTES

1. P. W. Anderson, *Phys. Rev. Lett.* **9**, 309 (1962).
2. G. J. Van Gurp, *Phys. Rev.* **166**, 436 (1968).
3. G. Blatter *et al.*, *Rev. Mod. Phys.* **66**, 1125 (1994).
4. S. Field *et al.*, *Phys. Rev. Lett.* **74**, 1206 (1995).
5. U. Yaron *et al.*, *Nature* **376**, 753 (1995).
6. K. Harada *et al.*, *ibid.* **360**, 51 (1992).
7. A. Tonomura, *Electron Holography* (Springer, Heidelberg, 1993).
8. A video clip is available at *Science*'s Beyond the Printed Page site on the World Wide Web: <http://www.aaas.org/science/beyond.htm>.
9. We thank H. Hirose for assistance in irradiating an ion beam to Nb samples and H. Ezawa, C. N. Yang, and Y. S. Ono for valuable discussions and comments.

30 August 1995; accepted 12 December 1995

APPLIED PHYSICS LETTERS   VOLUME 76, NUMBER 10   6 MARCH 2000

Fine crystal lattice fringes observed using a transmission electron microscope with 1 MeV coherent electron waves

T. Kawasaki,[a] T. Yoshida, T. Matsuda, N. Osakabe, and A. Tonomura
*Advanced Research Laboratory, Hitachi, Ltd., Hatoyama, Saitama 350-0395, Japan
and CREST, Japan Science and Technology Corporation (JST), Kawaguchi, Saitama 332-0012, Japan*

I. Matsui
Hitachi, Ltd. Instruments, Katsuta, 312-0033, Japan

K. Kitazawa
*Department of Applied Chemistry, University of Tokyo, Tokyo 113-8656, Japan
and CREST, Japan Science and Technology Corporation (JST), Kawaguchi, Saitama 332-0012, Japan*

(Received 16 December 1999; accepted for publication 10 January 2000)

A transmission electron microscope with a 1 MeV cold field-emission electron source has been developed for coherent and penetrating electron waves. We confirmed the coherence and overall stability of the microscope by observing Au($33\bar{7}$) lattice fringes. These fringes have a 0.498 Å spacing. © *2000 American Institute of Physics.* [S0003-6951(00)01110-4]

Bright-beam sources have been pivotal in opening new frontiers in science and technology. The invention and development of the laser revolutionized the communication and information industry. And, due to the advent of the field-emission (FE) source in electron microscopy,[1] we have been able to use subnanometer imaging and analysis and electron holographic interferometry[2] in fields ranging from fundamental physics[3] to biology. We can handle coherent electron waves and fine electron probes by using a monochromatic electron-beam tunneled from an apex of a sharpened metal tip (FE) instead of using thermally induced electrons from a broad area of a metal filament [thermionic emission (TE)].

The trend towards higher accelerating voltages in transmitsion electron microscopes (TEMs) has been driven in the quest for shorter wavelengths for high-resolution imaging and higher penetrability for observing practical thick specimens. A high-voltage TEM with a TE gun up to 3.5 MV has been constructed; however, the accelerating voltage of routinely used FE-TEMs has been limited to 350 kV.[4] A coherent electron wave with an ultrashort wavelength should accelerate the present efforts to push the point resolution beyond 1 Å by allowing us to enhance the information limit. To investigate the electromagnetic structure of practical materials, a penetrating and coherent electron wave is necessary. Magnetic vortices in high-T_c superconductors,[5] for example, have to be observed in thick films to reveal their three-dimensional structure.

Here, we report the development of a 1 MV TEM with a cold field-emission electron source. We observed a crystal lattice image formed from the interference of wave functions located in 0.498-Å-spacing gold lattice planes.[4] This lattice resolution limit is the most significant index of the overall performance of microscopes, namely temporal and spatial coherence, mechanical and electrical stability, and principal resolving power.

The configuration of our microscope is designed to pro-

duce highly coherent electron beams. The temporal coherence length[6] $l_t \sim \lambda^2/\Delta\lambda$ is related to the energy spread ΔE as $l_t \sim 2\lambda E/\Delta E$ (nonrelativistic). The intrinsic energy spread of the field emission is ~ 0.5 eV at a total emission current of 30 μA [W(310) emitter]. Therefore, the fluctuation of the accelerating potential should be comparable to or less than the intrinsic energy spread to fully use the monochromatic capability of the FE source. The FE gun system consists of three pressure vessels to avoid cross talk between the components (Fig. 1). The vessel on the right contains a Cockcroft–Walton high-voltage generator, the vessel in the middle contains a floating power supply that controls the field emission, and the vessel on the left contains an accelerator tube. The high-voltage cables ($C = 2$ nF) connecting the vessels serve as a ripple filter. The high voltage is then kept stable at 5×10^{-7} /min (0.5 V/min at 1 MV).

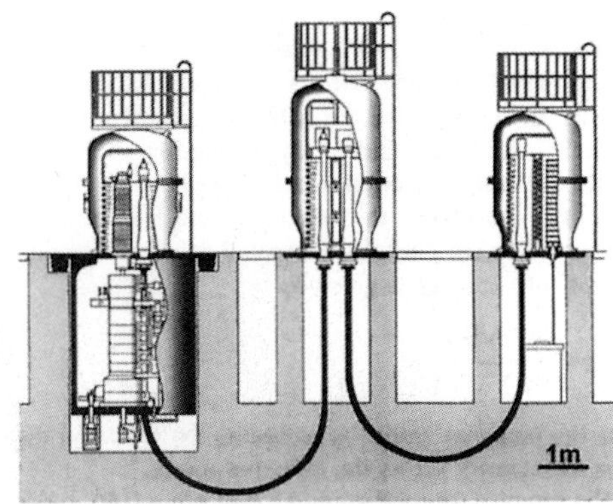

FIG. 1. Schematic diagram of the 1 MV field-emission electron microscope: The vessel on the right contains a 15-stage Cockcroft–Walton circuit, the vessel in the middle contains a floating power supply for the FE gun, and the vessel on the left contains a 36-stage accelerator tube. The microscope column is suspended from the accelerator vessel. Aberration coefficients of the objective lens are $C_s = 1.60$ mm and $C_c = 2.95$ mm. Operations can be performed via remote control.

[a]Electronic mail: tkawa@harl.hitachi.co.jp

Appl. Phys. Lett., Vol. 76, No. 10, 6 March 2000   Kawasaki *et al.*   1343

The spatial coherence length l_s is defined as $\lambda/2\beta$, where β is a semiangle of the beam divergence. From the point of the electron source, the spatial coherence is evaluated by using the average brightness B defined as $i/\pi\beta^2$, where i is the current density.[7] Though the inherent brightness of the FE source is three to four orders of magnitude larger than that of the TE source, the aberrations of the electron gun blur the crossover spot, causing the average brightness to deteriorate. Therefore, we designed the accelerating tube to achieve the optimum performance. Based on computer simulation, the brightness is expected to be 3×10^{10} A/cm^2 (total emission current$=30\,\mu$A, beam current $=3$ nA).[8] The measured value is 1.8×10^{10} A/cm^2 and the discrepancy of a factor 0.6 is considered to be due to the vibration of the gun and the disturbance of ac stray magnetic fields, which are now being improved.

Crystal lattice fringes are formed from the interference between Bragg-diffracted electron waves. The visibility of the fringes depends on the coherence of the electron wave and the overall stability of the microscope.

The temporal coherence directly limits the lattice resolution. In electron interferometry, the FE beam is actually considered to be monochromatic (quasimonochromatic[6]), however, in a lattice fringe observation, the dispersion of the lens action in the magnetic field of the objective lens deteriorates the visibility of the fringes. The effect is estimated using the chromatic aberration coefficient C_c. A defocus value spread of Δ occurs due to the aberration as $\Delta = C_c \Delta E/E$, blurring the image by an amount of $\alpha\Delta = (\lambda/d)\Delta$, where α is the scattering angle, and d is the spacing of the lattice. This sets the minimum resolvable lattice spacing d to be

$$d \approx \sqrt{\lambda\Delta}, \tag{1}$$

for the lattice fringe formed from the interference between the diffracted beam and the direct beam running along the optical axis. The instabilities of the accelerating voltage V and the excitation current of the objective lens I are equivalent to the energy spread. Thus, the defocus spread can be extended as $\Delta = C_c[(\Delta E/E)^2 + (\Delta V/V)^2 + 4(\Delta I/I)^2]^{1/2}$.

The spatial coherence affects the depth of focus, that is, the range of focus where the lattice fringes of a particular periodicity should be visible.[9] By defocusing an amount of Δz, the diffracted electron wave laterally shifts by $\alpha\Delta z = (\lambda/d)\Delta z$. And since the amount of the lateral displacement should be smaller than the spatial coherence length l_s, the depth of focus Δz is given by

$$\Delta z = \frac{(\ln 2)^{1/2}}{\pi} \frac{l_s d}{\lambda}, \tag{2}$$

where the prefactor enters by assuming the Gaussian distribution of intensity across the effective source.[9]

The samples we used were Au thin films (150 Å thick), prepared by successive vacuum deposition onto the Ag layer (2000 Å thick) using NaCl [for Au(001)] and mica [for Au(111)] as substrates. Figure 2 shows the lattice fringe images of a Au(100) film. The Au(200) fringe ($d=2.04$ Å) is often used as a benchmark for checking electron microscope resolution. A half-spacing fringe formed from $(\bar{2}00)$ and (200) reflections is often observed, however, there is a 1/3 spacing lattice fringe [Fig. 2(a)] formed from the interference

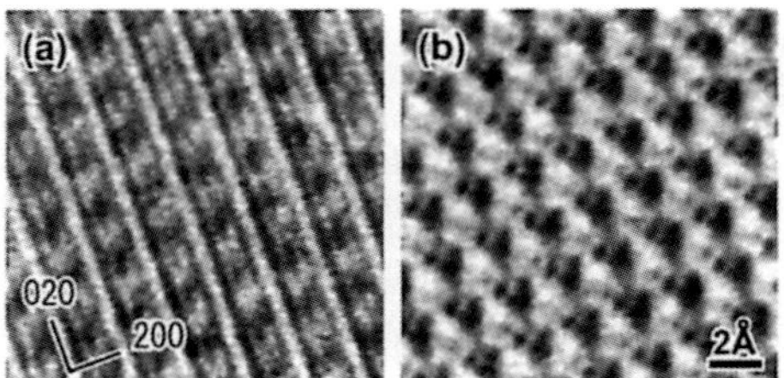

FIG. 2. (200) lattice fringe ($d=2.04$ Å) images of a Au(001) thin film: (a) 1/3 spacing fringes formed from $(\bar{2}00)$ and (400) reflections; and (b) fringes formed from many high-order reflections. The Fourier transform of the micrograph extends to ~ 0.5 Å

between $(\bar{2}00)$ and (400) reflections. This rather unexpected image is due to the coherence of the electron wave discussed above.

The coherence was quantitatively confirmed by observing the fine spacing lattice fringes. The lattice resolution limit of our microscope is calculated to be ~ 0.6 Å from Eq. 1 ($\Delta V/V = \Delta I/I = 5 \times 10^{-7}$). The value is better than that of conventional 1 MV TEM (~ 1 Å). We slightly tilted the illumination to partially escape from the chromatic effect and to take a shorter spacing lattice image. We have taken micrographs of a few fields of view in an Au(111) film, each field being taken by a ten-defocus-value series. Figure 3 shows one of the micrographs. Lattice fringes of 0.498 Å are clearly visible. We thus rather easily observed this spacing on behalf of the coherence. The depth of focus for a 0.5 Å lattice is calculated to be ~ 7400 Å from Eq. (2), assuming

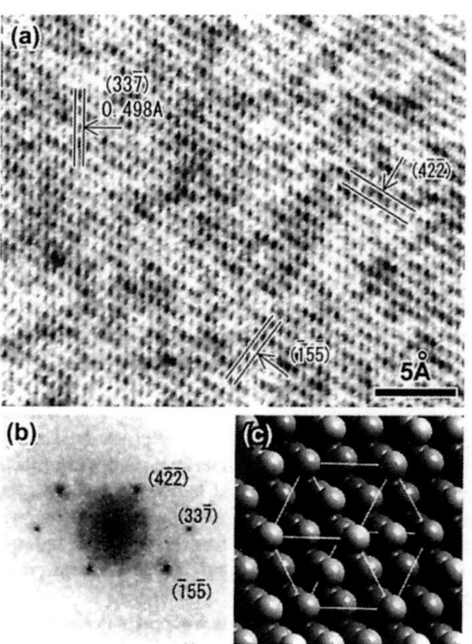

FIG. 3. (a) Electron micrograph of 0.498 Å spacing fringes in a Au(111) thin film; $(4\bar{2}\bar{2})(d=0.83$ Å) and $(\bar{1}5\bar{5})(d=0.57$ Å) fringes are observed simultaneously; (b) Fourier transform of the micrograph; and (c) schematic diagram of [10 11 9]-oriented Au crystal.

1344   Appl. Phys. Lett., Vol. 76, No. 10, 6 March 2000   Kawasaki *et al.*

the present condition (exposure time=1 s, magnification $=10^6$). The value is 10–30 times larger than those found in the previous microscopes, making it easier to observe fine fringes. Detailed crystallographic analysis confirmed that orientation of the crystal towards the beam was [10 11 9] and that the finest fringes we observed were (33$\overline{7}$) lattice fringes.

The atomic information, that is, the projected electric potential map, is obtained using a rather thin specimen. The diffracted beams should interfere with the unscattered wave with the phase shift of $\pi/2$ to satisfy the condition for the Zernike phase contrast microscope.[6] The phase shift is adjusted by defocusing, taking into account the spherical aberration coefficient C_s. The resolution (point resolution) is then written as $0.65C_s^{1/4}\lambda^{3/4}$. Although high-voltage microscopes with short wavelengths improve resolution, it is still restricted to ~ 1 Å due to the robust spherical aberration. This C_s-limited resolution is now aggressively challenged by several proposals[10] to push the resolution beyond 1 Å for better understanding of the detailed structure of materials. If the aberration effects are compensated, the resolution is determined by coherence [information limit, that is, the width of the damping envelope of a contrast transfer function,[9] which is almost determined by Eq. (1)]. We believe that a road map to sub-Å resolution is provided by the combination of coherent electron waves of ultrashort wavelength and the appropriate aberration correction.[11]

Another aspect of the coherent and penetrating electron wave is found in the investigation of electromagnetic structures of new materials. This is best illustrated in the study of the dynamics of magnetic vortices in high-T_c superconductors, which governs the critical current density (a key for power application). The magnetic field of individual vortices can be imaged in real time using a 350 kV FE-TEM,[5] however, a large magnetic penetration length of $\sim 10^3$ Å and a curved vortex line due to the layered structure require more penetrating and coherent electron waves. Given these considerations, we believe that coherent and penetrating electron waves in electron microscopy will help us to better understand the materials.

The authors thank T. Katsuta, T. Furutsu, S. Hayashi, T. Onai, K. Myochin, N. Ebisawa, M. Numata, H. Mogaki, S. Kokubun, M. Tate, H. Masuda, S. Kobayashi, S. Saito, S. Matsunami, N. Moriya, and S. Kubota for their technical support. The authors also thank T. Akashi and M. Gorai for assisting with the microscope operation and appreciate discussions with J. Endo.   We also acknowledge O. Kamimura for preparing samples.

[1] A. V. Crewe, D. N. Eggenberger, J. Wall, and L. M. Welter, Rev. Sci. Instrum. **39**, 576 (1968).

[2] A. Tonomura, *Electron Holography*, 2nd ed. (Springer, Heidelberg, 1999).

[3] A. Tonomura, N. Osakabe, T. Matsuda, T. Kawasaki, J. Endo, S. Yano, and H. Yamada, Phys. Rev. Lett. **56**, 792 (1986).

[4] T. Kawasaki, T. Matsuda, J. Endo, and A. Tonomura, Jpn. J. Appl. Phys., Part 2 **29**, L508 (1990). E. Zeitler and A. Crewe, in *Proceedings of the 5th International Conference on High Voltage Electron Microscopy*, edited by T. Imura and H. Hashimoto (Japanese Society of Electron Microscopy, Tokyo, 1977), p. 99; R. K. Garg, A. Seguela, and B. Jouffrey, in *Proceedings of the 3rd Asia-Pacific Conference on Electron Microscopy*, edited by C. M. Fatt (Applied Research Corporation, Singapore, 1984) p. 100; Nagoya University group constructed a 1 MV FE TEM [S. Maruse and H. Shimoyama, in *Proceedings of the 4th Asia-Pacific Conference on Electron Microscopy*, edited by V. Mangclavirsj, W. Banchorndhevakul, and P. Ingkaninun (Electron Microscopy Society of Thailand, Bangkok, 1988), p. 207] and have been improving the FE electron gun system [C. Morita, S. Arai, T. Kuroyanagi, K. Miyauchi, T. Onai, H. Shimoyama, and M. Hibino, in *Proceedings of the 14th International Conference on Electron Microscopy*, edited by H. Benavides and M. J. Yacaman (Insititute of Physics, Bristol, 1998), Vol. 1, p. 285].

[5] A. Tonomura, H. Kasai, O. Kamimura, T. Matsuda, K. Harada, J. Shimoyama, K. Kishio, and K. Kitazawa, Nature (London) **397**, 308 (1999).

[6] The term *coherence* in this letter refers to the concept of *partial coherence* [M. Born and E. Wolf, *Principles of Optics*, 5th ed. (Pergamon, New York, 1975)].

[7] P. W. Hawkes and E. Kasper, *Principles of Electron Optics* (Academic, London, 1989), Vol. 2, p. 971.

[8] T. Kawasaki *et al.*, J. Electron Microsc. (to be published).

[9] J. C. H. Spence, *Experimental High-Resolution Electron Microscopy* (Oxford University Press, Oxford, 1980).

[10] A. Orchowski, W. D. Raw, and H. Lichte, Phys. Rev. Lett. **74**, 399 (1995); T. Ikuta, J. Electron Microsc. **38**, 415 (1989); M. Haider, S. Uhlemann, E. Schwan, H. Rose, B. Kabius, and K. Ulban, Nature (London) **392**, 768 (1998).

[11] Another route to sub-Å resolution is to combine incoherent illumination and aberration correction: P. D. Nellist and S. J. Pennycook, Phys. Rev. Lett. **81**, 4156 (1998).

letters to nature

Observation of individual vortices trapped along columnar defects in high-temperature superconductors

A. Tonomura*†, H. Kasai*†, O. Kamimura*†, T. Matsuda*†, K. Harada*†,
Y. Nakayama†‡, J. Shimoyama†‡, K. Kishio†‡, T. Hanaguri†§,
K. Kitazawa†§, M. Sasase‖ & S. Okayasu‖

* *Advanced Research Laboratory, Hitachi Ltd, Hatoyama, Saitama 350-0395, Japan*
† *CREST, Japan Science and Technology Corporation (JST), Kawaguchi, Saitama 332-0012, Japan*
‡ *Department of Applied Chemistry, University of Tokyo, Tokyo 113-8656, Japan*
§ *Department of Advanced Materials Science, School of Frontier Sciences, University of Tokyo, Tokyo 113-0033, Japan*
‖ *Department of Material Science, Japan Atomic Energy Research Institute, Tokai, Naka-gun, Ibaraki 319-1195, Japan*

Many superconductors do not entirely expel magnetic flux—rather, magnetic flux can penetrate the superconducting state in the form of vortices. Moving vortices create resistance, so they must be 'pinned' to permit dissipationless current flow. This is a particularly important issue for the high-transition-temperature superconductors, in which the vortices move very easily[1]. Irradiation of superconducting samples by heavy ions produces columnar defects, which are considered[2] to be the optimal pinning traps when the orientation of the column coincides with that of the vortex line. Although columnar defect pinning has been investigated using macroscopic techniques[3,4], it has hitherto been impossible to resolve individual vortices intersecting with individual defects. Here we achieve the resolution required to image vortex lines and columnar defects in $Bi_2Sr_2CaCu_2O_{8+\delta}$ (Bi-2212) thin films, using a 1-MV field-emission electron microscope[5]. For our thin films, we find that the vortex lines at higher temperatures are trapped and oriented along tilted columnar defects, irrespective of the orientation of the applied magnetic field. At lower temperatures, however, vortex penetration always takes place perpendicular to the film plane, suggesting that intrinsic 'background' pinning in the material now dominates.

There are several methods of directly observing vortices, but none of them can determine the behaviour of individual vortex lines inside superconductors. This is because these methods detect vortices at the superconductor surfaces, even though an attempt has been made to obtain evidence of wandering vortex lines near material defects by using a two-sided Bitter decoration technique to detect the vortex positions on both sides of the film[6]. At present, the only methods which enable the observation of individual vortices and defects inside superconducting thin films are Lorentz microscopy[7] and interference microscopy[8], where the vortex magnetic fields are detected using a penetrating electron beam. However, owing to the low penetration power of the existing 300-kV field-emission electron beam, only a film thinner than the diameter of the vortex magnetic flux (that is, twice the magnetic penetration depth) has been observed. Vortex lines oriented in different directions cannot be distinguished in such a thin film, because the vortex magnetic fields inside it do not change very much.

To obtain clear images of vortices inside thicker films, we have developed a large electron microscope[5]. This 40-ton microscope has a 1-MV field-emission electron beam that has more than twice the penetration power of a 300-kV beam and also the highest brightness ($2 \times 10^{10}\,\mathrm{A\,cm^{-2}\,sr^{-1}}$) ever attained. These features of the electron beam have allowed us not only to observe vortices in Bi-2212 films 400 nm thick with high contrast, but also to

Reprinted with permission. doi: 10.1038/35088021

distinguish between two vortex lines[9], one perpendicular to the film plane and the other trapped along a tilted columnar defect.

A schematic of the observation principle is shown in Fig. 1. Film samples were prepared by cleaving a single crystal of Bi-2212 (transition temperature $T_c = 85$ K) grown by the floating-zone method[10]. The films were then obliquely irradiated with parallel 240-MeV Au[15+] ions (incidence angle $\theta_\phi = 70°$) as sparsely as 0.05 ions μm^{-2}, or with a matching magnetic field of 0.1 mT. We intentionally made the column density extremely sparse in our experiments so that we could use untrapped vortex lines perpendicular to the film plane as reference lines to unambiguously distinguish vortex lines trapped along tilted columns. A collimated electron beam was applied incident to the tilted film (tilting angle $\alpha = 30°$) and a magnetic field was applied in an arbitrary direction (incidence angle $\theta = (-70°) - (+70°)$) as shown in the schematic. We then observed the Lorentz images obtained by image-defocusing.

Examples of the observation results at $\theta_\phi = \theta = 70°$ are shown in Fig. 2. In the in-focus image (Fig. 2a), the projected images of tilted columnar defects are seen as short black lines. As the defocusing distance Δf increases, the column images first disappear completely by image-blurring. At larger Δf values, spots with bright-and-dark contrasting features appear instead, as shown in Fig. 2b. These spots appear because the phase change of the transmitted electron beam due to the vortex magnetic flux is transformed into the visible intensity variations by image-defocusing. They are Lorentz images of vortices.

There are two kinds of images in this Lorentz micrograph: circular spots and elongated spots with lower contrast. The elongated spots are indicated by the arrows in Fig. 2b. Image simulation[9] reveals that the circular images correspond to vortex lines penetrating the film perpendicularly to the film plane and that the elongated

images correspond to vortex lines trapped along tilted columns. The image elongation direction is not necessarily the same as that of the tilted columns because of the effect of the inclination of the sample film, and the image contrast is weakened because of the spread of the magnetic flux distribution owing to the tilting of vortex lines from the c axis.

The vortex line direction inside a superconductor can be more precisely determined by comparing its Lorentz images taken while the sample is rotated around an axis normal to the film surface with the corresponding simulated images. A perpendicular vortex line does not change its direction with this rotation, but a tilted vortex line does, owing to the precession. In addition, a comparison of the two images in Fig. 2a and b reveals that columnar defects can be found at the exact centres of the elongated vortex images; this indicates that the elongated images actually represent vortex lines trapped along tilted columnar defects. Vortex lines oriented in different directions can thus be observed as different Lorentz images.

We then investigated whether vortex lines remained trapped along columnar defects even when the direction of the magnetic field or the sample temperature changed. First, we rotated the magnetic field from the column direction ($\theta = 70°$) to the film normal direction ($\theta = 0°$). In so doing, however, the elongated vortex images did not change at all, indicating that the vortex lines at columns remained trapped along them. Even when we further tilted the magnetic field direction until $\theta = -70°$, the vortex lines remained trapped. Lorentz micrographs at $\theta = 70°$ and $\theta = -70°$ are shown in Fig. 3a and b. The elongated images are indicated by the arrows.

We also found that the elongated images remained unchanged when we increased T from 35 K to T_c. However, when we gradually decreased T, we found to our surprise that the trapped vortices at columnar defects began to "stand up" perpendicularly to the film plane between 12 and 14 K. Figure 3c shows a Lorentz micrograph of the vortices at $\theta = -70°$ when T was 10 K. The elongated vortex images with weak contrast at 35 K (Fig. 3b) had changed to circular vortex images (Fig. 3c), the same as those surrounding untrapped perpendicular vortices.

We also found that this change was not reversible; there was hysteresis when the temperature was increased or decreased. That

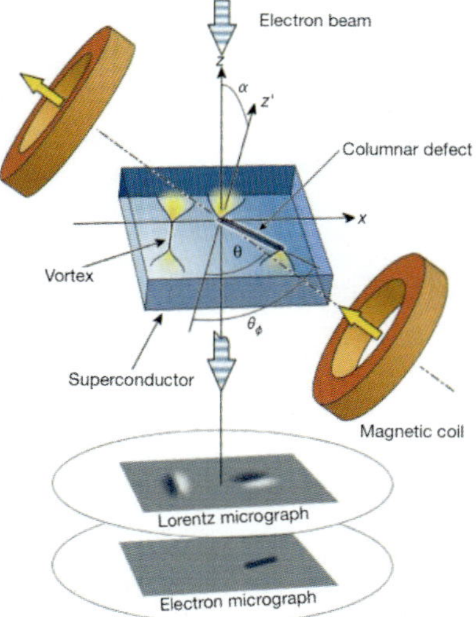

Figure 1 Lorentz microscopy of vortices in a superconducting Bi-2212 thin film with tilted columnar defects. The film is tilted at $\alpha = 30°$, α being the rotation angle around the x axis between the optical axis (z axis) and the film normal (z' axis). A magnetic field is applied to the film in the x–z' plane at an incidence angle of θ, thus producing vortices, whereas that of the column direction is θ_ϕ. An electron beam is vertically applied incident to the film, and the phase change of the transmitted beam that is due to the vortex magnetic flux inside the film is transformed into intensity variations as a Lorentz image by image-defocusing. The vortex images appear to be different for perpendicular and tilted vortex lines under appropriate defocusing distances.

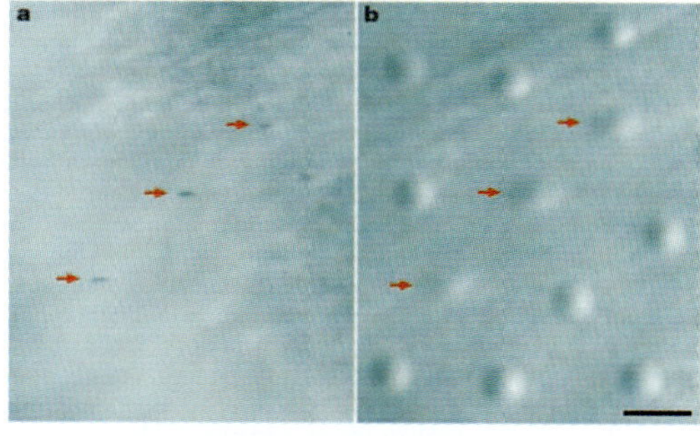

Figure 2 Vortices trapped and untrapped at columnar defects in Bi-2212 thin film ($H = 0.5$ mT, $\theta = \theta_\phi = 70°$). **a**, Electron micrograph. **b**, Lorentz micrograph ($\Delta f = 500$ mm, $T = 30$ K). Columnar defects can be seen as short black lines in the in-focus electron micrograph (**a**). When the micrograph is greatly defocused, the column images completely disappear and the vortex images appear instead as can be seen in the Lorentz micrograph (**b**): the images of the perpendicular vortex lines are circular, whereas those of the tilted vortex lines are elongated and have lower contrast. This was confirmed by image simulation and is illustrated in Fig. 1. It should be noted here that the elongated vortex images (indicated by the arrows in (**b**)) appear at the positions of the column images in (**a**). Owing to the low density of columnar defects, only some vortex lines are trapped along columnar defects; the others are untrapped and penetrate the film perpendicularly to the film plane. Scale bar, 2 μm.

　　　　621

letters to nature

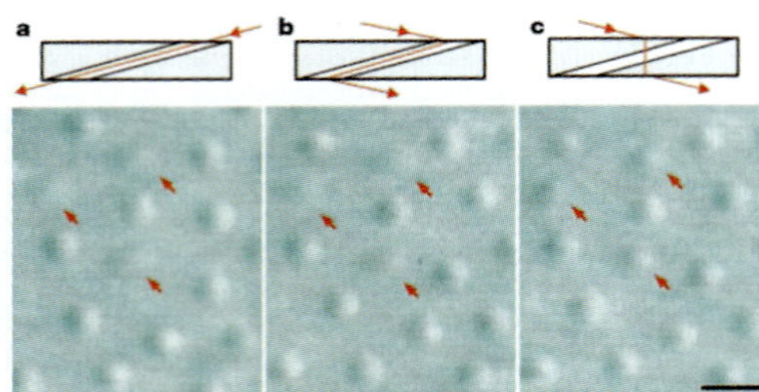

Figure 3 Lorentz micrographs showing vortex-line arrangements under various conditions. **a**, Magnetic field applied parallel to columns ($\theta = 70°$) at $T = 35$ K. **b**, Magnetic field applied in the direction of $\theta = -70°$ at $T = 35$ K. **c**, Magnetic field in the direction of $\theta = -70°$ at $T = 10$ K. The vortex lines trapped along columnar defects (in **a**) remained trapped even when the direction of the magnetic field was greatly tilted, as can be seen in **b** (indicated by the arrow). When T decreased, however, the vortex lines trapped along tilted columnar defects began to penetrate the film perpendicularly to the film plane at 12–14 K, even though the vortex lines were located at the positions of the columns, and therefore the vortex images became circular and their contrast became higher, as can be seen in **c**. Illustrations above each Lorentz micrograph indicate vortex-line arrangements. Scale bar, 2 μm. Video clips showing the behaviour of the vortices are available at http://www.hatoyama.hitachi.co.jp.

is, when T was increased, perpendicular vortex lines at 7 K began to tilt along columnar defects at 15–19 K, rather than at 12–14 K. The transition temperatures were rather scattered for the different vortex lines. For example, even at 7 K, we found vortex lines tilted at columns in very rare cases. This does not seem to originate from the different pinning forces of individual columns, because electron microscopy showed that column structures were fairly uniform in shape. Such tilting at 7 K may be due to the fact that very strongly pinned vortices are trapped simultaneously at multiple columns that gather by chance rather than at a single isolated column.

The observed behaviour of vortices can be interpreted by taking into consideration the temperature dependence of the pinning forces of columnar defects estimated from the observation of vortex movements. Above 25 K, changes in the magnetic field caused sparsely distributed vortices to hop from one columnar defect to another when a driving force was applied to them. Such a movement occurred regardless of the direction in which the magnetic field was applied. This can happen when columnar-defect pinning is dominant. These results are consistent with the previous observation that the vortex lines are firmly trapped even along greatly tilted columns irrespective of the direction of the applied magnetic field at $T > 19$ K.

When T was decreased below 25 K, the hopping movement gradually changed in such a manner that vortices intermittently stopped and migrated. This kind of movement can be interpreted to occur, because the background collective pinning due to more abundant smaller atomic-size defects of other types (for example, oxygen defects) is thought to increase relative to the columnar-defect pinning. The migration speed decreases at lower T; thus the migration is thought to be caused by thermally activated depinning of a vortex line from a large number of atomic-size defects one by one. The vortex lines, when stopped, were observed to be tilted along columns. When T was decreased below 12 K, however, the columnar-defect pinning became weaker relative to the background pinning, and trapped vortex lines at columnar defects were observed to stand up perpendicularly to the film plane. Around 7 K, all the vortices migrated or drifted uniformly when driven, as if the pinning force of columnar defects vanished and vortices were moving uniformly in a viscous medium. The pinning of columnar defects is almost completely hidden by the background pinning,

although there were still some exceptional vortices that were strongly pinned, possibly at multiple defects.

In this way, direct observation microscopically elucidated the behaviour of vortices in Bi-2212 thin films. The isotropic pinning of vortices by columnar defects at low magnetic fields indirectly inferred by macroscopic magnetization measurements[3,11] can be explained microscopically by our experiments as follows: under our conditions of thin films, vortex lines are trapped along columns at temperatures above 19 K irrespective of the direction of the applied magnetic field; and at temperatures below 12 K vortex lines always penetrate the film perpendicularly to the film plane owing to the increased background pinning compared to the columnar-defect pinning and also because of the demagnetization effect of thin films. □

Received 27 March; accepted 4 June 2001.

1. Crabtree, G. W. & Nelson, D. R. Vortex physics in high-temperature superconductors. *Phys. Today* **50**, 38–45 (1997).
2. Civale, L. *et al.* Vortex confinement by columnar defects in YBa₂Cu₃O₇ crystals: Enhanced pinning at high fields and temperatures. *Phys. Rev. Lett.* **67**, 648–651 (1991).
3. Hardy, V. *et al.* Accommodation of vortices to tilted line defects in very high-T_c superconductors with various electronic anisotropies. *Phys. Rev. B* **54**, 656–664 (1996).
4. Schuster, Th. *et al.* Observation of in-plane anisotropy of vortex pinning by inclined columnar defects. *Phys. Rev. B* **50**, 9499–9502 (1994).
5. Kawasaki, T. *et al.* Fine crystal lattice fringes observed using a transmission electron microscope with 1 MeV coherent electron waves. *Appl. Phys. Lett.* **76**, 1342–1344 (2000).
6. Yao, Z., Yoon, S., Dai, H., Fan, S. & Lieber, C. M. Path of magnetic flux lines through high-T_c copper oxide superconductors. *Nature* **371**, 777–779 (1994).
7. Harada, K. *et al.* Real-time observation of vortex lattices in a superconductor by electron microscopy. *Nature* **360**, 51–53 (1992).
8. Tonomura, A. *Electron Holography* 2nd edn (Springer, Heidelberg, 1999).
9. Fanesi, S. *et al.* Influence of core misalignment and distortion on the Fresnel and holographic images of superconducting fluxons. *Phys. Rev. B* **59**, 1426–1431 (1999).
10. Kotaka, Y. *et al.* Doping state and transport anisotropy in Bi2212 single crystals. *Physica C* **235–240**, 1529–1530 (1994).
11. Klein, L., Yacoby, E. R., Yeshurun, Y., Konczykowski, M. & Kishio, K. Evidence for line vortices in Bi₂Sr₂CaCu₂O₈. *Phys. Rev. B* **48**, 3523–3525 (1993).

Acknowledgements

We are grateful to N. Hatano for his discussions on tilted columnar defects, and to G. Pozzi, M. Beleggia, N. Osakabe, T. Yoshida and J. Masuko for their simulations of Lorentz micrographs of tilted vortex lines. We also thank F. Nori for discussions and T. Akashi and I. Matsui for help with the experiments.

Correspondence and requests for materials should be addressed to A.T. (e-mail: tonomura@harl.hitachi.co.jp).

RESEARCH ARTICLES

Oscillating Rows of Vortices in Superconductors

T. Matsuda,[1] O. Kamimura,[1] H. Kasai,[1] K. Harada,[1] T. Yoshida,[1] T. Akashi,[2] A. Tonomura,[1]* Y. Nakayama,[3] J. Shimoyama,[3] K. Kishio,[3] T. Hanaguri,[4] K. Kitazawa[4]

Superconductors can be used as dissipation-free electrical conductors as long as vortices are pinned. Vortices in high-temperature superconductors, however, behave anomalously, reflecting the anisotropic layered structure, and can move readily, thus preventing their practical use. Specifically, in a magnetic field tilted toward the layer plane, a special vortex arrangement (chain-lattice state) is formed. Real-time observation of vortices using high-resolution Lorentz microscopy revealed that the images of chain vortices begin to disappear at a much lower temperature, T_d, than the superconducting transition temperature, T_c. We attribute this image disappearance to the longitudinal oscillation of vortices along the chains.

When a magnetic field is applied to a type II superconductor, tiny magnetic vortices, typically arranged in a triangular lattice, form inside it (*1*) and play a crucial role in the practical use of superconductors as electrical conductors. Unless they are pinned, the vortices begin to flow when driven by an applied current, eventually suppressing superconductivity. In high-temperature superconductors, however, the vortex behaviors are more complicated because of their anisotropic layered

structure (*2*). For example, at high temperature T and magnetic field H, a single vortex line can split itself into vortex pancakes in each layer that can be displaced and move independently in a liquid-like state (*2–4*). In this regime, no pinning occurs and the critical current vanishes. The information about the microscopic behaviors of the vortices is therefore indispensable for developing high–critical current materials.

The effect of the anisotropic layered structures also appears in the static arrangement of vortices. When the field direction is tilted away from the c axis, a triangular vortex-lattice state is transformed into an unconventional arrangement of vortices consisting of alternating domains of linear chains and triangular lattices (*5*). This phenomenon has recently attracted much attention but has not yet been fully understood, despite both theoretical (*6, 7*) and experimental (*5, 8*) investigations.

[1]Advanced Research Laboratory, Hitachi Ltd., Hatoyama, Saitama 350-0395, Japan. [2]Hitachi Instruments Service Co. Ltd., 4-28-8 Yotsuya, Shinjuku-ku, Tokyo 160-0004, Japan. [3]Department of Applied Chemistry, University of Tokyo, Tokyo 113-8656, Japan. [4]Department of Advanced Materials Science, School of Frontier Sciences, University of Tokyo, Tokyo 113-0033, Japan.

*To whom correspondence should be addressed. E-mail: tonomura@harl.hitachi.co.jp

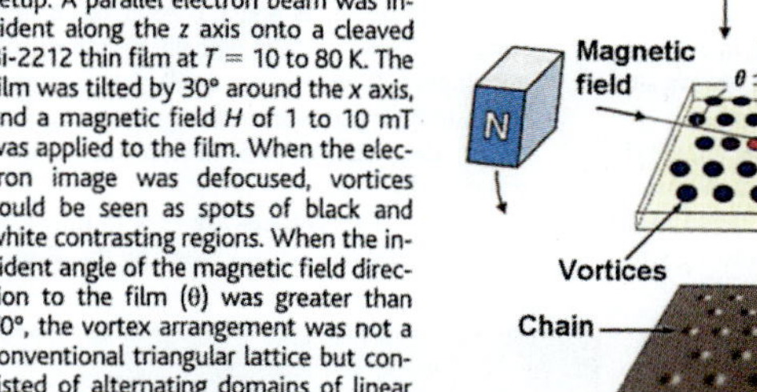

Fig. 1. Schematic of the experimental setup. A parallel electron beam was incident along the z axis onto a cleaved Bi-2212 thin film at $T = 10$ to 80 K. The film was tilted by 30° around the x axis, and a magnetic field H of 1 to 10 mT was applied to the film. When the electron image was defocused, vortices could be seen as spots of black and white contrasting regions. When the incident angle of the magnetic field direction to the film (θ) was greater than 70°, the vortex arrangement was not a conventional triangular lattice but consisted of alternating domains of linear chains and triangular lattices.

We report findings about the unusual disappearance of the images of chain vortices in $Bi_2Sr_2CaCu_2O_{8+\delta}$ (Bi-2212) that were found using high-resolution Lorentz microscopy (9, 10), and we explain the results in terms of mobile incommensurate chain vortices surrounded by less mobile commensurate ones. The chain vortices oscillate longitudinally even at temperatures sufficiently low that freezing of the vortex lattices would be expected. The oscillation of chain vortices that begins even at low T is of practical importance because such vortex motion will affect vortex pinning and melting. In addition, our direct observation of incommensurate vortices provides valuable microscopic information about similar phenomena in physical systems such as charge density waves and Josephson junction arrays.

In our experiment (Fig. 1), a 400-nm-thick cleaved single crystal of Bi-2212 was tilted at 30° around the x axis, and an electron beam was incident on the sample from above. A magnetic field H of 1 to 10 mT was applied to the film at incidence angles (θ) of 70° to 90°, and then the sample was cooled down to T in H so that individual vortices in the chain-lattice state could be observed in the out-of-focus plane as Lorentz micrographs.

In the Lorentz micrograph (Fig. 2), the spacing of chain vortices is narrower than, and therefore incommensurate with, that of the surrounding triangular-lattice vortices. The chain vortices appear to be constrained to arrange themselves along these lines. Some of these chains, like the rightmost chain in Fig. 2, appear to terminate or split into two branches when crossing a surface step, as indicated by the dark arrows. However, closer examination shows that the chain did not actually disappear, but only changed direction by bending to the left. Even when the vortex chains were moved by a driving force applied to the vortices by changing H, the line shape was retained even though it was not always straight but curved and changed over time.

Although all the vortex images in the chain-lattice state are clearly observed at low T, the images of some chain vortices begin to disappear when the temperature rises to T_d, which is

still considerably lower than T_c. The Lorentz micrograph at $T = 70$ K shows three such chains that have disappeared (Fig. 3A). The vortex images of the top and bottom chains have disappeared completely, whereas those of the middle chain are just beginning to disappear.

Why do the images of chain vortices disappear? It is evident from the contrast of the vortex images, which gradually weakens and completely vanishes when T increases, that the vortices themselves do not disappear, but rather the images do. Even among vortices belonging to the same chain, the vortex images are blurred gradually from one place to another, as in the case of the middle chain. Vortices a and b can clearly be seen, but the vortices between them cannot. Such image disappearance must be due to the movement of the vortices, or the spreading out of the magnetic flux of each vortex.

We attribute this vortex-image disappearance to the oscillations of vortices along the chain direction that are too fast to be observed by a detection system with a TV scanning rate. In a triangular lattice, all vortices are commensurate with the underlying potential energy landscape. There is no reason for only one group of vortices to start moving. In a chain-lattice state, however, chains of vortices become incommensurate with the potential distribution. Therefore, only incommensurate chains can move longitudinally at relatively low T, as is known from the theory of commensurate structures (11).

At low enough T, vortices in the equilibrium chain-lattice state receive no forces and therefore do not move. When T increases, however, locally unstable chain vortices such as those located between locally stable vortices a and b in Fig. 3A begin to vibrate back and forth along the chain direction, forming a coupled oscillation (12) (Fig. 3, B and C). As a result, the images of chain vortices partially disappear. At still higher T, the vortex oscillation becomes even stronger and whole chains of vortices begin to disappear, as did the top and bottom chains in Fig. 3A.

The fact that vortices are easy to move in the chain direction is supported by the obser-

vation that the vortices actually move in such a way that some sections of chain vortices reciprocate suddenly and simultaneously along the chain direction. This happens from time to time when the vortex configuration does not completely reach equilibrium, although this movement is not observed under the same conditions of the vortex-image disappearance, but rather under conditions where individual vortices are well separated and their movements can be clearly observed—for example, at $B_\perp$ (normal component of the magnetic flux density) = 0.2 mT, $T = 56$ K, and $\theta = 85°$. Furthermore, chain vortices are predicted to move freely along the chain when they become completely incommensurate (i.e., when the distance between the node vortices a and b increases infinitely). Indeed, we observed such a case

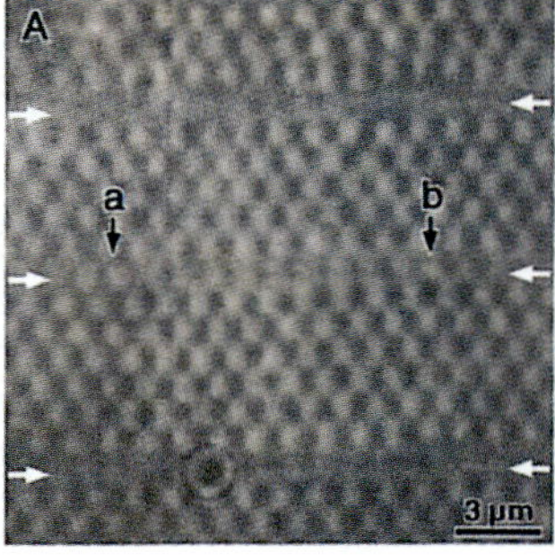

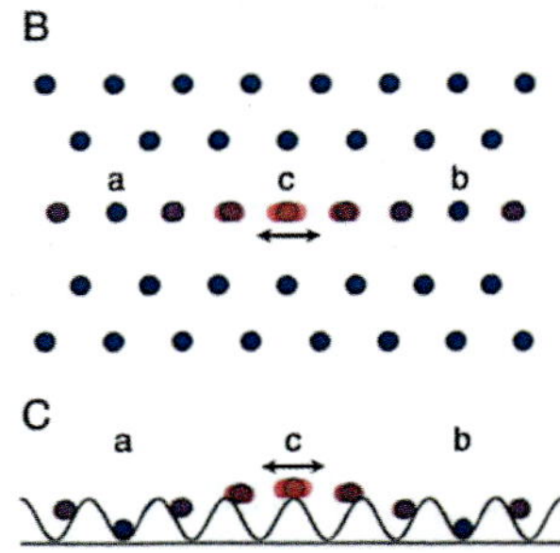

Fig. 3. Disappearing chain vortices. (**A**) Lorentz micrograph ($T = 70$ K, $B_\perp = 1.2$ mT, $\theta = 80°$): The image of the middle chain partially disappears except for vortices a and b, for example, which are stabilized by the surrounding triangular-lattice vortices. The images of the other two chains disappear along their whole lengths. (**B**) Vortex arrangement in the chain-lattice state: Red vortices easily oscillate along the chain direction when T increases because their positions are unstable. (**C**) Vortex arrangement situated on the vortex potential: Vortices a and b are stable while vortex c is unstable and tends to oscillate back and forth in the chain direction as a result of thermal fluctuations. When the distance between a and b becomes infinitely long, chain vortices become incommensurate and can move freely.

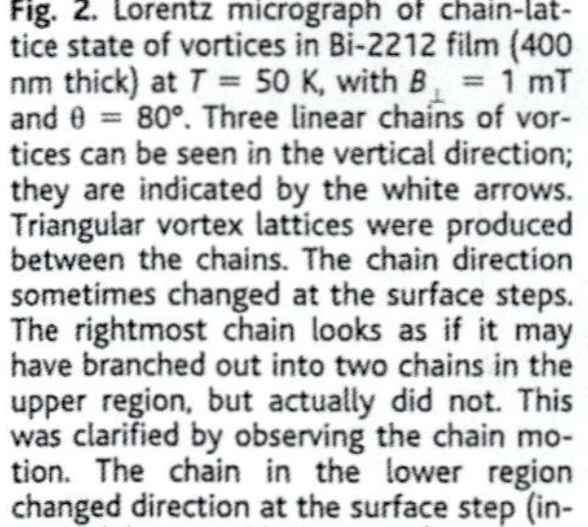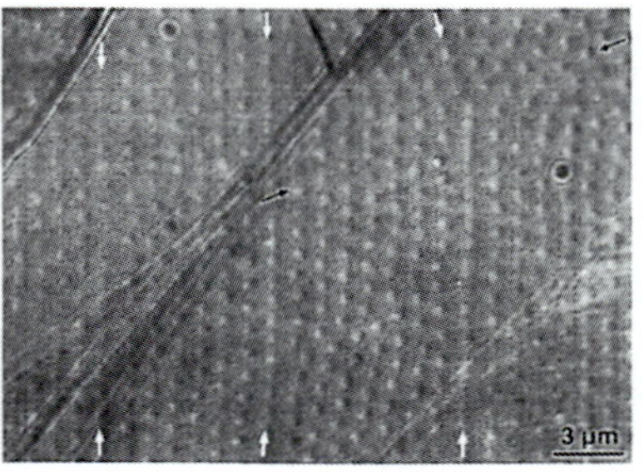

Fig. 2. Lorentz micrograph of chain-lattice state of vortices in Bi-2212 film (400 nm thick) at $T = 50$ K, with $B_\perp = 1$ mT and $\theta = 80°$. Three linear chains of vortices can be seen in the vertical direction; they are indicated by the white arrows. Triangular vortex lattices were produced between the chains. The chain direction sometimes changed at the surface steps. The rightmost chain looks as if it may have branched out into two chains in the upper region, but actually did not. This was clarified by observing the chain motion. The chain in the lower region changed direction at the surface step (indicated by two black arrows) and was connected to the left branch in the upper region.

Fig. 4. Disappearing images of more vortices in a single chain ($T = 57$ K, $B_\perp = 0.8$ mT, $\theta = 80°$). We attribute the disappearance of the vortex images to the synchronous oscillation of vortices between the node vortices a and b along the chain direction, just like a coupled oscillation.

where a much larger number of chain vortices disappeared (Fig. 4).

If thermal fluctuations excite the oscillations of chain vortices, the temperature at which the oscillation begins and the vortex image disappears (T_d) must differ depending on the vortex density, because the vortex density changes the spring constant of the chain vortices and also the periodic potential distribution for chain vortices. When we measured T_d as a function of vortex density or $B_\perp$, we found that the T_d values greatly depended on $B_\perp$ (Fig. 5). For example, T_d was as low as 20 K when $B_\perp = 1.7$ mT, whereas T_d was as high as 70 K when $B_\perp = 0.4$ mT.

The tendency for T_d to decrease when $B_\perp$ increases can be qualitatively understood as follows: When $B_\perp$ increases and the spacing a between vortices becomes narrower, the amplitude of the potential oscillation shown in Fig. 3C decreases as a result of the overlap of each vortex magnetic field. Lowering the potential barrier will naturally make chain vortices start oscillating at lower T. The spring constant k between closer chain vortices increases, thus making it easier for a more rigid incommensurate chain of vortices to slide freely as a whole. This explains why our experimental results show that T_d tends to decrease as $B_\perp$ increases (Fig. 5).

We expected to be able to detect the direct indication of the vortex movement at the very beginning of the oscillation under the assumption that vortex lines move as straight rods; hence, we attempted to observe the vortex motion while T gradually increased and crossed T_d. However, the individual vortex images were gradually blurred, and no sign of vortex movement was detected. As a result, our experiments cannot rule out the possibility that a chain vortex line was split into pancake vortices (4) that oscillated in each layer independently for different layers, thus blurring the averaged magnetic field of the vortex line. Even in this case, the partial disappearance of vortex images can be explained because the oscillation of pancake chain vortices in each layer can have the nodes fixed by the surrounding straight vortex lines belonging to the triangular lattices.

The disappearance of the vortex images

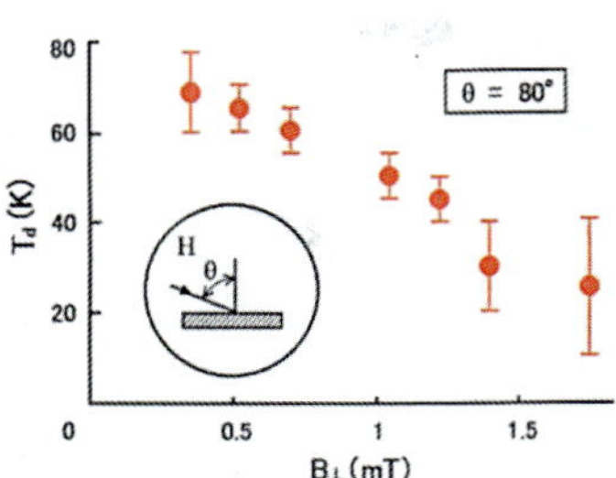

Fig. 5. Disappearance temperature T_d versus magnetic flux density perpendicular to the film $B_\perp$. T_d decreased greatly as $B_\perp$ increased. This tendency can be explained by assuming the movement of chain vortices as being that of a coupled oscillation of chain vortices situated on the periodic potential distribution (see Fig. 3C). When vortices are dense (large $B_\perp$ and small a), the periodic potential barrier becomes lower as a result of the overlap of the magnetic fields of surrounding triangular vortices, and chain vortices can more easily start oscillating at lower T, thus leading to lower T_d.

may be explained by assuming that the vortex lines are greatly tilted (7). Indeed, the image contrast was shown to decrease when vortex lines were trapped along columnar defects tilted by 70° (13). However, in the present experiment we were able to prove that the observed vortex image disappearance was not attributable to the vortex-line tilting, because the vortex images remained invisible even at a large defocusing distance Δf. The image contrast of tilted vortex lines, which is weak at a relatively short Δf, must become as vis-

ible as that of nontilted vortex lines at greater Δfs (13). Our experiments cannot rule out the possibility that the vortex lines do not simply tilt, but that their magnetic fluxes are somehow blurred in the chain direction (14).

The above discussion may be too simplified and may overlook several features of this phenomenon. For example, the adjacent lines of vortices on both sides of the disappearing vortex chains in Figs. 3A and 4 are not straight; they are slightly bent in such a way that the regions of the disappeared chains are wider at the vortex oscillation and narrower at the oscillation nodes. This may be due to the effect of the locally violent oscillation of vortices.

References and Notes
1. A. A. Abrikosov, *Sov. Phys. JETP* **5**, 1174 (1957).
2. D. J. Bishop, P. L. Gammel, D. A. Huse, C. A. Murray, *Science* **255**, 165 (1992).
3. A. Oral et al., *Phys. Rev. Lett.* **80**, 3610 (1998).
4. J. R. Clem, *Phys. Rev. B* **43**, 7837 (1991).
5. C. A. Bolle et al., *Phys. Rev. Lett.* **66**, 112 (1991).
6. A. I. Buzdin, A. Yu. Simonov, *JETP Lett.* **51**, 191 (1990).
7. A. E. Koshelev, *Phys. Rev. Lett.* **83**, 187 (1999).
8. I. V. Grigorieva, J. W. Steeds, G. Balakrishnan, D. M. Paul, *Physica C* **235–240**, 2705 (1994).
9. K. Harada et al., *Nature* **360**, 51 (1992).
10. T. Kawasaki et al., *Appl. Phys. Lett.* **76**, 1342 (2000).
11. L. M. Floria, J. J. Mazo, *Adv. Phys.* **45**, 505 (1996).
12. C. Reichhardt, C. J. Olson, F. Nori, *Phys. Rev. B* **58**, 6534 (1998).
13. A. Tonomura et al., *Nature* **412**, 620 (2001).
14. More detailed simulations and experiments to pinpoint the cause of the disappearance of chain-vortex images are now in progress (A. Tonomura et al., in preparation).
15. We thank F. Nori, A. Koshelev, T. Hashizume, S. Kagoshima, T. Onogi, P. Gammel, and D. Bishop for their useful suggestions. Supported by SORST, Japan Science and Technology Corporation (JST).

4 September 2001; accepted 1 November 2001

VOLUME 88, NUMBER 23   PHYSICAL REVIEW LETTERS   10 JUNE 2002

Observation of Structures of Chain Vortices Inside Anisotropic High-T_c Superconductors

A. Tonomura,[1,9,10,*] H. Kasai,[1,9,10] O. Kamimura,[1,9] T. Matsuda,[1,9,10] K. Harada,[1,9,10] T. Yoshida,[1,9] T. Akashi,[2]
J. Shimoyama,[3,9] K. Kishio,[3,9] T. Hanaguri,[4,9] K. Kitazawa,[4,9] T. Masui,[5] S. Tajima,[5] N. Koshizuka,[5]
P. L. Gammel,[6] D. Bishop,[7] M. Sasase,[8] and S. Okayasu[8]

[1]*Advanced Research Laboratory, Hitachi, Ltd., Hatoyama, Saitama 350-0395, Japan*
[2]*Hitachi Instruments Service Co., Ltd., 4-28-8 Yotsuya, Shinjuku-ku, Tokyo 160-0004, Japan*
[3]*Department of Applied Chemistry, University of Tokyo, Tokyo 113-8656, Japan*
[4]*Department of Advanced Materials Science, School of Frontier Sciences, University of Tokyo, Tokyo 113-0033, Japan*
[5]*International Superconductivity Technology Center (ISTEC), Shinonome, Koto-ku, Tokyo 135-0062, Japan*
[6]*Agere Systems, Murray Hill, New Jersey 07974-0636*
[7]*Bell Laboratories, Lucent Technologies, Murray Hill, New Jersey 07974-0636*
[8]*Japan Atomic Energy Research Institute (JAERI), Tokai, Ibaraki 319-1195, Japan*
[9]*SORST, Japan Science and Technology Corporation (JST), 3-4-15 Nihonbashi, Chuo-ku, Tokyo 103-0027, Japan*
[10]*Frontier Research System, The Institute of Chemical and Physical Research (RIKEN),
Wako, Hirosawa, Saitama 351-0198, Japan*
(Received 22 January 2002; published 23 May 2002)

In order to elucidate the formation mechanism of unconventional arrangements of vortices in high-T_c superconducting thin films at an inclined magnetic field to the layer plane, we investigated the structures of vortex lines inside the films by Lorentz microscopy using our 1-MV field-emission electron microscope. Our observation results concluded that vortex lines are tilted to form linear chains in $YBaCu_3O_{7.8}$. Vortex lines in the chain-lattice state in $Bi_2Sr_2CaCu_2O_{8+\delta}$, on the other hand, are all perpendicular to the layer plane, and therefore only vortices lined up along Josephson vortices form chains.

DOI: 10.1103/PhysRevLett.88.237001   PACS numbers: 74.60.Ge, 68.37.Lp, 74.72.Hs, 74.76.Bz

When a magnetic field is applied to a type-II superconductor, the field penetrates it in the form of flux lines, or magnetic vortices, which usually form a triangular lattice. This is the case even for anisotropic superconductors as long as the magnetic field is directed along the anisotropy c axis. When the magnetic field is greatly tilted away from the c axis, however, the Bitter images show that the vortices no longer form a triangular lattice, but instead form arrays of linear chains along the direction of the field tilting for $YBaCu_3O_{7.8}$ (YBCO) [1] and alternating domains of chains and triangular lattices for $Bi_2Sr_2CaCu_2O_{8+\delta}$ (Bi-2212) [2–4]. While the chain state in YBCO can be explained by the tilting of vortex lines within the framework of the anisotropic London theory [5–7], consistent with the experimental data [1], the chain-lattice state in Bi-2212 has long been an object of discussion. It has been attributed to two sets of vortex lines having different orientations [3,8–12], one set forming chains and the other set triangular lattices. Koshelev [12], for example, proposed that vortices that perpendicularly intersect Josephson vortices form chains and the rest of the vortices form triangular lattices. Recent investigations [4] report detailed observations of "pancake" vortices at a tilted magnetic field. No direct evidence for such mechanisms, however, was given experimentally due to the lack of methods to observe vortex lines inside superconductors.

Lorentz microscopy with our newly developed 1-MV electron microscope [13] has made it possible to distinguish between two vortex lines that are oriented in different directions [14], and was used for the present experiments

determining the tilting of vortex lines. The schematic of our experiments is illustrated in Fig. 1(a). Film samples, 300–400 nm thick, of single-crystalline YBCO ($T_c = $ 92 K) and Bi-2212 ($T_c = $ 85 K) were prepared by thinning a region 30 μm $\times$ 100 μm, of a YBCO single crystal with a focused ion beam machine (Hitachi FB-2000) and by cleaving a Bi-2212 single crystal, respectively. These samples, the surfaces of which were parallel to the layer plane, were tilted around the y' axis and an electron beam was incident onto them from above. A magnetic field of 0–10 mT was applied obliquely to the surface of the samples at incidence angles (θ) of 70°–90°, and vortices in arrangements reflecting the anisotropic layered structure of the materials were observed as Lorentz micrographs.

Typical examples of Lorentz micrograph and Bitter pattern of the chain-lattice state in Bi-2212 are shown in Figs. 1(b) and 1(c), respectively. In both pictures, one can see two chains of vortices in the vertical direction and more or less triangular lattices in between. In the case of YBCO, however, only chains were observed to be produced; there were no lattices [1]. These apparently similar pictures in Figs. 1(a) and 1(c) contain different information about vortices. While the Bitter pattern indicates fine ferromagnetic particles gathered at the locations of the vortex magnetic fields on the surface of the film, the Lorentz micrograph is formed by the electron phase shifts caused by vortex magnetic fields mainly inside the film due to the Aharonov-Bohm effect [15]. In the case of vortex lines that are perpendicular to the film plane, the two images

VOLUME 88, NUMBER 23   PHYSICAL REVIEW LETTERS   10 JUNE 2002

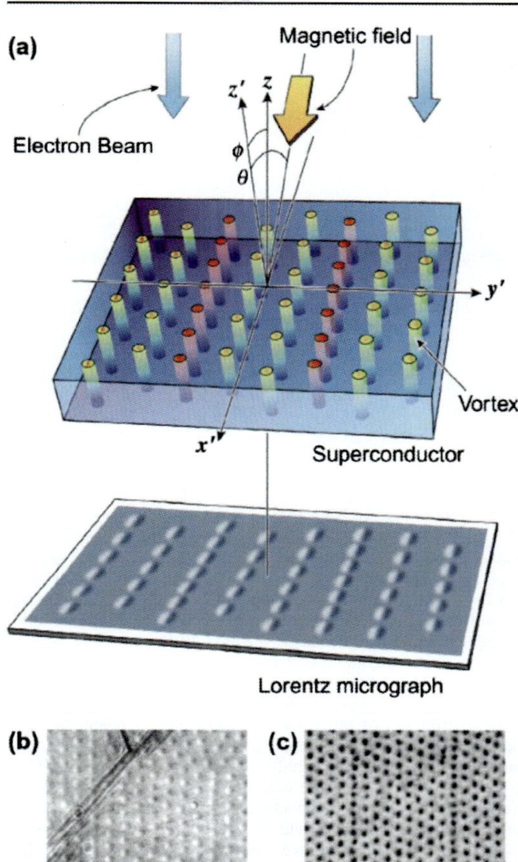

FIG. 1 (color). Observation of magnetic vortices in high-T_c superconductors at a tilted magnetic field. (a) Lorentz microscopy of vortices in a superconducting thin film. (b) Lorentz micrograph of vortices in Bi-2212 film. (c) Bitter pattern of vortices in Bi-2212. Similar chain-lattice patterns are observed in both (b) and (c). However, these two images are different in that Lorentz micrograph [14] is formed by the electron phase shifts caused mainly due to the vortex magnetic field inside the film and that Bitter pattern [1] is formed by the fine iron particles gathered at the locations where the vortex magnetic field appeared just outside the superconductor surface.

look very similar. The Lorentz images of vortex lines that are tilted considerably, however, become quite different from the Bitter images, i.e., elongated and weak in contrast [16], as demonstrated using vortex lines trapped along tilted columnar defects [14].

We first observed YBCO films using Lorentz microscopy to investigate whether vortex lines actually tilted to form

chains as predicted [1] when the magnetic field was tilted away from the c axis. Examples of the Lorentz micrographs obtained are shown in Fig. 2. The vortex images look circular when θ is less than 75° under the defocusing condition of $\Delta f = 300$ mm [Fig. 2(a)]. When θ increases above 80°, the vortices begin to form chains in the vertical direction in Figs. 2(b), 2(c), and 2(d), though the vortex chains are not so straight due to the pinning. At the same time, the vortex images become gradually elongated in the direction of the chains. The gradual image elongation with an increase in θ provides direct evidence that vortex lines tilt in the direction of the chains as illustrated below in Fig. 4(e). This confirms our previous understanding that the chain structure is formed by the attractive interaction between tilted vortex lines in YBCO. The tilt angles of these vortex lines cannot be quantitatively measured from these images without careful comparison between the observed images and the simulated images under various defocusing distances (Δf), and this work is now in progress.

The chain-lattice state in Bi-2212 cannot be explained by the same mechanism, i.e., the tilting of vortex lines [12]. We therefore investigated how the chain-lattice state is formed in Bi-2212 by observing the three-dimensional arrangement of vortex lines inside the superconductor by Lorentz microscopy. An example of a Lorentz micrograph of the chain-lattice state at a magnetic field tilted by 85° is shown in Fig. 3. All the images of chain and lattice vortices are the same and are circular. If vortex lines are tilted greatly just as the magnetic field, their images must be elongated and weak in contrast as we reported in Ref. [14]. We can thus conclude that neither chain vortices nor lattice vortices in Fig. 3 are tilted, but these two kinds of vortices are both perpendicular to the layer plane. We cannot exclude the small-angle tilting of the vortex lines, however,

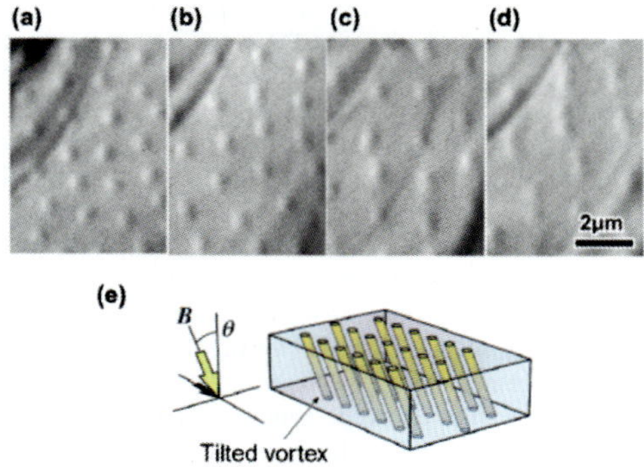

FIG. 2 (color). Lorentz micrographs of vortices in YBCO film sample at tilted magnetic fields ($T = 30$ K, $B_{z'} = 0.3$ mT); (a) $\theta = 75°$; (b) $\theta = 82°$; (c) $\theta = 83°$; (d) $\theta = 84°$. (e) Schematic of tilted vortex lines. When the tilt angle becomes larger than 80°, the vortex images begin to become elongated, and at the same time begin to form arrays of linear chains. This implies that chain vortices in YBCO are produced by some attractive force between tilted vortex lines.

VOLUME 88, NUMBER 23   PHYSICAL REVIEW LETTERS   10 JUNE 2002

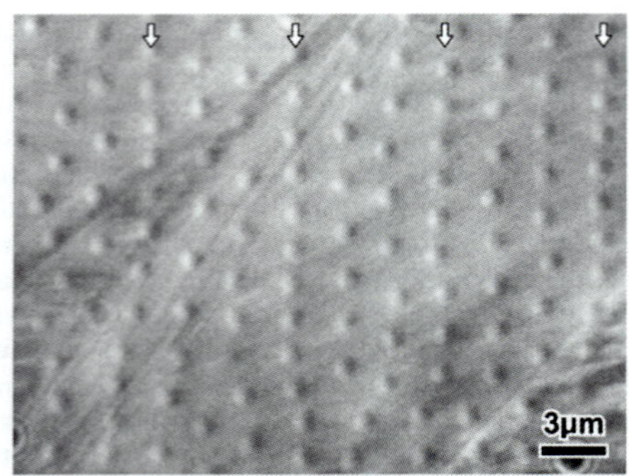

FIG. 3. Lorentz micrographs of vortices in Bi-2212. The chain-lattice structure is formed at a magnetic field tilted by $T = 50$ K, $B = 5$ mT. The chains are indicated by white arrows. If vortex lines are tilted at an angle comparable to the tilt angle of the magnetic field, 85°, the vortex images must be elongated and weak in contrast under this defocusing condition. Since the images of chain vortices as well as of lattices vortices are not elongated but circular, the vortex lines are not tilted but perpendicular to the layer plane.

if the tilt angle was much less than that of the magnetic field, 85°, since our method cannot detect the effect of such slight tilting.

We therefore carried out experiments to investigate whether or not the chain-lattice state was formed using the samples with vertical columnar defects much denser than vortices. The defect density corresponded to the vortex density of 20 mT. Since our previous experiments [14] assures us that vortex lines are trapped along tilted columnar defects regardless of the direction of the applied magnetic field as far as sample temperature T is above 20 K, the experiments were performed at $T = 50$ K, $B = 9$ mT, and $\theta = 85°$. Even under conditions where almost all the vortex lines were trapped along vertical columnar defects which were therefore perpendicular to the layer plane, we were able to observe the chain-lattice state, though the chain vortices were not arranged in such straight lines as those in samples free of columnar defects as shown in Figs. 1(b) and 1(c), but were zigzag a little bit due to the random distribution of columnar defects. This result provides evidence that the chain-lattice state can be formed even with nontilted, vertical vortices. If we accept that all the vortices are equally vertical, we can find no reason for specific vertical vortices to form chains. Therefore, we need some mechanism for specific vortices to form chains, such as the perpendicular crossing of these specific vortices with Josephson vortices. To confirm this possibility, we attempted to directly detect the Josephson vortices by Lorentz microscopy. The magnetic flux distribution of a Josephson vortex, however, was too thick, say, 50 μm wide in the layer plane to detect directly with our method. [See Fig. 4(e).]

Therefore, instead of directly observing Josephson vortices we carried out experiments to obtain evidence for the

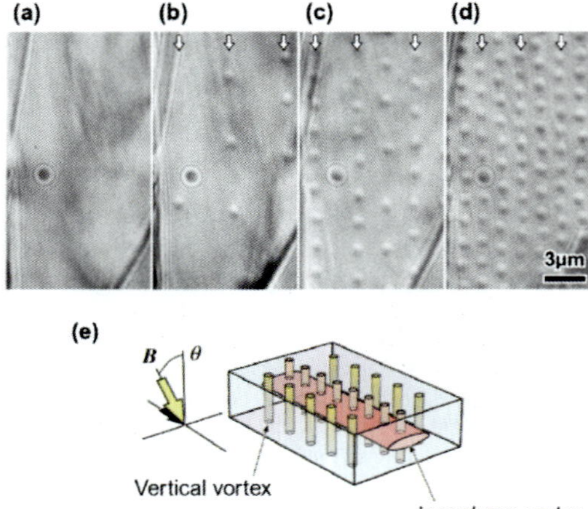

FIG. 4 (color). A series of Lorentz micrographs of vortices in a field-cooled Bi-2212 film sample when a magnetic field $B_{z'}$ perpendicular to the layer plane begins to be applied and increases at a fixed in-plane magnetic field $B_{x'}$ of 5 mT at $T = 50$ K. (a) $B_{z'} = 0$. (b) $B_{z'} = 0.02$ mT. (c) $B_{z'} = 0.1$ mT. (d) $B_{z'} = 0.17$ mT. (e) Schematic of the vortex lines consisting of vertical vortices and Josephson ones. When a magnetic field is applied parallel to the layer plane as in (a), no vertical vortices are produced, and therefore no vortex images can be seen, though Josephson vortices parallel to the layer plane should exist but cannot be observed by Lorentz microscopy due to the wide distribution of the vortex magnetic field. When the vertical magnetic field increases, vertical vortices at first begin to appear along straight lines indicated by white arrows as in (b), which are considered to be determined by Josephson vortices. Since vortices are arranged along straight lines even at large intervals between vortices and therefore no interaction between them seems to take place, we can find no other reason for the production of chain vortices than the assumption that vertical vortices crossing Josephson vortices form chains as illustrated in (e). Above $B_{z'} = 0.1$ mT, vertical vortices appear also between chain vortices as shown in (c) and (d).

existence of Josephson vortices as follows. In the first experiment, a magnetic field $B_{x'}$ was applied parallel to the layer plane, and then a magnetic field perpendicular to the layer plane $B_{z'}$ was applied and gradually increased. A series of resultant Lorentz micrographs during this process are shown in Fig. 4. At $B_{x'} = 5$ mT and $B_{z'} = 0$, no vortex images can be seen [Fig. 4(a)], since there are no vertical vortices, but only Josephson vortices if any. When $B_{z'}$ increased from zero, images of vertical vortices began to appear. The vortices were not randomly distributed, but arranged along horizontal straight lines as shown in Fig. 4(b) as if they favored falling into lines of straight ditches (indicated by white arrows) though these are not visible.

These experimental results showing that vortices are located along unseen straight lines, even when vortices are so sparse that the attractive interaction between vortices may not occur, cannot be explained without assuming the existence of some kind of ditches for vortices. When $B_{z'}$

VOLUME 88, NUMBER 23   PHYSICAL REVIEW LETTERS   10 JUNE 2002

further increases, the vertical vortices began to appear also between the chains, thus forming the chain-lattice state [Figs. 4(c) and 4(d)]. These results lead to the conclusion that these ditches must consist of the Josephson vortices [see Fig. 4(e)] that should be produced when a magnetic field $B_{x'}$ is applied parallel to the layer plane. These Josephson vortices were once thought to have no interaction with vertical vortices but were shown by Koshelev [12] to act as potential wells (ditches) for vertical vortices. One may notice in Figs. 4(b), 4(c), and 4(d) that vortex chains are not always fixed in their locations as well as in their intervals. The locations of Josephson vortices can be changed because each Lorentz micrograph was photographed every time under new field-cooled conditions: the sample temperature was first increased above T_c, then a magnetic field was applied, and finally the sample was cooled in the presence of an applied magnetic field. The spacing of the ditches indicated by white arrows in Fig. 4 is $3-5$ μm. This value is shorter than the estimated spacing 8 μm if gamma is assumed to be 200 [12]. This discrepancy may be due to the small thickness 300 nm of the film comparable to the magnetic vortex size.

We then carried out an observation of dynamics to see how vertical vortices begin to penetrate the film when $B_{z'}$ is increased from zero at a fixed $B_{x'}$ of 5 mT. Josephson vortices must have already been produced at $B_{z'} = 0$. When we gradually increased $B_{z'}$, vortex images began to enter our field of view one after another from one side along Josephson vortices which were fixed in their locations in this case, and then also between them. These static and dynamic behaviors of vortices confirm Koshelev's model [12] that vertical vortices can be stably located at Josephson vortices, thus forming chains even if they are extremely sparse, and can easily move along the directions of the Josephson vortices. Our previous finding [17] of the oscillation of chain vortices along the chain direction at much lower temperature than T_c also supports this conclusion.

The mechanisms that produce unconventional vortex arrangements at a tilted magnetic field reflecting the anisotropic layered structure of high-T_c superconductors were thus elucidated by direct observation of vortex lines inside the superconductors using our 1-MV holography electron microscope. While the chain structure in YBCO is formed by the attractive force between tilted vortex lines, the chain-lattice state in Bi-2212 is formed by the interaction of chain vortices with Josephson vortices as predicted by Koshelev [12].

We are very grateful to Dr. A. Koshelev of Argonne National Laboratory, Dr. F. Nori of University of Michigan, and Dr. G. Pozzi and Dr. M. Beleggia of University of Bologna for their useful suggestions and discussions. We also thank Dr. Y. Nakayama of University of Tokyo, Dr. T. Matsumoto, Dr. N. Osakabe, Dr. K. Takagi, and Dr. T. Fukazawa of Hitachi, Ltd., and Dr. Y. Shiohara of ISTEC for sample preparation. This work was partially supported by the New Energy and Industrial Technology Development Organization (NEDO) as Collaborative Research and Development of Fundamental Technologies for Superconductivity Applications.

---

*Email address: tonomura@harl.hitachi.co.jp

[1] P. L. Gammel, D. J. Bishop, J. P. Rice, and D. M. Ginsberg, Phys. Rev. Lett. **68**, 3343 (1992).
[2] C. A. Bolle *et al.*, Phys. Rev. Lett. **66**, 112 (1991).
[3] I. V. Grigorieva and J. W. Steeds, Phys. Rev. B **51**, 3765 (1995).
[4] A. Grigorenko *et al.*, Nature (London) **414**, 728 (2001).
[5] A. I. Buzdin and A. Y. Simonov, JETP Lett. **51**, 191 (1990).
[6] A. M. Grishin, A. Y. Martynovich, and S. V. Yampol'skii, Sov. Phys. JETP **70**, 1089 (1990).
[7] L. L. Daemen, L. J. Campbell, and V. G. Kogan, Phys. Rev. B **46**, 3631 (1992).
[8] D. A. Huse, Phys. Rev. B **46**, 8621 (1992).
[9] L. L. Daemen, L. J. Campbell, A. Y. Simonov, and V. G. Kogan, Phys. Rev. Lett. **70**, 2948 (1993).
[10] A. Sudbø, E. H. Brandt, and D. A. Huse, Phys. Rev. Lett. **71**, 1451 (1993).
[11] E. Sardella, Physica (Amsterdam) **275C**, 231 (1997).
[12] A. E. Koshelev, Phys. Rev. Lett. **83**, 187 (1999).
[13] T. Kawasaki *et al.*, Appl. Phys. Lett. **76**, 1342 (2000).
[14] A. Tonomura *et al.*, Nature (London) **412**, 620 (2001).
[15] A. Tonomura *et al.*, Phys. Rev. Lett. **56**, 792 (1986).
[16] S. Fanesi *et al.*, Phys. Rev. B **59**, 1426 (1999).
[17] T. Matsuda *et al.*, Science **294**, 2136 (2001).

Akira Tonomura's Books and Major Publications

(1) Books (in English)

1. M. Peshkin and A. Tonomura, *The Aharonov-Bohm Effect*, Lecture Notes in Physics, Springer-Verlag, Berlin, Heidelberg, 1989.
2. A. Tonomura, *Electron Holography* (Springer Series in Optical Sciences 70), Springer-Verlag, Berlin, Heidelberg, 1993.
3. A. Tonomura, L. F. Allard, G. Pozzi, D. C. Joy, and Y. A. Ono (Eds). *Electron Holography*, North-Holland, Amsterdam, 1995.
4. A. Tonomura, *The Quantum World Unveiled by Electron Waves*, World Scientific, Singapore, 1998.
5. A. Tonomura, *Electron Holography*, 2nd ed. (Springer Series in Optical Sciences 70) Springer-Verlag, Berlin, Heidelberg, 1999.

(2) Major Publications in Scientific Journals (in English)

1. A. Tonomura, A. Fukuhara, H. Watanabe, and T. Komoda, Optical reconstruction of image from Fraunhofer electron-hologram, *Jpn. J. Appl. Phys.* **7**(3), 295, (1968).
2. A. Tonomura, T. Matsuda, and J. Endo, High resolution electron holography with field emission electron microscope, *Jpn. J. Appl. Phys.* **18**(1), 9-14, (1979).
3. A. Tonomura, T. Matsuda, J. Endo, H. Todokoro, and T. Komoda, Development of a field emission electron microscope, *J. Electron Microsc.* **28**(1), 1-11, (1979).
4. J. Endo, T. Matsuda, and A. Tonomura, Interference electron microscopy by means of holography, *Jpn. J. Appl. Phys.* **18**(12), 2291-2294, (1979).
5. A. Tonomura, T. Matsuda, J. Endo, T. Arii, and K. Mihama, Direct observation of fine structure of magnetic domain walls by electron holography, *Phys. Rev. Lett.* **44**(21), 1430-1433, (1980).
6. A. Tonomura, T. Matsuda, H. Tanabe, N. Osakabe, J. Endo, A. Fukuhara, K. Shinagawa, and H. Fujiwara, Electron holography technique for investigating thin ferromagnetic films, *Phys. Rev.* B**25**(11), 6799-6804, (1982).
7. A. Tonomura, T. Matsuda, R. Suzuki, A. Fukuhara, N. Osakabe, H. Umezaki, J. Endo, K. Shinagawa, Y. Sugita, and H. Fujiwara, Observation of Aharonov-Bohm effect by electron holography, *Phys. Rev. Lett.* **48**(21), 1443-1446, (1982).

8. N. Osakabe, K. Yoshida, Y. Horiuchi, T. Matsuda, H. Tanabe, T. Okuwaki, J. Endo, H. Fujiwara, and A. Tonomura, Observation of recorded magnetization pattern by electron holography, *Appl. Phys. Lett.* **42**(8), 746-748, (1983).

9. A. Tonomura, H. Umezaki, T. Matsuda, N. Osakabe, J. Endo, and Y. Sugita, Is magnetic flux quantized in a toroidal ferromagnet? *Phys. Rev. Lett.* **51**(5), 331-334, (1983).

10. A. Tonomura, T. Matsuda, T. Kawasaki, J. Endo, and N. Osakabe, Sensitivity-enhanced electron-holographic interferometry and thickness-measurement applications at atomic scale, *Phys. Rev. Lett.* **54**(1), 60-62, (1985).

11. A. Tonomura, N. Osakabe, T. Matsuda, T. Kawasaki, J. Endo, S. Yano, and H. Yamada, Evidence for Aharonov-Bohm effect with magnetic field completely shielded from electron wave, *Phys. Rev. Lett.* **56**(8), 792-795, (1986).

12. N. Osakabe, T. Matsuda, T. Kawasaki, J. Endo, A. Tonomura, S. Yano, and H. Yamada, Experimental confirmation of Aharonov-Bohm effect using a toroidal magnetic field confined by a superconductor, *Phys. Rev.* A**34**(2), 815-822, (1986).

13. A. Tonomura, T. Matsuda, J. Endo, T. Arii, and K. Mihama, Holographic interference electron microscopy for determining specimen magnetic structure and thickness distribution, *Phys. Rev.* B**34**(5), 3397-3402, (1986).

14. A. Tonomura, Applications of electron holography, *Rev. Mod. Phys.* **59**(3), 639-669, (1987).

15. Y. Honda, M. Futamoto, T. Kawasaki, K. Yoshida, M. Koizumi, F. Kugiya, and A. Tonomura, Observation of magnetization structure on Co-Cr perpendicular magnetic recording media by Bitter and electron holography methods, *Jpn. J. Appl. Phys.* **26**(6), L923-L925, (1987).

16. N. Osakabe, T. Matsuda, J. Endo, and A. Tonomura, Observation of atomic steps by reflection electron holography, *Jpn. J. Appl. Phys.* **27**(9), L1772-L1774, (1988).

17. A. Tonomura, J. Endo, T. Matsuda, T. Kawasaki, and H. Ezawa, Demonstration of single-electron buildup of interference pattern, *Amer. J. Phys.* **57**(2), 117-120, (1989).

18. S. Hasegawa, T. Kawasaki, J. Endo, A. Tonomura, Y. Honda, M. Futamoto, K. Yoshida, F. Kugiya, and M. Koizumi, Sensitivity-enhanced electron holography and its applications to magnetic recording investigations, *J. Appl. Phys.* **65**(5), 2000-2004, (1989).

19. T. Matsuda, S. Hasegawa, M. Igarashi, T. Kobayashi, M. Naito, H. Kajiyama, J. J. Endo, N. Osakabe, A. Tonomura, and R. Aoki, Magnetic field observation of a single flux quantum by electron-holographic interferometry, *Phys. Rev. Lett.* **62**(21), 2519-2522, (1989).

20. N. Osakabe, J. Endo, T. Matsuda, A. Tonomura, and A. Fukuhara, Observation of surface undulation due to single-atomic shear of a dislocation by reflection-electron holography, *Phys. Rev. Lett.* **62**(25), 2969-2972, (1989).

21. T. Matsuda, A. Fukuhara, T. Yoshida, S. Hasegawa, A. Tonomura, and Q. Ru, Computer reconstruction from electron holograms and observation of fluxon dynamics, *Phys. Rev. Lett.* **66**(4), 457-460, (1991).

22. Q. Ru, J. Endo, T. Tanji, and A. Tonomura, Phase-shifting electron holography by beam tilting, *Appl. Phys. Lett.* **59**(19), 2372-2374, (1991).

23. A. Tonomura, Electron-holographic interference microscopy, *Adv. Phys.* **41**(1), 59-103, (1992).

24. K. Harada, T. Matsuda, J. Bonevich, M. Igarashi, S. Kondo, G. Pozzi, U. Kawabe, and A. Tonomura, Real-time observation of vortex lattices in a superconductor by electron microscopy, *Nature* **360**, 51-53, (5 November 1992).

25. J. E. Bonevich, K. Harada, M. Matsuda, H. Kasai, T. Yoshida, G. Pozzi, and A. Tonomura, Electron holography observation of vortex lattices in a superconductor, *Phys. Rev. Lett.* **70**(19), 2952-2955, (1993).

26. T. Hirayama, Q. Ru, T. Tanji, and A. Tonomura, Observation of magnetic-domain states of barium ferrite particles by electron Holography, *Appl. Phys. Lett.* **63**(3), 418-420, (1993).

27. J. Chen, T. Hirayama, G. Lai, T. Tanji, K. Ishizuka, and A. Tonomura, Real-time electron-holographic interference microscopy with a liquid-crystal spatial light modulator, *Opt. Lett.* **18**(22), 1887-1889, (1993)

28. K. Harada, T. Matsuda, H. Kasai, J. E. Bonevich, T. Yoshida, and A. Tonomura, Vortex configuration and dynamics in $Bi_2Sr_{1.8}$ $CaCu_2O_x$ thin film by Lorentz microscopy, *Phys. Rev. Lett.* **71**(20), 3371-3374, (1993).

29. G. Lai, T. Hirayama, K. Ishizuka, T. Tanji, and A. Tonomura, Three-dimensional reconstruction of electric-potential distribution in electron-holographic interferometry, *Appl. Opt.* **33**(5), 829-833, (1994).

30. G. Lai, T. Hirayama, A. Fukuhara, K. Ishizuka, T. Tanji, and A. Tonomura, Three-dimensional reconstruction of magnetic vector fields using electron-holographic interferometry, *J. Appl. Phys.* **75**(9), 4593-4598, (1994).

31. K. Harada, H. Kasai, T. Matsuda, M. Yamasaki, and A. Tonomura, Real-time observation of the interaction between flux-lines and defects in a superconductor by Lorentz microscopy, *Jpn. J. Appl. Phys.* **33**(5A), 2534-2540, (1994).

32. Q. Ru, G. Lai, K. Aoyama, J. Endo, and A. Tonomura, Principle and application of phase-shifting electron holography, *Ultramicroscopy* **55**(2), 209-220, (1994).

33. T. Kawasaki, J. Endo, T. Matsuda, and A. Tonomura, Development and application of a 350 kV transmission electron microscope with a magnetic field superimposed field emission gun, *Microbeam Analysis* **3**, 287-291, (1994).

34. N. Osakabe, T. Kodama, J. Endo, A. Tonomura, K. Ohbayashi, T. Urakami, H. Tsuchiya, and Y. Tsuchiya, Fast and precise electron counting system for the observation of quantum mechanical electron intensity correlation, *Nucl. Instrum. & Methods* A**365**, 585-587, (1995).

35. T. Matsuda, K. Harada, H. Kasai, O. Kamimura, and A. Tonomura, Observation of dynamic interaction of vortices with pinning centers by Lorentz microscopy, *Science* **271**(5254), 1393-1395, (1996).

36. K. Harada, O. Kamimura, H. Kasai, T. Matsuda, and A. Tonomura, Direct observation of vortex dynamics in superconducting films with regular arrays of defects, *Science* **274**(5290), 1167-1170, (1996).

37. N. Osakabe, H. Kasai, T. Kodama, and A. Tonomura, Time-resolved analysis in transmission electron microscopy and its application to the study of the dynamics of vortices, *Phys. Rev. Lett.* **78**(9), 1711-1714, (1997).

38. K. Harada, H. Kasai, T. Matsuda, M. Yamasaki, and A. Tonomura, Direct observation of interaction of vortices and antivortices in a superconductor by Lorentz microscopy, *J. Electron Microsc.* **46**(3), 227-232, (1997).

39. N. Osakabe, T. Kodama, and A. Tonomura, Time-resolved electron microscopy by means of electron counting, *Phys. Rev.* B**56**(9), 5156-5163, (1997).

40. C. Sow, K. Harada, A. Tonomura, G. W. Crabtree, and D. G. Grier, Measurement of the vortex pair interaction potential in type-II superconductors, *Phys. Rev. Lett.* **80**(12), 2693-2696, (1998).

41. A. Tonomura, H. Kasai, O. Kamimura, T. Matsuda, K. Harada, J. Shimoyama, K. Kishio, and K. Kitazawa, Motion of vortices in superconductors, *Nature* **397**(6717), 308-309, (1999).

42. T. Kawasaki, T. Yoshida, T. Matsuda, N. Osakabe, A. Tonomura, I. Matsui, and K. Kitazawa, Fine crystal lattice fringes observed using a transmission electron microscope with 1-MeV coherent electron waves, *Appl. Phys. Lett.* **76**(9), 1342-1344, (2000).

43. T. Kawasaki, I. Matsui, T. Yoshida, T. Katsuta, S. Hayashi, T. Onai, T. Furutsu, K. Myochin, M. Numata, H. Mogaki, M. Gorai, T. Akashi, O. Kamimura, T. Matsuda, N. Osakabe, A. Tonomura, and K. Kitazawa, Development of 1 MV field-emission transmission electron microscope, *J. Electron Microsc.* **49**(6), 711-718, (2000).

44. A. Tonomura, H. Kasai, O. Kamimura, T. Matsuda, K. Harada, Y. Nakayama, J. Shimoyama, K. Kishio, T. Hanaguri, K. Kitazawa, M. Sasase, and S. Okayasu, Observation of individual vortices trapped along columnar defects in high-temperature superconductors, *Nature* **412**(6847), 620-622, (2001).

45. T. Matsuda, O. Kamimura, H. Kasai, K. Harada, T. Yoshida, T. Asahi, A. Tonomura, Y. Nakayama, J. Shimoyama, K. Kishio, T. Hanaguri, and K.

Kitazawa, Oscillating rows of vortices in superconductors, *Science* **294**(5551), 2136-2138, (2001).

46. A. Tonomura, H. Kasai, O. Kamimura, T. Matsuda, K. Harada, T. Yoshida, T. Akashi, J. Shimoyama, K. Kishio, T. Hanaguri, K. Kitazawa, T. Masui, S. Tajima, N. Koshizuka, P. L. Gammel, D. Bishop, M. Sasase, and S. Okayasu, Observation of structures of chain vortices inside anisotropic high-T_c superconductors, *Phys. Rev. Lett.* **88**(23), 237001, (2002).

47. O. Kamimura, H. Kasai, T. Akashi, T. Matsuda, K. Harada, J. Masuko, T. Yoshida, N. Osakabe, A. Tonomura, M. Beleggia, G. Pozzi, J. Shimoyama, K. Kishio, T. Hanaguri, K. Kitazawa, M. Sasase, and S. Okayasu, Direct evidence of the anisotropic structure of vortices interacting with columnar defects in high-temperature superconductors through the analysis of Lorentz images, *J. Phys. Soc. Jpn.* **71**(8), 1840-1843, (2002).

48. T. Akashi, K. Harada, T. Matsuda, H. Kasai, A. Tonomura, T. Furutsu, N. Moriya, T. Yoshida, T. Kawasaki, K. Kitazawa, and H. Koinuma, Record number (11 000) of interference fringes obtained by a 1 MV field-emission electron microscope *Appl. Phys. Lett.* **81**(10), 1922-1924, (2002).

49. M. Beleggia, G. Pozzi, J. Masuko, N. Osakabe, K. Harada, T. Yoshida, O. Kamimura, H. Kasai, T. Matsuda, and A. Tonomura, Interpretation of Lorentz microscopy observations of vortices in high-temperature superconductors with columnar defects, *Phys. Rev.* **B66**(17), 174518, (2002).

50. K. Harada, A. Tonomura, Y. Togawa, T. Akashi, and T. Matsuda, Double-biprism electron interferometry, *Appl. Phys. Lett.* **84**(17), 3229-3231, (2004).

51. M. Beleggia, G. Pozzi, A. Tonomura, H. Kasai, T. Matsuda, K. Harada, T. Akashi, T. Matsui, and S. Tajima, Model of superconducting vortices in layered materials for the interpretation of transmission electron microscopy images, *Phys. Rev.* **B70**(18), 184518, (2004).

52. K. Harada, A. Tonomura, T. Matsuda, T. Akashi, and Y. Togawa, High-resolution observation by double-biprism electron holography, *J. Appl. Phys.* **96**(11), 6097-6102, (2004).

53. K. Harada, T. Akashi, Y. Togawa, T. Matsuda, and A. Tonomura, Optical system for double-biprism electron holography, *J. Electron Microsc.* **54**(1), 19-27, (January 2005).

54. Y. Togawa, K. Harada, T. Akashi, H. Kasai, T. Matsuda, F. Nori, A. Maeda, and A. Tonomura, Direct observation of rectified motion of vortices in a niobium superconductor, *Phys. Rev. Lett.* **95**(8), 087002, (2005).

55. A. Tonomura, Direct observation of thitherto unobservable quantum phenomena by using electrons, *Proc. Natl. Acad. Sci. USA* **102**(42), 14952-14959, (2005).

56. F. Nori and A. Tonomura, Helical spin order on the move, *Science* **311**(5759), 344-345, (2006).

57. A. Tonomura, The Aharonov-Bohm effect and its applications to electron phase microscopy, *Japan Academy Ser. B* **82**(2), 45-48, (2006).

58. A. Tonomura, Conveyor belts for magnetic flux quanta, *Nature Materials* **5**, 257-258, (2006).

59. J. J. Kim, K. Hirata, Y. Ishida, D. Shindo, M. Takahashi, and A. Tonomura, Magnetic domain observation in writer pole tip for perpendicular recording head by electron holography, *Appl. Phys. Lett.* **92**(16), 162501, (2008).

60. A. Tonomura and F. Nori, Disturbance without the force, *Nature* **452**, 298-299, (20 March 2008).

61. A. Tonomura, Development of electron phase microscopes, *Nuclear Instruments and Methods in Physics Research* A601, 203-212, (2009).

62. K. Hirata, Y. Ishida, J. J. Kim, H. Kasai, D. Shindo, M. Takahashi, and A. Tonomura, Electron holography observation of in-plane domain structure in writer pole for perpendicular recording heads, *J. Appl. Phys.* **105**(7), 07D538, (2009).

63. H. Batelaan and A. Tonomura, The Aharonov-Bohm effects: Variations on a subtle theme, *Physics Today*, **62**(9), 38-43, (September 2009).

64. Y. Murakami, H. Kasai, J. J. Kim, S. Mamishin, D. Shindo, S. Mori, and A. Tonomura, Ferromagnetic domain nucleation and growth in colossal magneto-resistive manganite, *Nature Nanotech.* **5**(1), 37-41, (2010).

65. M. Uchida and A. Tonomura, Generation of electron beams carrying orbital angular momentum, *Nature* **464**(8904), 737-739, (2010).

66. Y. Murakami, Y. Nii, T. Arima, D. Shindo, K. Yanagisawa, and A. Tonomura, Magnetic domain structure in the orbital-spin-coupled system MnV_2O_4, *Phys. Rev.* B**84**(5), 054421, (2011).

67. T. Kodama, N. Osakabe, and A. Tonomura, Correlation in a coherent electron beam, *Phys. Rev.* A**83**(6), 063616, (2011).

68. A. Tonomura, X. Z. Yu, K. Yanagisawa, T. Matsuda, Y. Onose, N. Kanazawa, H. S. Park, and Y. Tokura, Real-space observation of skyrmion lattice in a helimagnet MnSi thin sample, *Nano Letters* **12**, 1673-1677, (2012).

69. T. Tanigaki, Y. Inada, S. Aizawa, T. Suzuki, H. S. Park, T. Matsuda, A. Taniyama, D. Shindo, and A. Tonomura, Split-illumination electron holography, *Appl. Phys. Lett.* **101**(4), 043101-1 – 043101-4, (2012).

70. T. Suzuki, S. Aizawa, T. Tanigaki, K. Ota, T. Matsuda, and A. Tonomura, Improvement of the accuracy of phase observation by modification of phase-shifting electron holography, *Ultramicroscopy* **118**, 21-25, (2012).

71. H. S. Park, Y. Murakami, K. Yanagisawa, T. Matsuda, R. Kainuma, D. Shindo, and A. Tonomura, Electron holography studies on narrow magnetic domain walls observed in a Heusler alloy $Ni_{50}Mn_{25}Al_{12.5}Ga_{12.5}$, *Adv. Funct. Mater.* **22**, 3434-3437, (2012).

72. K. Hirata, Y. Ishida, T. Akashi, D. Shindo, and A. Tonomura, Electron holography study of magnetization behavior in the writer pole of a perpendicular magnetic recording head by a 1 MV transmission electron microscope, *J. Electron Microscopy* **61**(5), 305-308, (2012).

73. H. S. Park, K. Hirata, K. Yanagisawa, Y. Ishida, T. Matsuda, D. Shindo, and A. Tonomura, Nanoscale magnetic characterization of tunneling magnetoresistance spin valve head by electron holography, *Small* **8**(23), 3640-3646, (2012).

Video Clips of Tonomura's Experiments

The attached DVD-ROM contains the following four video clips of Tonomura's experiments with his narrations. (courtesy of Hitachi, Ltd.)

1. Build-up of single electron events to form an interference pattern in the double-slit experiment. (doubleslit-n.mpeg)

2. Real-time observation of movements of vortices (magnetic flux lines) in niobium thin films by Lorentz microscopy. (vortices1-n.mpeg)

3. Movements of vortices penetrating into a niobium thin film with a square array of artificial defects when the applied magnetic field increases. (vortices2-n.mpeg)

4. Movements of vortices in a high-Tc $Bi_2Sr_2CaCu_2O_{8+\delta}$ thin film when the largely tilted magnetic field increases or decreases. (vortices3-n.mpeg)

Acknowledgment: These video clips are copyrighted by Hitachi, Ltd. The editors would like to thank Research and Development Group of Hitachi, Ltd. for permitting us to include these video files in the book.